T0135617

Geochemical investigations on rocks from the Ophiolite Zone of Zermatt-Saas Fee (Western Alps, Switzerland) with special emphasis on the potential of Aitchison's log-ratio method

Dissertation
zur Erlangung des Doktorgrades
der Mathematisch-Naturwissenschaftlichen Fakultäten
der Georg-August-Universität zu Göttingen

vorgelegt von
Wendelin Himmelheber*
aus Stuttgart (Geburtsort)

Göttingen 2003

* Author's address:
W. Himmelheber
Bischof-Meinwerk-Str. 6c
33719 Bielefeld
Germany
himmelheber@bitel.net

D 7

Referent:	Prof. Dr. J. Hoefs
Korreferent:	Prof. Dr. G. Wörner
Tag der mündlichen Prüfung:	27.1.3

Bibliografische Information Der Deutschen Bibliothek

Die Deutsche Bibliothek verzeichnet diese Publikation in der Deutschen Nationalbibliografie; detaillierte bibliografische Daten sind im Internet über http://dnb.ddb.de abrufbar.

ISBN 3-8325-0263-7

Logos Verlag Berlin
Comeniushof, Gubener Str. 47,
10243 Berlin
Tel.: +49 030 42 85 10 90
Fax: +49 030 42 85 10 92
INTERNET: http://www.logos-verlag.de

Table of contents

4

Table of figures

Table of tables

Abbreviations

b.d.	Below detection limit
cps	Counts per second
HFSE	high field strength elements
HREE	heavy rare earth elements
ICP-MS	inductively coupled plasma mass spectrometer
LILE	large ion lithophile element
LREE	light rare earth elements
MORB	middle oceanic ridge basalt
MREE	middle rare earth elements
pca	principle component analysis
PTt	Pressure-Temperature-Time
REE	rare earth elements
XRF	X-ray fluorescence
ZSF	ophiolite zone of Zermatt - Saas Fee

Mineral abbreviations are after Kretz (1983)

0. Introduction

This study is concerned with the geochemistry of the Zermatt-Saas Fee Zone (ZSF hereafter). This ophiolitic suite marks the suture between the European and the Adriatic continental plates, together with the other ophiolites from the Western Alps. Its rocks, polymetamorphic mafic and ultramafic extrusives and intrusives and minor metasediments, are supposed to represent oceanic crust from the middle Jurassic to middle Cretaceous Ligurian-Piemonte ocean, a small basin that opened up in response to movements related to the opening of the central Atlantic ocean (Gealey 1988, Dercourt et al. 1986, Pfiffner 1992).

The three major rock types present in the ZSF, namely serpentinites, metagabbros and metabasalts, will be studied in this paper from a geochemical point of view.

The main focus is on the magmatic genesis of the ZSF rocks and on their relationship. The peculiar difficulties of doing this kind of work in a polymetamorphic terrain are dealt with. In particular it is tried to shed some light on the geochemical effects of the metamorphic overprint. It is considered whether statistical or other methods can help to evaluate the relative mobility and immobility of diagnostic trace elements.

The serpentinites are the rocks in the ZSF that attracted least attention. In particular the exact nature of their protoliths is not known: have they been restites or cumulates, are they of continental or oceanic provenance? These questions will be addressed. Also the time of their serpentinization and the nature of the serpentinizing fluid are not well constrained. Ocean floor metamorphism seems to be the most plausible candidate, but not necessarily the only possible one. Now whereas stable isotope investigations would be most promising to tackle this question, some hints may perhaps also be gained from major- and trace element geochemical results.

Furthermore it seems an interesting peculiarity that the former ultramafics have been thoroughly serpentinized, whereas the other rocks have undergone a water-deficient metamorphism, as described by many authors (see below). This raises doubts as to whether the standard model of ophiolite formation is appropriate for the ZSF.

The metagabbros and the metabasalts supposedly underwent ocean floor metamorphism too. The effects of this on their geochemical composition is also a focus of this work.

For this study a total of 85 rock specimens were collected. These were studied petrographically and a subset of 64 specimens and subspecimens were selected for chemical analysis.

Collecting specimens at Allalinpass: Robert Schönhofer

1. Geological Background

Since the days of Argand in the beginning of the last century, the Zermatt area has been studied in great depth by a large number of geologists. The following paragraphs are a short introduction to the regional geology of the Zermatt-Saas Fee ophiolite zone; for further information the reader is referred to the maps and publications of P. Bearth (1953, 1954a, 1954b, 1959, 1963, 1967, 1973), to the publications of Milnes (1974), Milnes et al. (1981), Steck (1989, 1990), Barnicoat & Fry (1986) and to the papers mentioned below.

1.1 Generalities

The Ophiolite Zone of Zermatt-Saas Fee (as well as the adjacent nappes) belongs to the Penninic realm of the Alps. This paragraph will deal with the relationships between these nappes.

Fig. 1.1 gives a tectonic profile through the area; the nappes depicted will be shortly described from bottom to top (cf. table 1.1).

Nappe terminology underwent some revisions since the late 1980ies; herein the names as defined by Sartori (1987) will be used, with due regard given to the older terminology.

The basic geometry of the regional nappes has been correctly described already by Argand (1911). Large extrapolations from surface geology down to depths of ca. 12 km are made possible by the strong relief of the area and by the axial plunge of the major structures of ca. 20° towards the west.

The ZSF is underlain by the Monte Rosa nappe, made up of a pre-Alpine basement (para- and orthogneisses) and its sedimentary cover (Gornergratzone). The Monte Rosa nappe is usually interpreted as part of the European continental margin, more specifically, the Briançonais (e.g. Stampfli & Marthaler 1990), but it might also have been a microplate of its own (e.g. Platt 1986). The Briançonais has recently been interpreted as a terrain of Iberic affinity moved eastward (Stampfli 1993, Stampfli & Marchant 1997), but this has been questioned by Trümpy (1998). The Gornergratzone is parallelized by Sartori (1987) to the Frilihorn series of the Mt. Fort nappe (see below).

Below the Monte Rosa nappe a second ophiolitic sliver is to be found, the Antrona Zone. Whether this comes from a separate ocean basin or from the same one as the ZSF is presently still a matter of debate (cf. Pfeiffer et al. 1989, Ballèvre & Merle 1993).

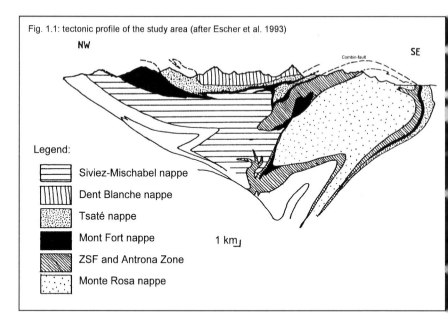

Fig. 1.1: tectonic profile of the study area (after Escher et al. 1993)

NW

Combin-fault

SE

Legend:

Siviez-Mischabel nappe

Dent Blanche nappe

Tsaté nappe

Mont Fort nappe   1 km⌐

ZSF and Antrona Zone

Monte Rosa nappe

Above the ZSF lies the metasedimentary Mt. Fort nappe, made up of two units: the Frilihorn series, a sequence of middle to upper Triassic sediments ("Würmlizug" of the former Combin Zone) and the Série Rousse, siliceous and phyllitic marbles from the upper Cretaceous (Sartori 1987, Escher et al. 1993).

This, in turn, is capped by the Tsaté nappe, interpreted as a former accretionary prism (Stampfli & Marthaler 1990). Again a subdivision in two units is possible: the Zone de Tracuit, made up mainly of ophiolitic material, and the Série Grise, a meta-flysch made of calcschists and marbles with a few marker beds of other materials. Mt. Fort nappe and Tsaté nappe together make up the former Combine zone.

Above these thin metasedimentary cover nappes follows the Siviez-Mischabel nappe. It consists again of pre-Alpine basement with some sedimentary cover (the Barrhorn Series). If the Monte Rosa nappe was not a independent microcontinent of its own, it was probably continuous with the Siviez-Mischabel nappe, up to the moment when they were affected by a late folding episode taking place under the orogenic lid of the Dent-Blanche nappe, a unit of southern Alpine affinities.

Concerning models of continental breakup and ocean formation, there have recently been some revisions. Wernicke (1981, 1985) proposed the so-called simple shear model. According to him, rifting takes place essentially along a huge listric normal fault penetrating down into the mantle. In that way subcontinental litho-

Table 1.1
Nappes of the Zermatt area

Dent Blanche nappe	
Siviez-Mischabel nappe	Barrhorn series: metasediments
	Basement
Tsaté nappe	Série grise: meta-flysch: calcschist and marble
	Zone de Tracuit: ophiolites
Mont Fort nappe	Série rousse: upper cretaceous marbles
	Frilihorn series: middle to upper triassic sediments
Zermatt-Saas Fee zone	ophiolites
Monte Rosa nappe	Gornergrat zone: metasediments
	Basement: para- and orthogneisses
Antrona-zone	ophiolites

sphere can become denuded, which leads to the formation of a "lithospheric ocean". This model has been applied by Lemoine et al. (1987) to the alpine ophiolites as well as to those from Corsica and the Apennines. According to them, the alpine serpentinites and peridotites all belong to the former lithospheric ocean. This model is supported and refined by Stampfli & Marthaler (1990). It is also strengthened by geochemical results from the Ligurian Apennines, an other part of the Piemont-Ligurian ocean, which is in many respects very similar to the ZSF. Here Rampone et al. (1998) found an extreme isotopic depletion in the peridotites, not consistent with an origin as oceanic lithosphere.

Such a model has of course also geochemical implications, or, geochemical data may be useful in deciding on the feasibility of the model.

1.2 Internal Structure of the ZSF

The internal structural relationships of the ZSF are controversial. Some authors (e.g. Argand 1911, Steck 1989, Steck et al. 1989) favor the theory of fold nappes, inspired by the marble-cake like geometry of the nappes and perhaps by memories of the "Glarner Doppelfalte". The advocates of this theory interpret the ZSF as a lying fold, the so-called Mittaghorn Synform. The inverted limb would be found in the Saas area and along the Gornergrat, the upright limb in the Täsch valley and in the Zermatt area.

According to others the theory of fold nappes is obsolete, and the internal structure of the ZSF is interpreted as a stack of thrust slices (e.g. Barnicoat & Fry 1986). P. Bearth, probably the authority with the best knowledge of the field relationships, has published diverging opinions on this question. In Bearth (1967) the fold nappe interpretation is disclaimed (p. 8); the Mittaghorn synform is interpreted as just an upbending of the frontal parts of the nappe, due to the late alpine backfolding (p. 14). According to Bearth (1952) also the Monte Rosa nappe cannot be interpreted as a fold nappe, as its internal structure is at variance with this; there is no inverted fold limb and no frontal bending. On the other hand, Bearth & Schwander (1981) claim (without very much documentation), that the metasedimentary sequence, which can be found in places at the outer rims of the ZSF and which is believed to be (at least in some places) in primary contact with the underlying metamafic rocks, is showing different polarities on both sides of the Zone, thus lending support to the fold nappe interpretation. Also the inverted polarity of the Allalin Gabbro (Meyer 1983b) could be fitted into that picture.

However this may be, fig. 1.2 sums up the structural and lithological information to be gained from published sources in the form of a horizontal tectonic section, which has been constructed after the precedence of Steck et al. (1989). The favorable exposures in the Täsch valley make it possible to draw a very detailed picture of the relationships there, whereas in the Zermatt and Saas areas much less structural and lithological detail can be observed. Therefore the section contains a good deal of interpretation. For example, it is tentatively proposed that the Allalin Gabbro and the gabbro sheet at Mellichen are linked, as the structural trends favor such an alignment. Meyer (1983a), Bearth (1967) and Ganguin (1988) however, on petrographical grounds, believed this not to be the case. Also the step-like geometry of the Allalin Gabbro, suggestive of an imbrication, can not be completely justified from exposed relationships, although the ultramafic bands shown in the map (Bearth 1954b) would tend to encourage such a view. Attention is drawn to the fact that the different lithological units do not seem to wrap around the Allalin body, as would be necessary if the ZSF was a fold nappe. Particularly two metasedimentary bands to the northeast are striking in the wrong direction. Another peculiarity of the Allalin body is the unconformity towards metabasalt and serpentinite units at its lower (southern) fringe.

The thrust sheets in the Täsch area are very well documented due to the favorable exposures there.

The "upper tectonic sheet" in the south western (Italian) part of the section is adopted from the little map in Dal Piaz & Ernst (1978). According to these authors it consists of all the different lithologies present.

14

The continuation of the ZSF into the Valtournanche and further south towards and beyond the Aosta fault can not be detected at the altitude of the section. The only trace of it is possibly the small "nose" at the southwestern end of the area displayed.

To sum up, the geometry is most clear in the northeastern part (with the exception of the very frontal parts around Mittaghorn), where there is found some imbrication in the structurally higher parts of the ZSF, to be seen in the Täsch area in metabasalts and metagabbros, as well as in the lower parts with a repetition of the sequence serpentinite – metabasalt at a larger scale, although there much detail remains hidden below glaciers. The intermediate parts in the Zermatt region are badly exposed. The southwestern parts again look very orderly, although the shape of the huge serpentinite body seems somewhat enigmatic. It is however well documented despite much glaciation in these parts. At this point attention is drawn to the little metasedimentary band folded into this body at Lichenbretter as a testimony to its very variable and intensely folded internal structure. Perhaps it would be a rewarding exercise in structural mapping to try to unravel the internal structure of this favorably exposed serpentinite body. Also renewed mapping of the frontal nappe parts around Mittaghorn might help settle the open question of fold nappes.

1.3 Rock types and metamorphism

All rock types of the oceanic lithosphere have been present and can now be found as their metamorphic equivalents. The main rock types are, of course, metabasalts, metagabbros and serpentinites, but there are also minor metasediments, massive sulfides (Castello 1981, Reinecke 1991) and epidosites (Barnicoat & Bowtell 1995).

The metabasalts in some places display pillow structures (Bearth 1959). The existence of a sheeted dike complex has been claimed by Barnicoat and Fry (1986), but field evidence seems inconclusive. What can be discerned at the place of the purported dike complex (Swiss reference grid 631.8/98.1) are boudinaged dikes of metabasalt in a metagabbro (flasergabbro) matrix. Also at the eastern ridge of Spitzi Flue (631.2/96.3) a complex of metabasalt and metagabbro layers can be observed. The Garten-Riffelberg formation (cf. fig. 1.2) is a peculiar melange of metasedimentary and ophiolitic components, probably an olistolith (Bearth 1967).

The common genetic lineage of serpentinites, metagabbros and metabasalts as former residual mantle, magma chamber cumulate and extrusive volcanics has been questioned by Lemoine et al. (1987) on the basis of field evidence for other alpine-suture ophiolites from the Western Alps, the Apennines and Corsica. They also proposed a genetic model based on the simple shear model of continental rifting (cf.

Fig 1.2: Horizontal tectonic section, altitude 3000 m, through the Zermatt-Saas Fee ophiolite zone.

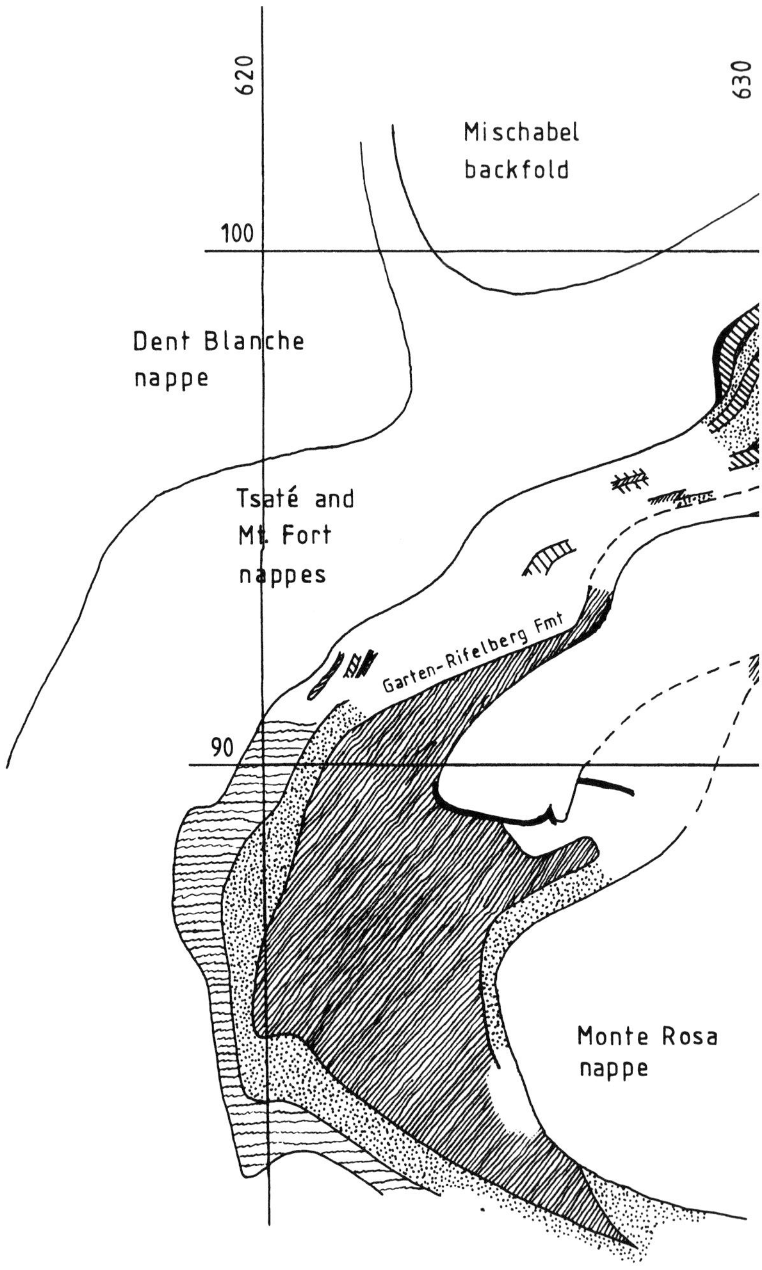

620

630

Mischabel
backfold

100

Dent Blanche
nappe

Tsaté and
Mt. Fort
nappes

Garten-Rifelberg Fmt

90

Monte Rosa
nappe

16

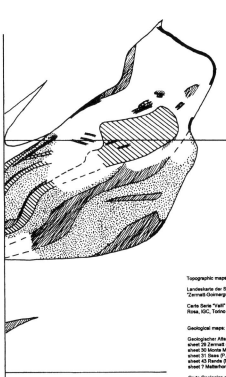

 Metasediments

Metabasalts

Metagabbros

Serpentinites

"Upper tectonic sheet"
of DalPiaz 1965

Topographic maps:

Landeskarte der Schweiz 1:25000, Zusammensetzung 2515
"Zermatt-Goirnergrat", Wabern 1988

Carte Serie "Valli" Nr. 5, 1:50000, Cervino-Matterhorn e Monte
Rosa, IGC, Torino 1989

Geological maps:

Geologischer Atlas Schweiz 1:25000:
sheet 29 Zermatt (P. Bearth 1953)
sheet 30 Monte Moro (P. Bearth 1954)
sheet 31 Saas (P. Bearth 1954)
sheet 43 Randa (P. Bearth 1963)
sheet 7 Matterhorn (K. Bucher & R. Oberhänsli, forthcoming)

Carta Geologica d'Italia 1:100000, Foglio Monte Rosa

A. Steck, J.-L.- Epard, A. Escher, R. Marchand, H. Masson, L.
Spring (1989): Coupe tectonique horizontal des Alpes
Centrales. Mém. Géol. Lausanne 5

P. Bearth (1967): Die Ophiolithe der Zone von Zermatt-Saas
Fee. Beiträge geol. Karte Schweiz NF 132

P. Bearth & Schwander (1981) The post-triassic sediments of
the ophiolite zone of Zermatt-Saas Fee and the associated
manganese mineralisations. Eclogae geol. Helv. 74, 189-205

Ganguin J (1988) Contribution a la charactérisation du
métamorphisme polyphasé de la zone de Zermatt-Saas Fee.
Ph.D. thesis No. 8731, ETH Zürich (map on p. 24),

Dal Piaz & G.V., Ernst, W.G. (1978) Areal geology and
petrology of eclogites and associated metabasites of the
Piemont ophiolite nappe, Breuil-St. Jaques area, Italian
western Alps. Tectonophysics 51, 99-126 (containing a map at
a very reduced scale, the only published outcome of a major
collaborative mapping effort in the scale 1:10000)

Additional material used:

B.J.Davidson &R.P. Metcalfe, unpublished maps, as published
in Fry, N. & Barnicoat, A.C. (1987) The tectonic implications of
high-pressure metamorphism in the western Alps. J Geol Soc
London 144, 653-659. Similar to the map by Ganguin 1988 but
less precise

WIDMER, T.W.: Entwässerung ozeanisch alterierter Basalte in
Subduktionszonen (Zone von Zermatt-Saas Fee). PhD-thesis,
Basel, 308pp (1996). Profile on p. 11q

Escher, J., Masson, H., Steck, A.(1993) Nappe geometry in the
Western Swiss Alps. J Struct Geol 15, 501-509

A.C. Ellis, Profile of the Täsch-area, in: A.C. Barnicoat (1988),
the mechanism of veining and retrograde alteration of Alpine
eclogites. J. metam. Geol. 6, 545-558

Stampfli & Marthaler 1990) to account for the different genetic provenance of these rock types. The question of the three main rock types' genetic lineage will be a major topic of this study.

The metamorphic history of the area has been complex. Relic minerals from the magmatic stage can be found rarely as plagioclase in the metagabbros (Meyer 1983a, Wayte et al. 1989) and as cpx and chromian spinel, sometimes perhaps also olivine, in the serpentinites (Bearth 1967, Ganguin 1988). The protoliths have been subjected

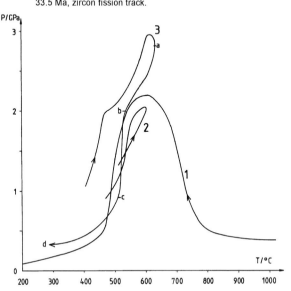

Fig. 1.3: PTt paths proposed in the literature for the ZSF. 1: Meyer (1983a); 2: Fry & Barnicoat (1987); 3: Reinecke (1998). The points marked on Reinecke's PTt-path are related to geochronological data given by Amato et al. (1999): a: 40,6 Ma, eclogite Sm-Nd; between b and c: 38 Ma, metasediment Rb-Sr; d: 33.5 Ma, zircon fission track.

to a polyphase metamorphic history, comprising ocean floor metamorphism (Barnicoat & Cartwright 1995, 1997, Widmer 1996, Barnicoat & Bowtell 1995), an eoalpine HP metamorphic event with at least two stages (Meyer 1983a, b, Barnicoat and Fry 1987) and a retrograde overprint in the greenschist facies with again two stages (Meyer 1983a, b, Ganguin 1988, Barnicoat & Fry (1986), Pfeifer et al. 1989). PT-Paths have been constructed by Meyer (1983a), Fry & Barnicoat (1987) and Reinecke (1998), see fig. 1.3. The path given by Oberhänsli (1980) comes from a time when the scientific community was reluctant to accept high pressures and is not included here. The somewhat exotic PT path of Meyer (1983a, b) seems well documented and has even gained access into textbooks (Bucher & Frey 1994). It is however challenged by the finding of Barnicoat & Cartwright (1997) that the Allalin Gabbro underwent ocean floor metamorphism.

As for dating the metamorphic events, there is an older and a younger school. The older school used to date the HP-event at around 100-65 Ma (eoalpine stage) and the younger greenschist-facies overprint at around 45-30 (mesoalpine stage) and 30-10 Ma (neoalpine stage; Hunziker 1974, Monié 1985). The younger school now puts the HP-event into the eocene, around 52 to 40 Ma (Barnicoat et al 1991, Bowtell et al.

1994, Rubatto et al. 1997). A recent paper by Amato et al. (1999) presents new evidence for the late-HP theory and also sums up previous findings; the geochronological information from this publication is also displayed in fig. 1.3. According to Ballèvre & Merle (1993) the conflicting evidence can be reconciled if it is assumed that the European and oceanic units underwent two events of HP-metamorphism, one eoalpine, one mesoalpine. The two events are separated by a major episode of divergence. The later event then partly reset the clocks.

The area from which specimens were collected can be delimited by the following landmarks: Zermatt – Hirli – trockener Steg – Kleines Matterhorn – Riffelhorn – Fluealp – Pfulwe – Allalinpass – Täschalp – Zermatt, between 1600 and 3560 m altitude (see fig. 1.4). Some specimens were also picked from the moraine of the Allalin Glacier in the Saas valley at Mattmark. Some specimens were taken from the outcrop, some from recent moraines. The latter is legitimate, as due to the geographic situation the provenance of the specimens can never be in doubt. Specimen locations are marked on the map fig. 1.4.

Fig 1.4: Topographic overview of the study area with specimen locations

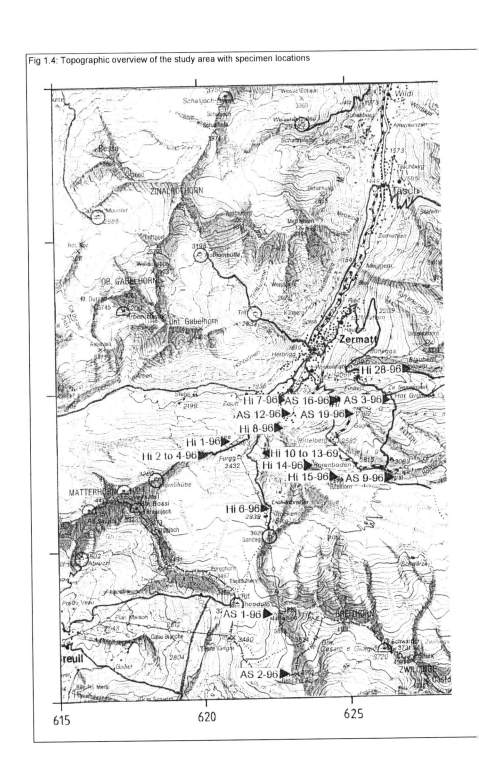

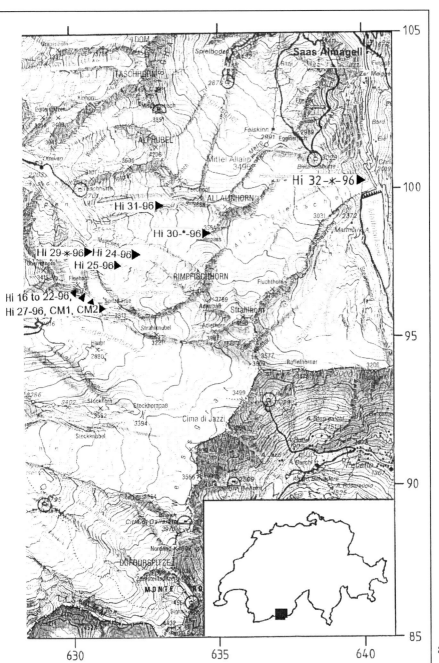

2.   Petrographic description

There are many excellent descriptions of the petrography of the metabasalts and metagabbros to be found in the literature, notably Bearth (1967), Meyer (1983a), Ganguin (1988), Barnicoat & Fry (1986). The seemingly monotonous serpentinites attracted less interest.

Detailed descriptions of the individual specimens are deferred to appendix 1. The following general descriptions are only aimed at giving a broad overview.

2.1 Metabasalts

Macroscopically, the metabasalts can be distinguished from the metagabbros by their usually darker color and by their texture. Whereas the metagabbros have a coarse cm-size flaser structure, the metabasalts are fine-grained except for garnet porphyroblasts and occasional pseudomorphs after medium-grained lawsonite or after plagioclase.

Due to the complicated metamorphic history and to differing degrees and modes of ocean-floor hydration, prograde dehydration and retrograde rehydration, there is to be found quite some diversity among this group of rocks. The basalts seem to have been largely hydrated during ocean-floor metamorphism, as no true magmatic relics (except for textural relics like pillow structures and possibly plagioclase pseudomorphs) are observed. The ensuing HP-metamorphism then resulted in dehydration. During greenschist facies metamorphism, the rehydration was patchy and incomplete and thus many relics of the older metamorphic stages survived. I will proceed by describing some idealized types.

There are two types of HP-metamorphic basalts to be found in the ZSF, the bluish glaucophanites and the eclogites. The question whether they represent differences in metamorphic grade or in protolith chemistry has been discussed controversially. Bearth & Stern (1971), Oberhänsli (1982), Widmer (1996) and Barnicoat (1988) argue that the two types represent differences in the alteration of the protoliths during ocean floor metamorphism. The glaucophanites then are believed to be chemically altered in different ways, viz. they are either richer in Mg (Widmer 1996), in Na (Bearth & Stern 1971) or in H_2O (Barnicoat 1988). Fry & Barnicoat (1987), on the contrary, argue that glaucophanites are products of the early retrograde stages (still HP). Perhaps both theories apply, but to different rocks.

The HP-paragenesis of the glaucophanites consists of glaucophane, garnet, epidote, paragonite, phengite, rutile, pyrite, apatite, sometimes also Mg-chloritoid, omphacite and zoisite. According to the exact paragenesis several subtypes can be distinguished (Ganguin 1988, Widmer 1996). Garnet usually contains inclusions, among others sometimes glaucophane and pseudomorphs after lawsonite (the latter occur also outside garnet). These inclusions are interpreted as relics from a prograde blueschist stage (Fry & Barnicoat 1986). The texture usually is schistose, showing preferred shape orientation of glaucophane crystals.

Eclogitic metabasalts consist of omphacite, garnet, epidote, quartz, rutile, phengite, paragonite, glaucophane and apatite. Garnets contain the same inclusions as in glaucophanites. Usually schistosity is weak. Often square white patches, consisting of zoisite, micas, quartz and other minerals are observed. There is some confusion in the literature as to whether these are pseudomorphs after lawsonite or after magmatic plagioclase. In metamorphic pillows they are interpreted by Bearth (1967:20) as pseudomorphs after plagioclase; Barnicoat & Fry (1986) and Ganguin (1988) however only mention lawsonite pseudomorphs.

Of course the HP-basalts all show signs of retrogression, but these were not mentioned in the idealized descriptions just given, as the kind of alteration that occurs will become clear from the description of the totally retrogressed metabasalts below.

The typical retrogressed metabasalts occur as either prasinites or ovardites. The prasinites' paragenesis consists of albite plus some of chlorite, hornblende and epidote/zoisite. Petrographically, ovardites are a subtype of prasinites, consisting of chlorite and large round blasts of albite. This poikiloblastic albite contains abundant inclusions of the matrix minerals.

Due to their large amount of chlorite and to the fact that it was deformation in the first place that made the retrogression possible by facilitating access of water, the retrogressed metabasalts usually have a pronounced schistosity. Sometimes prasinites and ovardites still contain garnet as a relic, with rims of barroisitic hornblende or chlorite, sometimes also of biotite. If eclogites and blueschists are believed to be compositionally different, it might be possible that the same holds also for the prasinites and ovardites. According to Ganguin (1988, p. 56) it is not always possible to find out from petrographical observations, whether a retrogressed specimen has been an eclogite or a glaucophanite before. According to him, symplectitic intergrowths of albite with blue-green hornblende and the appearance of chlorite are diagnostic of former eclogites. This is at variance with Widmer's (1996) contention that prasinites develop from eclogites, and ovardites from glaucophanites, as ovardites (sensu Bearth 1967, p. 59) are composed of albite plus chlorite.

Between the greenschists and the eclogite-facies metabasalts there are found intermediate stages such as garnet-amphibolites and albitic amphibolites. They are characterized by the progressive replacement of HP-minerals by mineral parageneses of lower grade. Thus glaucophane is replaced by barroisitic hornblende, omphacite by symplectitic intergrowth of albite and hornblende, chloritoid by white mica and chlorite, rutile by titanite or ilmenite. These changes were interpreted in detail by Ganguin (1988) and also related to the two basic basalt types, the type that underwent ocean floor metamorphism and the type that did not.

2.2 Metagabbros

The diversity of metagabbroic rock types is even greater than that of the metabasalts, due to a greater diversity of protoliths. Also magmatic relics are preserved more frequently with the metagabbros, enhancing again the diversity. These relics caused Meyer (1983a) to claim that the Allalingabbro had not been hydrated prior to his subduction and HP metamorphism. This has been questioned recently by Barnicoat & Cartwright (1997) on the basis of O-isotope data. They found the $\delta^{18}O$ values of metamafics derived from little-altered protoliths in the range 5.1 to 6.0 ‰, whereas those derived from low-T altered rocks are in the range 5.9 to 6.6 ‰. However this may be, the fact of abundant magmatic relics remains. In very rare cases even unaltered plagioclase has been found (Bearth 1967, p. 79, Meyer 1983a, Wayte et al. 1989).

Macroscopically the most abundant type of metagabbro has a flaser texture of whitish and greenish domains (developed in the places of former magmatic crystals), which have been elongated to some extent producing a weak schistosity. In specimens with a low degree of retrograde overprint and deformation, the green domains can be recognized as large omphacite porphyroclasts. With the development of retrograde minerals, the rock became more ductile and schistosity can increase considerably, up to the development of fuchsite schists (Ganguin 1988).

Of the three major primary magmatic minerals plagioclase has been always (with the above mentioned rare exceptions) altered to saussurite. According to Meyer (1983a) an early (pre-peak-HP) saussurite, consisting of zoisite, jadeite, quartz and some kyanite, can be distinguished from a later (peak-HP-) saussurite, made up of zoisite, omphacite, kyanite and sometimes quartz. In the greenschist stage, saussurite is further altered to zoisite/epidote, white mica or sometimes green fuchsite, chlorite and albite (Bearth 1967, p. 46, 52). Magmatic clinopyroxene has changed to omphacite containing inclusions of rutile. Sometimes additional talc develops, thus forming light green smaragdite. Olivine at first develops coronitic textures, of which the most obvious one is a rim of garnet. In later HP-stages olivine is changed to intergrowths of talc, chloritoid, kyanite and chlorite.

The later retrograde development is analogous to the metabasalts.

According to Ganguin (1988), three types of gabbros can be distinguished, differing chemically and also according to their parageneses. Very Mg-rich (olivine rich) gabbros are characterized by the occurrence of Mg-chloritoid as a HP-phase. The typical paragenesis is omp - grt -cld - tlc - zo - rt. This type is found only in the Allalin-gabbro and in some moraine blocks in the Täsch valley area. Normal magnesian gabbros form the majority of rocks; they are characterized by the paragenesis omp - zo ± grt ± paragonite. The Fe-Ti gabbro type, representing highly evolved protoliths, abound in rutile and garnet, and often host omphacite veins. The typical paragenesis is omp - grt ± glc - ep - rt ± phengite. Upon retrogression, these rock types all converge towards hbl - ab - ep/zo - paragonite - phengite - chl - tit. The Fe-Ti gabbros seem to be lacking in white mica however. This grouping into different gabbro types has been confirmed by the specimens studied here (see also chapter 6.1).

2.3 Serpentinites

In thin section I could not observe in my serpentinite specimens anything beyond those minerals mentioned already by Bearth (1967), i.e. antigorite (confirmed by XRD), magnetite, olivine, relics of clinopyroxene (augite) and sometimes chlorite and hornblende. Rahn & Bucher (1998) report Ti-clinohumite which they claim on textural reasons to be formed during greenschist facies ocean floor metamorphism. Ganguin (1988) reports rare relics of magmatic Cr-spinel. According to the same, olivine (forsterite) is not a magmatic relic, but newly formed, whereas according to Bearth (1967) it occurs both as magmatic relic and as metamorphic mineral. Some serpentinites were changed to schists dominated by chlorite.

Usually the schistosity is quite strong.

Associated to the serpentinites are some rocks of metasomatic origin, namely rodingites (former mafic dikes, now rich in Ca-silicates) and rocks developed at the contacts between serpentinites and metabasites or metasediments. Rodingites have a very variable mineralogy; the specimens sampled contain chlorite, vesuvianite, amphibole, pyroxene and opaques. Contact metasomatites contain talc, chlorite, amphibole, opaques and in one case (Hi 12-2-96) green garnet (uvarovite).

2.4 Veins

The veins sampled come from the HP metamorphic stage of the Allalin Gabbro. It is one omphacite vein (Hi 32-4-96), one chloritoid-talc vein (Hi 32-10-96) and one talc-kyanite-glaucophane vein (Hi 32-8-96). The former two have been analyzed, whereas the latter one is too small for separation.

3 Analytical procedure

Specimen size was around 1 kg for most of the fine grained serpentinites, around 2 kg for all other rock types. Due to the alpine character of the sampled area, the collection of larger sized specimens was not feasible. However, most of the material is rather fine grained, so samples can be confidently regarded as representative of some rock volume of homogeneous petrogenetic conditions (e.g. part of a single layer of a intrusion, subjected to the same metamorphic and retrometamorphic intensive and extensive conditions).

Weathering rinds usually were not present. After removing a slab for documentation purposes, specimens were crushed. An aliquot of ca. 100 g of crushed rock was ground down to silt grain size in preconditioned agate mills. This powder was used for all subsequent analytical procedures.

3.1 XRF

Glass discs were prepared by fusing 700 mg of rock powder with 4200 mg of Li-tetraborate in platinum crucibles at ca. 1100 $^{\circ}$C. Major and trace elements were measured in a routine procedure with the Phillips PW 1480 automated sequential spectrometer at Geochemisches Institut der Universität Göttingen with data processing by the Phillips X40 software package.

To check analytical accuracy, several discs of 2 in-house standards (BB and PB), which are similar in composition to the analyzed specimens, were prepared and measured. The results are very close to recommended values (Tab. 3.1). Except for the standards, no replicate analyses were carried out, as the method is routine and previous experience with similar specimens from the ZSF had shown reproducibility to be very good (Himmelheber 1996). Limits of detection were not determined specifically, but taken from the experience of daily routine operation for the trace elements (G. Hartmann, written communication 1996) and for the oxides from the literature (Weiss 1997, p. E2) as recorded in Tab. 3.1.

3.2 Ferric/ferrous Iron

Contents of ferrous iron were determined by wet chemical methods; ferric iron then was calculated by subtracting ferrous iron from total iron as determined by XRF.

Tab. 3.1: reproducibility, accuracy and limit of detection of XRF

| | BB | | | PB | | | |
	average	recomm.	Stdev	average	recomm.	Stdev	l.o.d.
SiO2	48.7	48.68	0.071	37.67		0.058	1.26
TiO2	2.351	2.330	0.011	0.573		0.011	0.01
Al2O3	13.54	13.50	0.055	6.10		0.000	0.29
Fe2O3	10.99	11.20	0.011	15.22		0.029	0.04
MnO	0.17	0.18	0.000	0.18		0.001	0.01
MgO	8.07	8.10	0.043	26.68		0.190	0.18
CaO	8.28	8.20	0.012	3.88		0.030	0.04
Na2O	3.45	3.45	0.047	0.33		0.012	0.16
K2O	1.73	1.75	0.004	0.18		0.000	0.06
P2O5	0.512	0.500	0.006	0.059		0.002	0.02
Nb	56	60	0.707	4	5.3	0.577	5
Zr	195	185	0.707	34	24	0.000	10
Y	23	21	1.225	5	6	1.528	5
Sr	929.4	930	2.881	102	97	0.577	5
Rb	43	44	0.707	11	10	1.732	5
Pb	5.2	4.2	0.837	3	1	1.155	10
Ga	19.4	19	1.342	7	6.6	0.577	5
Zn	118	125	3.162	91	94	1.000	5
Ni	182.6	180	4.099	1204	1200	6.658	10
Co	43.4	43	0.894	127	124	0.577	5
Cr	264.8	265	0.837	1486	1500	8.963	5
V	192.4	190	3.782	105	100	3.512	5
Ba	653.4	650	6.656	37	44	1.000	10
Sc	18.4	18	0.894	11	12	0.000	5

To determine ferrous iron, a titration method as described by Heinrichs & Herrmann (1990) was used. Around 200 to 400 mg of rock powder were heated with each 5 ml of HF (40 %), H_2SO_4 (96 %) and H_2O to 170 $^{\circ}$C for ca. 10 minutes in platinum crucibles with lids. The acids serve to dissolve most of the silicates and other minerals, whereas the water serves (as vapor) to expel oxygen from the crucibles. The resulting solution is poured into a mixture of dilute boric and phosphoric acid and immediately titrated with $KMnO_4$.

Four replicate analyses were carried out. The calculated standard deviation was 0,32 wt % absolute or 11 % relative.

Magnetite, ilmenite, pyrite and some garnets are not completely digested by this method. With some specimens there could be observed a reddish residue, sometimes also a whitish one. As magnetite is an important phase in the serpentinites, as are ilmenite and garnet in the metabasites, it has to be conceded that the determined FeO contents are too low. The error can be in the range of 1 wt % absolute, as can be judged from estimated magnetite and ilmenite contents.

Pyrite, although frequently present, is of less concern, as it is shown by the S analyses to be a very minor component.

3.3 C/S

Analyses for total carbon and sulfur were carried out on a Metalyt CS 100/1000 RF from Eltra company. This device works by inductive heating of a mixture of ca. 100 mg of rock powder with some iron and tungsten in a current of pure O_2. The sulfur and carbon compounds are completely volatilized and oxidized. The IR-absorption of the SO_2 and CO_2 gas produced is measured in the flow and calibrated against appropriate mixtures of elemental sulfur, carbonate, and quartz.

All specimens were measured at least twice. Reproducibility is only at the level of ca. 10 % relative. In accordance with the petrographic observations, S was supposed to be present as sulfides, C as carbonates. Contents were calculated accordingly and are reported in the table in appendix 2.

3.4 H_2O

Water content was determined from LOI, correcting for the S and CO_2 determined as described above and for Fe^{II} oxidized in the process. No effort was made to differentiate between H_2O^+ and H_2O^-. At any rate, with the rocks studied, H_2O^- is of very minor amount and of no petrogenetic significance.

LOI determination is subject only to minor sources of error, but nevertheless the H_2O contents must be regarded as some-

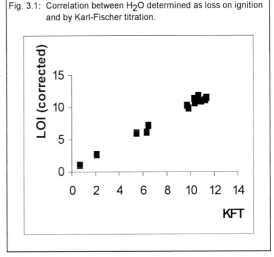

Fig. 3.1: Correlation between H_2O determined as loss on ignition and by Karl-Fischer titration.

what lacking in precision due to the many steps involved. As an independent control, with 14 specimens a H_2O-determination by Karl-Fischer titration was carried out. This method is very precise, but the device used for it (684 KF Coulomat by Metrohm, furnace by Haereus) is rather difficult to handle. The correlation between the two methods was very good (see fig. 3.1).

The analytical data were not recalculated water-free, except for the REE diagrams and the spidergrams.

For ICP-MS analysis, rock powder had to be digested. Ca. 100 mg specimen powder and 3 ml each of HF (40 %, ultapure, commercially available) and $HClO_4$ (70 %, sub-boiling distilled) were put in teflon autoclaves. Autoclaves reached 200 $^{\circ}C$ after 10 hours of stepwise heating and were kept at this temperature for 4 hours. After cooling down, lids were removed and the solution evaporated to dryness.

The specimens were taken up in 2 ml HNO3 (67 %, ultrapure) to which for the metagabbro and metabasalt specimens 2 ml HCl (6N, 4fold distilled) were added. Also an internal standard (In and Re) was added. The mixture was reheated to just below the boiling point for complete dissolution and then cooled down and diluted with ultrapure water to 100 ml (dilution factor ca. 1:1000). Upon dilution a few solutions developed a milky appearance due probably to some precipitation of Al-compounds. According to the results of the analysis discussed below this did not seem to affect the analytical results; probably the precipitate had redissolved in the time between solution preparation and analysis.

Specimen solutions were transferred to preconditioned PE-bottles and stored there for one respectively two weeks until the first two measuring sessions, and for 21 resp. 54 weeks until the last two sessions.

Specimen solutions were analyzed by ICP-MS for the 36 Elements given in Table 3.2. Determinations were performed on a VG Plasma Quad II+. Three consecutive measurements are taken for each specimen in the routine procedure of the Geochemisches Institut der Universität Göttingen. The measuring queue started with 4 calibration solutions (blank; Cl; BA; Gr) and contained 4 further measurements of the calibration solution "BA" after regular intervals as a control of the devices' performance ("quality control"). It also contained blanks and international standards (JB 3 and JGb-2).

A total of 106 measurements of 68 different specimens (plus 12 measurements of international standards) was performed in 6 different measuring sessions.

In the first two measuring sessions there was a hardware malfunction of the ICP-MS, resulting in an erratic switching between the PC and analog detection modes. As a consequence, normalizing to the internal standards was not possible. However, drift and matrix effects were subordinate (so much at least could be seen from the raw counts of the internal standards In and Re), and thus a direct calibration was carried out. Cps were two-point calibrated to the calibration solutions blank and "BA". For those pieces of data collected in analogue mode calibration to sltn "Gr" was tried whenever this had been measured in analogue mode. However, these calibrations were found to be not consistent with the other ones and thus were not used except for testing the correlation between ICP-MS and XRF as described below.

After calibration, drift correction was effected by linear interpolation between the 4 quality control measurements.

With the four other measuring sessions normalization to the internal standards was tried, as well as 3- and 4-point calibration. With the third session the best results were obtained by not normalizing, a two point calibration and QC drift correction. With the fourth and sixth sessions, the best results were obtained by normalizing to In, a 2-point calibration and a drift correction by QC. With the fifth session normalization to In was not possible, as In had been measured mostly in analogue mode. A two point calibration and QC drift correction was applied. The values from this last session were only used for some REE values.

Finally, blank contents were subtracted and the values recalculated to the weight of the specimen powder digested.

The reproducibility of the results was good, within measuring sessions as well as between. Even aging of the solutions by one year did not deteriorate the results, not even with elements like Nb, which have been reported to show aging behavior (Münker 1998) (see fig. 3.2).

Table 3.2 gives the isotopes used, limits of detection (as 3 of blanks) and the from QC and from the analyzed international standard. As the reproducibility of the QC is only a function of the performance of the ICP-MS, whereas the reproducibility of the international standard

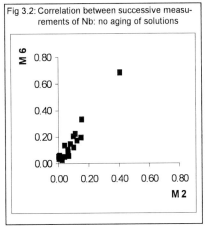

Fig 3.2: Correlation between successive measurements of Nb: no aging of solutions

ards contains also all kinds of deviation caused by the previous steps of the analytical procedure, it is clear that the latter is usually much larger than the former. Except for some elements which were discarded, the standard deviation is in the range of 10 % (relative).

The data themselves are to be found in appendix 2, some typical analyses are given in Tab. 3.3.

Tab 3.2: Elements analyzed by ICP-MS

Element	Isotope	Stdev(QC) N=8		Stdev (JB) N=5		L.O.D. N=8
		ppm	%	ppm	%	ppm
Li	7	0.96	9.60	1.07	14.83	0.011
Sc	45	0.34	3.36	2.72	8.05	0.053
Cr	52	6.42	6.42			3.927
Co	59	2.09	8.35	5.74	16.72	0.243
Ni	60	5.40	10.81	5.94	16.40	0.409
Cu	65			0.039		2.384
Zn	66	2.82	11.27	18.96	18.96	0.384
Ga	71	(only measuring run 4)				
Rb	85	14.75	14.75	0.91	6.02	0.668
Sr	88	6.07	6.07	42.04	10.43	0.084
Y	89	0.87	8.73	1.95	7.24	0.036
Zr	91	2.81	7.03	5.82	5.95	0.102
Nb	93	0.84	8.38	0.09	3.68	0.059
Mo	95	0.05	5.13	0.06	5.56	0.048
Sn	120	0.28	13.78	0.39	20.78	0.045
Ba	137	6.62	6.62	25.66	10.47	0.116
La	139	0.63	5.04	0.56	6.34	0.037
Ce	140	1.57	6.28	1.73	8.06	0.046
Pr	141	0.13	5.06	0.29	9.37	0.004
Nd	146	0.33	3.32	1.49	9.52	0.025
Sm	147	0.13	5.37	0.54	12.69	0.038
Eu	151	0.18	17.51	0.18	13.80	0.153
Gd	157	0.33	6.62	0.55	11.80	0.045
Tb	159	0.03	5.42	0.07	9.05	0.071
Dy	163	0.10	3.87	0.46	10.23	0.032
Ho	165	0.03	6.41	0.09	11.64	0.002
Er	166	0.10	10.50	0.33	13.15	0.010
Tm	169	0.01	4.29	0.04	9.82	0.003
Yb	174	0.09	8.63	0.30	11.94	0.016
Lu	175	0.03	10.10	0.04	9.00	0.006
Ta	181	0.07	6.79	0.14	10.17	0.015
Pb	208	1.28	6.39	0.59	10.66	0.200
Th	232	0.47	4.66	0.05	4.21	0.104
U	238	0.18	8.96	0.06	12.60	0.008

3.6 Reliability of the ICP-MS data

As the reproducibility of the data is found to be good within the limitations of the method, there remains the question of accuracy. In this respect there are three major sources of error: firstly there are systematic errors, e.g. the problem of interferences. Secondly, some elements are known to pose problems in the ICP-MS apparatus, especially the HFSE, and among these particularly Nb and Ta. Thirdly, some elements are fraught with problems of the digestion process, be it that they are preferentially contained in refractory phases, like Zr and Hf, be it that oxidation and precipitation processes can occur, as with Cr.

Accuracy was checked by three methods: scrutiny of calibration measurements; analyzing international reference standards; checking one analytical method by another.

As for reference standards, JB-3 and JGb-2 were used. The recommended values could be reproduced with sufficient success. For trace elements with contents above 1 ppm the errors are in the range of 1 to 4 ppm.

The problems of interferences can be tackled also from another corner: they should show up in the response curve of the ICP-MS. In fig. 3.3 some typical response curves are shown. These have been constructed by taking the raw cps (minus the raw solution blank cps) of the measurements of the calibration solution "BA" and dividing them by the isotopic abundance of each element and by its concentration in the calibration solution. It is seen that there are only minor interferences on Gd and Yb and some on Cr and Ni.

There follows a discussion of the analytical accuracy of several element groups.

Tab 3.3: Some typical analyses

	Metagabbro Hi 32-13-96	Metabasalt Hi 22-96	Serpentinite AS 16-96
Li	1.94	22.46	0.46
Sc	18.00	22.21	14.38
Cr	52.76	152.61	922.66
Co	44.07	41.60	85.32
Ni	194.85	340.55	1036.07
Zn	32.99	91.73	41.37
Rb	b.d.	0.71	b.d.
Sr	294.12	173.09	0.32
Y		21.63	0.78
Zr	7.96	22.67	0.94
Nb	0.33	3.16	0.10
Mo	0.42	0.12	b.d.
Sn	0.48	2.74	0.07
Ba	2.73	12.89	0.20
La	0.28	1.80	0.04
Ce	1.95	14.51	0.12
Pr	0.34	2.30	b.d.
Nd	1.68	10.59	0.11
Sm	0.65	3.11	b.d.
Eu	0.44	1.01	b.d.
Gd	0.63	3.51	0.08
Tb	0.11	0.60	0.02
Dy	0.72	4.14	0.12
Ho	0.15	0.93	0.03
Er	0.41	2.65	0.10
Tm	0.06	0.38	0.02
Yb	0.34	2.44	0.12
Lu	0.05	0.38	0.02
Ta	0.15	0.70	b.d.
Pb	0.24	0.63	b.d.
Th	b.d.	0.16	b.d.
U	b.d.	0.12	b.d.

REE and Y

These are regarded as reliable, as usual for this analytical method. Measured values for the international standard JB-3 are in very good agreement with recommended values. Y is slightly lower than the recommended value.

Correlation between ICP-MS and XRF values for Y is excellent, however RFA values are systematically higher by about 20 %.

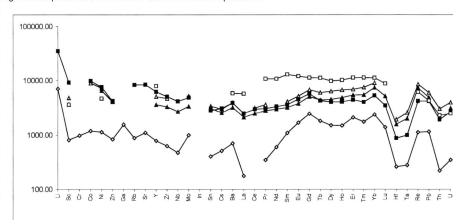
Fig. 3.3: Response curves of the ICP-MS. See text for explanation.

Alkali and alkaline earth elements (Li, Rb, Cs, Sr, Ba)

Except for Li, which did show poor reproducibility, these again seem basically reliable. There were some problems concerning calibration in the first two measuring sessions, where Li, Rb and Sr had to be calibrated on the "Cl" sltn, whose low contents result in a large, though systematic error for specimens with higher contents. Similarly Sr in measuring session 6 had to be calibrated to the Cl solution. This problem of extrapolating beyond the calibration does not concern the Rb values, which were (except for one specimen) all below the content of the Cl calibration solution.

For Sr there is a good correlation between ICP-MS and XRF values. However, there is some systematic error, ICP-MS being somewhat lower than XRF . This is due to the calibration problem stated above. Thus for the metabasalts and metagab-

34

bros the XRF-values have been chosen. As the Sr contents of the serpentinites and some other specimens (some rodingites, veins) were too low to give reliable XRF values, the ICP-MS values were adopted for these.

With Ba, most of the XRF values are below this method's l.o.d. For specimens with Ba contents high enough for reliable XRF-values the agreement between both methods is good and thus the ICP-MS values are thought to be accurate.

Rb contents were mostly below or close to both method's l.o.d. and accordingly the correlation diagram is showing much scatter. However some correlation can be observed.

The values for Cs have been discarded altogether, as they were mostly below l.o.d.

HFSE

For Zr values a distinction has to be made between specimens low in Zr, like the serpentinites, and specimens high in Zr, like the metagabbros and metabasalts. For the latter, Zr values obtained by ICP-MS are only 1/3 to 1/5 of those obtained by RFA. The possibility of defective calibration solutions can be excluded, because the response curves are smooth in the neighborhood of Zr (fig. 3.3). The remaining possibilities are problems with refractory minerals or precipitation/sorption phenomena. On the one hand, the fact that the ICP-MS measurements of the standard JB-3 were much closer to recommended values (between 60 and 76 ppm, recommended value: 98 ppm) could easily be ascribed to differences in the mineralogical composition between samples and standards. On the other hand, this seems doubtful for the metabasalts, as there has never been any zircon observed in them, not even by people explicitly searching for it (oral communication D. Rubatto, 1996). A similar discrepancy between values obtained for a standard and for the metamorphic rocks under study was observed in Himmelheber (1996).

For the serpentinites it was found that Zr values determined by XRF are not reliable, because they are too close to this method's l.o.d. However, as there was found a very good reproducibility of the ICP-MS data between the second measuring session and the sixth one, which was done over one year later, it is believed that with Zr contents below ca. 10 ppm no problems of precipitation or polymerization occurred. As it is further to be assumed that Zr is not stored in refractory minerals in the serpentinites, the ICP-MS values were chosen for this group of rocks, as well as for the rodingites and veins.

With Nb and Ta the case is somewhat different. These elements have a strong tendency for polymerization and for sticking to the sample solution conduits of the

ICP-MS apparatus. This behavior should result in memory effects (e.g. a slow decrease in counts during the three measuring runs of a blank). However, this could not be observed in the raw data.

It has been reported that Nb and Ta measurements depend on the age of the solution (Münker 1998). Usually a decrease in measured contents with age is observed, but also sometimes some kind of "recovery". As yet the factors responsible for this behavior have not been sorted out. With the solutions at hand, these effects could not be observed. Probably the Nb contents of all specimens were low enough to stay in solution. However, the ratio of Nb to Ta is not close to a chondritic ratio, except for measuring run 6 (geometric mean 3,65, chondritic value around 18). This seems due to a deficiency of the calibration solutions (cf. also the response curves fig. 3.3). The same holds for Hf. Therefore Ta and Hf values are not recorded. For Nb the ICP-MS values were chosen, as they correlate very well with XRF values, but have a lower l.o.d than the latter.

Cr, Co, Ni

These metals suffer from the calibration problems dealt with above. Therefore the correlation diagrams show a large systematic error for the measuring sessions 1 to 3, although a good correlation. A second problem lies in the high contents of some specimens that lay beyond the contents of the calibration solutions, again resulting in a systematic error. Therefore the values used are those from XRF.

All other elements measured

Ga, Cu, Sc, Zn, Mo, Sn, Pb, Th, U were measured as well (each Ga and Cu only in one of the 6 measuring sessions). Most of them show a good agreement with recommended values of the standards and also mostly good correlation with XRF-values. For Pb correlation is bad due to the low contents close to or below both method's l.o.d. As a consequence, Pb values are discarded altogether. For Sc and Zn correlations between XRF and ICP-MS is only moderate. Sc suffers from interferences by HCO_2 and $^{13}CO_2$, Zn from CrN and TiO. Therefore XRF values are used for these elements.

As no particular problems related to digestion or analytical procedure are known for Cu, Ga, Mo, Sn, Th and U, data are thought to be reliable, as far as they are not below l.o.d.

4   Statistical methods used

4.1 The log-ratio method

In this paper, some emphasis is put on a sound statistical treatment of the geochemical data. There is a ongoing debate about what methods are appropriate in this
respect, as it has been known to statisticians since Pearson (1896) that classical
tools like the product moment correlation coefficient are not suitable for closed datasets. A closed data set is one where the different properties that have been determined for the specimens sum up to 1 (or 100 %), as it is the case with geochemical
data. Standard statistical methods were developed for a sample space equivalent to
$\mathbf{R}^n$, that is, each of the n properties determined can be mapped onto the real numbers. This is not the case with closed data sets, where the mapping is only to the
interval [0,1] and, moreover, there are only n-1 independent variables, as all n variables sum up to 1. Thus by closure correlation is enforced on the data. This problem
concerns also the Spearman rank correlation coefficient.

Nevertheless frequently correlation matrices (arrays of product-moment correlation coefficients for all element or oxide pairs) are used in the geologic literature as
a means to interpret geochemical data (e.g. Beach & Tarney 1978, Kerr et al.
1999). To give some appreciation of the problems involved, following Aitchison
(1986), three difficulties associated with this approach can be distinguished:
– the "subcomposition difficulty": when only a subset of the elements/oxides analyzed is used and this is being renormalized to 100 %, as for example in the familiar petrographic triangles, the correlation coefficients will not remain the same,
they can even change their sign;
– the "negative bias difficulty": there must be at least one negative entry in every
row and column of the correlation matrix (think of a composition of only two parts:
they will always have a correlation coefficient of -1);
– the "null correlation difficulty": lack of correlation (independence of variables) is
not indicated by a correlation coefficient of zero; indeed there is now way of telling
which value would correspond to a null correlation.

As a solution to this "constant sum problem", the method of log-ratios was proposed by J. Aitchison in a series of papers (e.g. 1984), culminating in his mono-
graph (1986). In Aitchison (1997) a group-theoretical foundation for the log-ratio
method is given. In addition to the following paragraphs, the reader not willing to
read Aitchison's very mathematical papers and in search of a perhaps more acces-
sible treatment is also referred to Rollinson (1992, 1993) and Reyment (1989).

The method of log-ratios is not contained in standard statistics program packages.
There exists a computer program by Aitchison himself, available from Chapmann &

Hall, written in Basic, containing a lot of advanced features not used here. Another program comes from Marcus (1993), written in the rather inaccessible Matlab. For the purposes of the present study I have written a program in Turbopascal, later translated to VBA (to work in conjunction with Excel), available from me as source code or as compiled run-time version (Appendix 3). For very small datasets the logratio method can also be implemented on a spreadsheet, but it takes a lot of work and of memory space and zero treatment is quite tedious.

Aitchison's method hinges on the observation that the ratio of two components of a composition remains unaffected by closure. Indeed, by their property of being closed data, geochemical data are essentially ratios. Therefore it is proposed that element ratios are used as the basic entity. Because these can vary over a range of 12 and more orders of magnitude (ratios formed from contents between 100 % and <1 ppm) the ratios are logarithmized. Also by taking logarithms a certain simplicity of the overall mathematical structure is obtained (invariance of distance measures under the group of perturbations, Aitchison 1997). A further benefit of logarithmizing is that scale invariance is achieved. That is, phenomena taking place at the percent level can be meaningfully related to phenomena at the ppm level. Gaining scale invariance is another benefit of the log-ratio method, not shared by the standard correlation coefficient.

Aitchison proposes three mathematically equivalent ways for data treatment, of which conceptually the most simple one is the matrix of log-ratio variances. In this matrix for any pair of elements/oxides the value of

$$\mathrm{var}(\log(c_i / c_j)) = \sum_{k=1}^{N} \left(\ln(\frac{c_{ik}}{c_{jk}}) - \sum_{k=1}^{N} \ln(\frac{c_{ik}}{c_{jk}}) / N \right)^2 / (N-1)$$

[c_{ik}: concentration of component i in specimen k; N: number of specimens]

is given. This can be visualized as a measure of the degree to which the data points, in a plot of c_i against c_j, deviate from a straight line through the origin. Thus, a small log-ratio variance of two elements/oxides would indicate some coherence of behaviour and is somewhat equivalent to a high positive correlation coefficient.

Another way of describing the distribution of the data is the centered logratio covariance matrix. This is a covariance matrix, computed from the data after they have been subjected to the so called logistic transformation, that is, the element/oxide contents of a specimen are divided by the geometric mean of all the element/oxide contents of that specimen and the ratio is logarithmized. This covariance matrix and the logratio variance matrix can be transformed into each other and thus are mathematically equivalent (Aitchison 1986):

$$cov(c_i,c_j) = 1/2 * \left(\sum_{k=1}^{m} (var(c_i, c_k))/m + \sum_{k=1}^{m} (var(c_k, c_j))/m - \sum_{k=1}^{m}\sum_{l=1}^{m} (var(c_k, c_l))/m^2 - var(c_i, c_j) \right)$$

[m: number of elements/oxides]

The centered log-ratio covariance matrix is intuitively less appealing than the log-ratio variance matrix, as normalizing to the geometric mean of a specimen's contents of all elements/oxides doesn't seem to make any geochemical sense. Nevertheless, the two forms are mathematically equivalent, and the covariances can be used as a basis for all methods of multivariate analysis.

The third way of describing the variance structure, not used in this paper, is acchieved by dividing all element/oxide contents by the content of one element/oxide singled out and calculating the covariance matrix for the resulting ratios. Again a one to one matrix transformation is possible.

From the centered logratio covariance matrix it is obvious to arrive at the definition of a logratio correlation coefficient:

$\rho(c_i,c_j) = cov(c_i,c_j) / sqrt(cov(c_i,c_i) * cov(c_j,c_j))$.

This correlation coefficient, to my knowledge introduced to the literature by Marcus (1993), isn't mentioned by Aitchison, probably because it suffers to a certain extent from the same problems as the "crude" correlation coefficient. It is dependent on the elements/oxides selected for analysis and on their behavior (the "subcomposition difficulty") and also is subject to the "negative bias difficulty" and to the "null correlation difficulty". As can be gained from the formulas given, these difficulties are inherited from the centered logratio covariance matrix.

However, this log-ratio correlation coefficient also seems to have some advantages over the log-ratio variance as a measure of correlation: it is insensitive to degenerate cases (e.g. very small absolute variation leads to a small log-ratio variance); values range in the familiar $[-1,+1]$ interval; there are meaningful negative values; standard testing procedures are possible. Also it should be stressed that the subcompositional difficulty is also shared by other methods explicitly advocated by Aitchison, as principal component analysis (see below). In practical work, it seems that a careful selection of elements/oxides, adapted to the problem studied, leads to sensible results.

A way out of the subcompositional difficulty might be to use always the totality of all components. With trace element studies this is however obviously not possible, at least at the present state of analytical technology.

A major problem associated with the log-ratio method is that it cannot deal with zero contents, as is obvious from the above formulas. With analytic data there can be two kinds of zeros: contents below the limit of detection and contents that were not determined for one reason or another. With contents below l.o.d. some kind of replacement might be feasible, e.g. 1/2 of the l.o.d. This could be accompanied by some kind of sensitivity analysis. This solution is not possible with the other case of zeros. Therefore I have chosen in my computer program just to pass by all zeros, that is, they are just not used for calculation of averages, variances etc. in all those element pairs that are affected.

Aitchisons method has not been greeted with unanimous enthusiasm. One major critic is A. Woronow (Woronow et al 1989, Woronow 1997a, 1997b). In Woronow's opinion the fact that closed data are not independent does not pose any problems. As an example he proposes to look at a set of olivine analyses. These would give, upon standard statistical analysis, a correlation coefficient of -1 between Mg and Fe, mirroring the exchange relationship. "The prepared mind would be capable of interpreting the physical causes recorded by such data." As Woronow further justly remarks, many geochemical relationships have to do with mixing or with chemical reactions, where mass conservation holds, and these relationships are linear in percentage space, but not in the space of logistically transformed data.

Although there exist also other geochemical relationships which are not linear in percentage space, like those that have to do with the distribution of trace elements, this criticism sounds rather convincing. However, things are not quite as positive for the crude correlation coefficient and not as bad for logratios as claimed in this statement. When a set of idealized olivine compositions is subjected to statistical analysis, the crude correlation matrix is the following:

	SiO_2	MgO	FeO
SiO_2			
MgO	1		
FeO	−1	−1	

The coefficient between Mg and Fe of -1 is alright, but what about the other two correlation coefficients? They are certainly misleading, as there is no connection between SiO_2 and MgO respectively FeO provided by the model. They are caused by the slight decrease in wt % of SiO_2 which is brought about by exchanging the heavier FeO for lighter MgO (a typical effect of closure). Incidentally, using mol % instead of wt % makes those correlation coefficients = 0.

The same matrices for log-ratio variances and log-ratio correlation coefficients are the following:

40

	SiO_2	MgO	FeO		SiO_2	MgO	FeO
SiO_2				SiO_2			
MgO	0.518			MgO	−0.097		
FeO	0.518	1.909		FeO	−0.096	−0.981	

The log-ratio correlation coefficient seems to perform much better than the crude correlation coefficient, and even the log-ratio variance performs rather well: all variances are rather high, indicating that there is no positive correlation, but highest is the MgO-FeO variance, reflecting the antipathetic relationship of the two oxides.

Besides Woronow's strategy of sticking to the product-moment correlation coefficient (and his second strategy of canceling one element/oxide in an effort to circumvent closure (Woronow 1997b), not dealt with here) there have been also other proposals. Hall (1990) proposes a complicated data transformation, where first percentages are transformed to ranks and these in turn are transformed to a quantity with standard normal distribution (for those who want to try and reproduce Hall's method: the formula to calculate a specimen's "normal score" reads in EXCEL like this:

"=Norminv(Rank(<specimen>;<array of specimens>;0)/(<N>+1);0;1)").

This method certainly is scale invariant, but it has to be remarked that ranks are affected by closure. Also it can be observed that results depend slightly on the standard deviation of the normal scores that are assigned. Finally a lot of information is lost by this transformation. In the above example of olivine analyses the correlation matrix obtained by this method would be the same as with the product moment correlation coefficient.

4.2 R-mode principal component analysis (pca)

A few words of explanation for this technique may be due for the less statistically minded reader. Whereas matrices of correlation coefficients, be it product moment or log-ratio variance, although straightforward to comprehend, are in essence bivariate, principal component analysis is a truly multivariate method. Imagine a space of n dimensions, n being the number of elements /oxides under consideration. Then any of the analyzed specimens will be represented by one single point in this space, independently of whether percentages, ranks or centered logratios are used. A set of specimens then will form a cloud of points in this space. This cloud can have any shape, depending on the character of the specimens and on any transformations performed on the data (like the logistic transformation). For example, with raw percent data the cloud would be rather flat in minor oxides like P_2O_5. Statisticians of course always hope that the cloud can be described by some kind of parametric distribution, preferably the (multi-)normal distribution. In this case the cloud would have the shape of a n-dimensional ellipsoid. Principal component analysis then is a method which determines the principal axes of this ellipsoid.

Usually it is found that these axes have very different lengths, that is, the ellipsoid is rather "flat". This means that most of the variability of the specimens is contained in just a few dimensions (hence the name "principal components"). So instead of working in a space of 40 to 50 dimensions (number of oxides and trace elements typically analyzed), for most purposes a reduction to 5 or less dimensions is possible. This is to be expected, as many oxides behave more or less coherently during petrogenetic processes (think e.g. of Zr and Hf, but also the classical categories of siderophile, chalcophile and lithophile elements do embrace sets of elements behaving coherently). Of course these principal components are oblique to the coordinates of the space we started with. The new axes are described in terms of the old ones by their "factor loadings", that is, the coordinates of the unit vectors in the directions of the principal axes of the ellipsoid. The relative lengths of these axes are taken as a measure of the variability contained in this component.

As yet, there are no statistically well founded criteria as to how many dimensions should be retained. Usually a cutoff at the 5 % level is recommended (5 % of total variance), or a cutoff, when variances retained by consecutive factors do not drop any more significantly (Reyment & Jöreskog 1996)

Principal component analysis is performed by the mathematical technique of extracting the eigenvectors and eigenvalues from a covariance or a correlation matrix. In the computer program written for this study a numerical method due to Jacobi was used for this.

A major problem with pca is that it is usually very scale sensitive (think of the "flatness" of the sample cloud in minor or trace elements). Hall (1990) used principal component analysis on data transformed as described above, thus circumventing this problem. With centered logratios scale invariance is also ensured.

After reducing the dimensions sometimes a second step of rotating the axes towards a configuration where variances along the primary coordinates are maximized (varimax rotation) is performed (e.g. Hall 1990). This is supposed to shift the focus from a description of sample variability towards element behavior. In my experience however the unrotated principal components do lend themselves more easily to interpretation. This is to be expected for the petrogenetic problems under consideration in this paper. Think for example of a batch of magma undergoing differentiation. The primitive magma would be represented by one point in logistic space. Differentiation then would produce other points, thus forming a elongated cloud. Any petrogenetic process can thus be viewed as a vector operating on a starting composition, thus forming a cloud of points. The longest principal axis of this cloud would be related to the most important process of differentiation, say, fractionation of olivine. Rotating axes would only rotate them away from this main direction.

Of course there is no necessity that other petrogenetic processes operating on this suite of rocks (other minerals fractionating, later alteration) operate in a direction orthogonal to the main process and to each other. The mathematical method however enforces orthogonality on the principal components. Yet there is always some angle other than zero between different petrogenetic vectors and thus they must have orthogonal components. So by applying some petrological common sense, often a second, third... process can be disentangled from the 2nd and further principal components.

In this study, the number of analyzed elements and oxides used in pca is larger then the number of specimens. It is clear that thus the sample space already has a reduced dimensionality. I do not think, however, that this does invalidate the results obtained by this method. It is clear from the outset that the number of relevant dimensions is much smaller than the number of elements and also of specimens used. If this was the point to prove, reasoning would indeed be faulty. But if this is taken as a point of departure, one can be assured that the few (4 to 5) dimensions that are supposed to emerge, are well defined by the specimen sets used. In another study with a large set of specimens (Himmelheber and Sheraton in prep.), the components emerging were quite similar from the point of view of petrogenetic interpretation.

Another potential criticism would be with sampling. Sampling in the field was not done according to a statistical scheme, but rather having in mind a coverage of the petrographic diversity. This probably results in an overrepresentation of exotic rock types. The statistical methods used are, however, not applied with the objective of giving a well founded overview over the variability of the rocks of the ZSF and their mean composition, but rather in a heuristic way to disentangle the multitude of processes that have been operating. Under this perspective a seeming outlier can be very significant.

To help understand pca it will be applied to the olivine example from above, with the following results:

pca applied to raw data
1st component contains 100 % of total variability
(reduction from 3 to 1 dimension)
factor loadings:

SiO_2	0.144
MgO	0.624
FeO	−0.768

pca used on centered logratios

	1st component	2nd component	contains
	97.26 %	2.74 %	of total variance
factor loadings:			
SiO_2	0	0.816	
MgO	0.707	−0.408	
FeO	−0.707	−0.408	

pca used on the Hall transform of the data
1st component contains 100 % of total variability
factor loadings:

SiO_2	0.577
MgO	0.577
FeO	−0.577

In the case of logratio pca the result does not depend on whether wt % or mole % are used (scale invariance). It is seen that in this particular case raw data pca performs rather well. With log-ratio analysis the first factor gives a very clear picture, the second factor however is a bit misleading. On the other hand, it would be usually be disregarded on account of its low variability content. The reason for the appearance of this component lies in the fact that ratios change when one element (SiO_2) stays constant while another (MgO or FeO) changes.

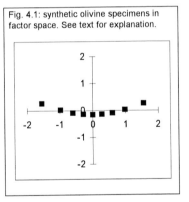

Fig. 4.1: synthetic olivine specimens in factor space. See text for explanation.

A way to judge the success of a pca is to plot the coordinates of the data points in the planes formed by any two components. For the olivine examples the result is displayed in fig 4.1. It appears that the data are certainly not normally distributed. Nevertheless it is clearly borne out that most of the variability is contained in one direction, which can be clearly related to Mg-Fe exchange by the factor loadings.

5 Geochemistry of the Serpentinites

5.1 General framework of the discussion

The geochemistry of the serpentinites is discussed in the following framework:
The protolith is a peridotite. It might be either part of former earth mantle or of cumulate origin. In the first case, the protolith might have been part of the subcontinental or of the oceanic mantle. As already mentioned in chapter 1, according to Lemoine et al. (1987) and Stampfli & Marthaler (1990), the alpine serpentinites and peridotites probably belong to a former "lithospheric ocean", formed by denudation of subcontinental lithosphere through rifting along a huge listric normal fault penetrating into the mantle, according to the model of Wernicke (1981, 1985). Geological considerations thus speak in favor of subcontinental mantle.

The second possibility, a cumulate origin, seems improbable considering the large volume of serpentinites in the ZSF (cf. fig. 1.2), but cannot be excluded a priori.

Again, the former peridotites might or might not have a magmatic relationship with the other rock types of the ZSF. According to Lemoine et al. (1987), for the Alpine ophiolites as well as for those from Corsica and the Appenines, a relationship of melt and residue is postulated for the metagabbros and the metaperidotites on field evidence, whereas the metabasalts are supposed to be of separate origin. Trommsdorff et al. (1993) have applied the same model to the Malenco area. By a careful interpretation of the field relationships they have shown that there simple shear type rifting has lead to the exposure of subcontinental mantle and to the intrusion of gabbros derived from underlying upwelling asthenosphere. Both rock types thus are not strictly consanguineous, albeit they are both coupled to one tectonic event. Basaltic extrusives found in that same area are of later origin.

Similar field evidence is lacking in the Zermatt area. The primary contact between serpentinites and metagabbros postulated by Bearth (1967), despite the admitted lack of really conclusive evidence, and construed as two horizontal sheetlike bodies seems not well documented, and at any rate is not what would be expected if a magmatic relationship exists. For the Internal Ligurides, Rampone et al. (1998), contrary to the proposal of Lemoine et al (1987), postulate a genetic link between metagabbros and metabasalts on the basis of a common isochron.

The provenance of the serpentinite specimens is briefly reviewed here, as it bears on the discussion of the geochemistry. About one half of the specimens comes from the large serpentinite body making up the Lichenbretter and the Breithorn (AS 1-96, AS 2-96, AS 3-96, AS9-96, AS 12-96, AS 16-96, AS 19-96, Hi 6-96, Hi 7-96, Hi 13-5-96, Hi 14-96, Hi 15-96). Some of these were collected close to the contacts to other units like the Garten-Riffelberg unit or the Monte-Rosa nappe. Other specimens are from the thin slivers sandwiched between metagabbroic and metabasaltic thrust sheets in the Täsch valley area and elsewhere, which have obviously served as lubricant in the imbrication process (Hi 3-96, Hi 17-96, Hi 21-96, Hi 27-3-96, Hi 28-96, Hi 28-2-96, Hi 32-9-96). One sample (Hi 1-96) is from a small (decameter) lens enclosed in calcschists.

5.2 General description of the serpentinites' geochemistry

Most major elements vary over a rather restricted range. SiO_2 ranges between 38,5 % and 41,2 % with the exception of one specimen having 44,5 % SiO_2.

The MgO contents of the majority of the specimens are lying between 38 % and 40 %. The most silica-rich sample however is as low in MgO as 29,5 %, and also some other rather SiO_2-rich specimens are slightly lower than 38 %.

Total iron (calculated as Fe_2O_3) ranges between 6 % and 10 %. The defgree of oxidation of the iron has to be compared to magnetite, as this is the mineral that stores Fe in serpentinites. With most specimens, Fe is more strongly oxidized than in mgt. In the 5 specimens AS 19-96, Hi 1-96, Hi 6-96, Hi 15-96, Hi 21-96 Hi 27-3-96 and Hi 32-9-96 the opposite is the case.

Aluminum is quite high for a serpentinite; contents between 1 % and 2 %, occasionally even almost 4 % point to a lherzolitic protolith (cf. Nicolas 1969).

All other oxides are low to very low. K_2O and Na_2O are below or close to the l.o.d. CaO varies between 0.01 % and 1.65 %. In one exceptional specimen (Hi 1-96) even 7.46 % are reached. This specimen is the same one already found above deviant with respect to its SiO_2 and MgO contents. It comes from a small lens of serpentinite enclosed in calcschists and thus surely metasomatism is responsible for its deviant character. All its other element contents however (including trace elements) are in the normal range.

MnO is very constant around 0.1 %. P_2O_5 contents are below the l.o.d, with one exception (AS 19-96) containing 0,142 %. TiO_2 varies between 0.02 % and 0.22 %.

H_2O contents are, as expected, high and range between 7.7 % and 12.2 %. Other volatiles (S, CO_2) are very low (mostly around 0.1 %).

Among the trace elements only the ferromagnesians are present in higher amounts, particularly Cr (between 1300 ppm and 4400 ppm) and Ni (between 1250 ppm and 2500 ppm). LIL elements (Sr, Ba, Rb) are very low, as are the HFSE and REE (Sr: 0 ppm to 2.2 ppm, Ba: mostly below 1 ppm, up to ~2 ppm, Rb: below l.o.d, Zr: 0.5 to 6 ppm, Nb: 0 to 0.6 ppm, REE: below 1 ppm).

5.3   REE diagrams and spidergrams

Fig 5.1 to 5.3 display some of the trace element data in the graphical form of the well-known normalized spidergrams. These diagrams have been calculated on a water-free basis. The normalization factors are to be found in Tab. 5.1. For the REE diagram, values below the l.o.d. have been arbitrarily set at 1/2 of the l.o.d. in order to obtain a better graphical representation. The l.o.d. itself is marked with a heavy line.

The curves in the REE diagram (fig. 5.1) are lacking in smoothness, due to low concentrations close to the l.o.d.. Nevertheless certain tendencies can be observed: The curves are rather flat, rising slightly from the LREE to the MREE and, even less

Table 5.1: Normalizing coefficients of the REE diagrams and spidergrams

REE diagram (after Boynton 1984)

La	Ce	Pr	Nd	Sm	Eu	Gd	Tb	Dy	Ho	Er	Tm	Yb	Lu
0.31	0.81	0.12	0.6	0.2	0.07	0.26	0.05	0.32	0.07	0.21	0.03	0.21	0.03

MORB-normalized spidergram (after Pearce 1983)

Sr	%K$_2$O	Rb	Ba	Th	Ta	Nb	Ce	%P$_2$O$_5$	Zr	Hf	Sm	%TiO$_2$	Y	Yb	Sc	Cr
120	0.15	2	20	0.2	0.18	3.5	10	0.12	90	2.4	3.3	1.5	30	3.4	40	250

Mantle-normalized spidergram (after Thompson 1982)

Ba	Rb	Th	K	Nb	Ta	La	Ce	Sr	Nd	P	Sm	Zr	Hf	Ti	Tb	Y	Tm	Yb
6900	0.35	0.04	120	0.35	0.02	0.33	0.87	11.8	0.63	46	0.2	6.84	0.2	620	0.05	2	0.03	0,22

steeply, on to the HREE. This LREE depletion is stronger with those specimens that have lower total REE contents. Thus the spread of the LREE is larger than the spread in HREE contents.

Some specimens show a negative Ce-anomaly. This is particularly evident in those specimens having low total REE contents. As some doubts remain about the reality of this anomaly, due to contents close to the l.o.d., there will be presented below (in chapter 5.5) an additional argument pointing to disturbed Ce-contents. One specimen (AS 19-96) shows a distinct negative Eu-anomaly, even though its Eu content is below the l.o.d., as its Sm_N and Gd_N are much higher than Eu's l.o.d.$_N$. This same specimen is also the only one showing a decline from Gd_N to Lu_N. Due to the high l.o.d. for Eu, only for two other specimens a conclusive statement on Eu can be made; with these no Eu anomaly is present. For all other specimens an anomaly does not seem likely. For specimen AS 19-96 there are no peculiarities of petrography or provenance to be found, which might explain its deviant character. It is found deviant also in other respects (see below).

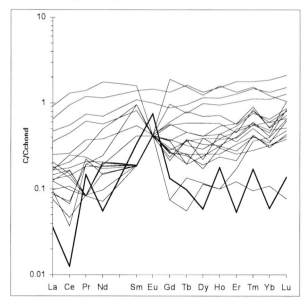

Fig. 5.1: REE diagram of serpentinites. Bold line: l.o.d. Eu values below l.o.d are set to ½ of l.o.d.

Fig. 5.2: mantle-normalized spidergram of serpentinites after Thompson (1982). Bold line: l.o.d. Values below the l.o.d are set to ½ l.o.d.

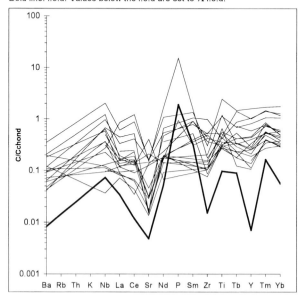

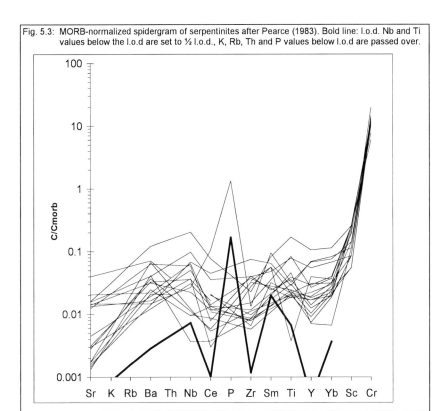

Fig. 5.3: MORB-normalized spidergram of serpentinites after Pearce (1983). Bold line: l.o.d. Nb and Ti values below the l.o.d are set to ½ l.o.d., K, Rb, Th and P values below l.o.d are passed over.

The mantle-normalized spidergram after Thompson (1982), Thompson et al. (1984) is shown in fig. 5.2. In its left part, supposed to represent the most incompatible elements, Rb, Th and K values are not represented at all, because they are below the l.o.d. The general aspect of the diagram is again, as with the REE, one of subhorizontal, subparallel curves. These curves are punctuated by some positive as well as negative peaks. Nb and Ti mostly are forming small positive anomalies. Negative anomalies are formed by Zr and, very conspicuously, by Sr. Also, at Sr the general subparallel character of the curves is most strongly disturbed. Not all specimens fit into that picture; for example, some show a positive Sr anomaly and some negative Nb and Ti anomalies. Some specimens displaying a positive Sr peak contain some amphibole, but the other peculiarities cannot be related unambiguously to any petrographic features of these specimens.

The picture is completed by the MORB-normalized spidergram after Peirce (1983). The only notable additional feature to be mentioned is the enrichment in Cr, which is quite typical for peridotites and points to some primary spinel content.

I will proceed with some preliminary interpretative efforts on these findings. The large, for the most part negative Sr anomaly is most easily explained by leaching during serpentinization. The agent for this may well have been sea water, as modern marine hydrothermal solutions usually are having Sr contents slightly in excess of the mean seawater content of around 7 - 8 ppm (Teagle et al. 1998, Charlou et al. 2000, Gieskes 1978), which shows that seawater is capable of leaching Sr from rocks. This assumption is strengthened by the correlation of Sr with Ca (fig. 5.4), an element usually lost upon serpentinization (see below).

The negative Zr anomaly as well as the positive anomalies at Nb and Ti seem not so easily explained by mobilization. For one, in contrast to Sr, these elements are usually deemed immobile. Furthermore, outside the Sr-anomaly, the curves in the spidergrams remain roughly parallel. This is also observed in the correlation diagrams (fig 5.4). Whereas the incompatible low-mobility elements Zr, Nb, Ti and the REE (with some outliers) are obviously, if loosely, correlated with one another, Sr does not show any correlation with them. Thus it seems that the contents of these elements are still close to the protolith and are governed by magmatic processes.

This might have been a melt extraction event, where Ti and perhaps Nb were buffered in the residue by some phase wherein they are compatible. If the serpentinites have been cumulates, an analogous argument applies. The differential REE depletion (La/Lu growing with increasing depletion, that is, with increasing amount of melt extraction) then would be easily explained by the presence of orthopyroxene and clinopyroxene in the residue. Garnet can be excluded, as this would result in much higher HREE values.

The Zr anomaly however is not easily explained by magmatic processes. Rampone et al. (1996) found similar Zr anomalies in peridotites from the N-Italian Internal Ligurides and attributed them to changes in the relative magnitude of partition coefficients, depending on the specific melting conditions. Whereas this seems hardly convincing, discussion is deferred chapter 9, where some magmatic modeling is performed.

What minerals might accomplish the buffering of Ti and Nb in the mantle during melt extraction? According to Green (1981), rutile might be a candidate. On the other hand, Ryerson & Watson (1987) have argued that rutile would dissolve easily in any basaltic magma extracted from such mantle. Still, this would depend on the amount of basaltic melt actually extracted. Also rutile is not known to concentrate Nb. An other candidate could be titanian clinohumite. This mineral is often found in serpentinites and peridotites, in particular also in serpentinites from the ZSF (Rahn

& Bucher 1998, Weiss 1997), and is supposed to be stable to great depths (Scambelluri et al. 1995, Okay 1993). It has also been found in kimberlites (McGetchin et al. 1970). In the case at hand, Ti-clinohumite is particularly attractive as a possible Ti-carrier as it also accommodates significant amounts of Nb and Zr, usually in the range of 5 ppm (Weiss 1997, Weiss & Müntener 1996) and around 35000 ppm Ti. A serpentinite containing 1 % of Ti-clinohumite (corresponding to ca. 0.05 % TiO_2 in the serpentinite, around 0.5 chondritic concentration) then would have Nb and Zr contents of around 0.1 and 0.01 chondritic concentration respectively. These contents compare favorably with the Nb, Zr and Ti contents of the measured serpentinites, thus also offering an explanation for the negative Zr anomaly.

However, according to Rahn & Bucher (1998), in the ZSF-serpentinites Ti-clinohumite was formed during serpentinization of the protolith in a greenschist-facies environment of ocean floor metamorphism. The Ti-clinohumite then would have accommodated Ti and HFSE from primary clinopyroxene, whereas the REE did not find any appropriate phase and were perhaps partly removed. The positive Ti and Nb anomaly then would be a relic of the primary element contents. Of course, the present Ti-clinohumite's being secondary does not exclude a primary content.

In addition to Ti-clinohumite, there exist even more phases in the serpentinites into which Ti and Nb can easily enter. According to Sokolov (1977) magnetite can contain up to 3-20 % TiO_2. Nb is preferentially contained in sheet silicates. It seems pertinent that the most Nb-rich serpentinite specimens are also rich in modal chlorite. They also deviate from the correlation line Nb-Zr towards higher Nb contents (fig 5.4).

This then would mean, that upon serpentinization, REE and perhaps Zr were leached. REE loss upon serpentinization has been argued for already by Ottonello et al. (1979) on thermodynamic grounds. Growing La/Lu with increasing depletion, as found in the specimens studied, is just what is to be expected according to these authors. There is general agreement in the literature that LREE are usually more mobile than HREE (Hajash 1984, Cullars & Medaris 1973, Mysen 1983, Flynn & Burnham 1978, Wood et al. 1976). The chlorite-rich serpentinites are those most rich in Al and thus the most fertile ones, which might explain their high Nb content.

5.4 Results of the log-ratio method 1: model

Before the results of the application of the log-ratio method to the serpentinites are discussed, a model is presented, dealing with the depletion of a 4-phase lherzolite by partial melting. The log-ratio method will then be applied to this model as a kind of calibration of the empirical findings. Table 5.2 gives the parameters of the

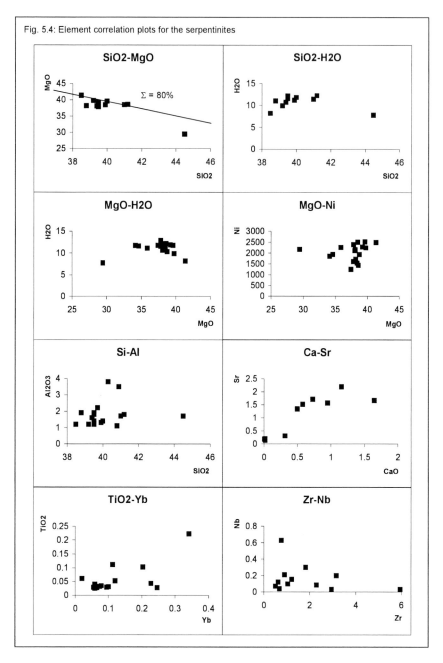

Fig. 5.4: Element correlation plots for the serpentinites

model in terms of source composition, melting modalities and distribution coefficients. The chemical composition of the model lherzolite is greatly simplified; only 4 major elements (oxides) are taken into account, viz. SiO_2, MgO, CaO and Al_2O_3. The compositions of the 4 phases olivine, orthopyroxene, clinopyroxene and spinel are idealized accordingly. In addition to these 4 oxides 3 trace elements, labeled Cr, Y, Zr, are introduced into the model as representative compatible respectively more or less incompatible elements.

Tab 5.2: Restite model parameters

phase proportions in the source (modified after Kinzler 1997)

ol	opx	cpx	spl
55	24.5	18	2.5

Phase proportions in melt (after Johnson et al. 1990)

ol	opx	cpx	spl
10	20	68	2

Distribution coefficients (after Bédard 2000)

	Ol	opx	cpx	spl
Cr	1.25	1.9	3.8	77
Y	0.007	0.2	0.412	0.002
Zr	0.02	0.021	0.26	0.015

Applying the familiar formulas for batch melting to this model results in the series of restite compositions figuring in Table 5.3. Fig. 5.5 shows how the element concentrations are developing. Applying the log-ratio method to this dataset leads to the results displayed in table 5.4. They are not significantly different when fractional melting is used for the modeling instead.

Tab 5.3: Compositions of model restites

extraction	1.0 %	2.5 %	4.0 %	5.5 %	7.0 %	8.5 %	10.0 %	11.5 %
SiO2	46.05	46.09	46.14	46.19	46.24	46.29	46.34	46.40
MgO	46.19	46.41	46.64	46.87	47.11	47.36	47.62	47.89
Al2O3	1.60	1.61	1.62	1.62	1.63	1.63	1.64	1.65
CaO	6.16	5.89	5.61	5.32	5.02	4.72	4.40	4.07
Cr	2710	2740	2771	2803	2835	2868	2901	2935
Y	2.90	2.63	2.39	2.18	2.00	1.83	1.68	1.54
Zr	6.95	5.76	4.86	4.16	3.60	3.14	2.76	2.43

Tab 5.4: log-ratio results for model restite

log-ratio variances (lower left) and log-ratio correlation coefficients (upper right)

	SiO2	MgO	Al2O3	CaO	Cr	Y	Zr
SiO2		1.000	1.000	-0.943	1.000	-1.000	-0.998
MgO	0.000		1.000	-0.944	1.000	-1.000	-0.998
Al2O3	0.000	0.000		-0.943	1.000	-1.000	-0.998
CaO	0.022	0.025	0.024		-0.945	0.944	0.922
Cr	0.001	0.000	0.000	0.030		-1.000	-0.998
Y	0.050	0.054	0.053	0.006	0.062		0.998
Zr	0.135	0.143	0.140	0.050	0.154	0.021	

Principal components

	percent of variance explained	
	99.67%	0.32%
SiO2	-0.2669	-0.1101
MgO	-0.2936	-0.1457
Al2O3	-0.2859	-0.1207
CaO	0.1255	0.8362
Cr	-0.3348	-0.1720
Y	0.3333	0.1576
Zr	0.7224	-0.4454

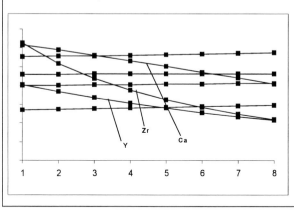

Fig. 5.5: development of element concentrations in model restite. Cf. text for detailed explanation.

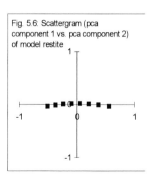

Fig. 5.6: Scattergram (pca component 1 vs. pca component 2) of model restite

According to the log-ratio variances and to the log-ratio correlation coefficients Si and Mg correlate perfectly, and Cr, due to its compatible character, joins this group of elements. This group of elements will be called the compatible elements. Also Al joins this group, as spl is going into the melt rather slowly in this model.

Ca is removed by the melting and behaves antipathetically to this group, showing very large log-ratio variances and high negative log-ratio correlation coefficients to the compatible elements.

"Y" and "Zr" both behave incompatibly, although "Y" is distinctly less incompatible than "Zr" (the difference is smaller, if fractional melting is used). This incompatible behavior is reflected in a moderately good correlation between each other and insignificant correlations with the other elements, except with Ca. The log-ratio variance of Ca with Zr however is rather high. This is caused by the stronger and not linear decrease Zr (cf. fig. 5.5).

Applying principal component analysis to these data results in two significant factors. The first one reflects the opposition between those elements concentrating in the restite and those elements most strongly removed from it: the incompatible and the compatible elements. This opposition results in positive loading on Y and particularly on the strongly incompatible Zr, as opposed to negative loadings of almost identical magnitude on the three compatible elements.

The second factor reflects the strong non-linearity of Zr's behavior, resulting in a curvature of the data point cloud in component space (fig 5.6). As the opposition between compatible and incompatible elements is already taken care of in component 1, component 2 is basically reflecting the difference of Ca's and Zr's behaviour. The data cloud's bending up on both ends does not mean that the depletion trend of Ca is somehow reversed; rather it has to be viewed as an artifact of the factorization algorithm; the true "depletion direction" would be slightly oblique to F1. It would be impossible to detect this kind of slight curvature in data from natural rocks; therefore it is not discussed any further.

The behavior of the oxides entering preferentially the melt is quite sensitive to the model parameters. When the starting amount of spinel is made smaller and the amount entering the melt made larger, Al will also enter the melt preferentially, resulting in somewhat different log-ratio results. With model parameters as proposed by Beccaluva et al. (1984), Al_2O_3 contents are even dropping much faster than CaO contents. Then Al contents drop about as fast as Y and Zr, resulting in good correlations among this group, whereas Ca joins the compatible group. These two groups are again reflected in the factor loadings of the first pca component.

5.5 Results of the log-ratio method 2: Correlation structure

How now do the log-ratio variances of the real serpentinites compare to this model? Table 5.5 contains the log-ratio variances and the log-ratio correlation coefficients. Two specimens (Hi 1-96 and AS 19-96) have been excluded from the sample as outliers (for the justification see below ch. 5.6); their inclusion however does not result in very different log-ratio variances (with the exception of the pair Ca-Sr). The ordering of the elements/oxides is different from the familiar one; it has been set up in an effort to produce coherent blocks of elements correlated with each other. Not all elements analyzed were used for the log-ratio analysis. Some elements were dropped because they contain to many values below the l.o.d.; one other (Li), because its reliability was not certain enough. For reconnaissance purposes log-ratio analysis including these elements was tried; no basically different results were obtained.

Si, Fe, Mg, Mn, Cr, Co, Ni, V, Zn and H_2O are forming one coherent block, to which also Al and S are weakly affiliated, the REE + Y are forming another one (although the correlations are impaired by the relatively large analytical error in the latter case). Ga is correlated to Sc. Ca shows the highest log-ratio variances of all elements; the only element to which it is at least loosely correlated is Sr. These findings are also confirmed by the element-element plots in fig. 5.4.

In some respects the log-ratio variance matrix compares well with the model discussed above. There are two groups of elements, compatible and incompatible ones. There are however also notable differences. The most striking one is the behavior of Ca. In the model Ca joins the compatible elements; conceivably it might also join the incompatible elements, if clinopyroxene would go much faster into the melt than spinel. In reality however it seems to belong to neither group. Also if pla-

Tab. 5.5: matrix of the serpentinites' log-ratio variances and log-ratio correlation coefficients. Upper right: log-ratio correlation coefficients; lower left: log ratios. Correlation coefficients significant at the 5 % level and log-ratios < 0.1 are in bold print.

	SiO2	MgO	H2O(t)	Cr	Co	Ni	Fe2O3(t)	MnO	Zn	V	S	CO2	Sc	Ga	Al2O3	TiO2	Zr	Y
SiO2		**0.991**	**0.962**	**0.934**	**0.909**	**0.837**	**0.931**	**0.892**	**0.864**	**0.822**	**0.755**	0.369	0.360	0.307	0.295	-0.370	-0.285	**-0.735**
MgO	**0.004**		**0.930**	**0.939**	**0.939**	**0.930**	**0.914**	**0.885**	**0.812**	**0.769**	0.409	0.326	0.281	0.198	-0.413	-0.210		**-0.718**
H2O(t)	**0.011**	**0.022**		**0.904**	**0.806**	**0.737**	**0.852**	**0.760**	**0.803**	**0.820**	**0.674**	0.250	0.445	0.450	0.400	-0.382	-0.356	**-0.655**
Cr	**0.025**	**0.022**	**0.033**		**0.866**	**0.845**	**0.938**	**0.861**	**0.895**	**0.803**	**0.658**	0.277	0.261	0.236	0.130	-0.359	-0.147	**-0.704**
Co	**0.024**	**0.019**	**0.054**	**0.046**		**0.978**	**0.855**	**0.876**	**0.843**	**0.637**	**0.823**	0.416	0.096	0.062	-0.003	-0.473	-0.157	**-0.606**
Ni	**0.047**	**0.039**	**0.079**	**0.053**	**0.007**		**0.803**	**0.814**	**0.815**	**0.551**	**0.775**	0.334	0.023	-0.127	-0.480	-0.079		**-0.529**
Fe2O3(t)	**0.024**	**0.024**	**0.048**	**0.022**	**0.047**	**0.065**		**0.928**	**0.827**	**0.858**	**0.702**	0.300	0.316	0.207	0.217	-0.202	-0.163	**-0.735**
MnO	**0.033**	**0.028**	**0.074**	**0.048**	**0.038**	**0.059**	**0.024**		**0.860**	**0.715**	**0.684**	0.386	0.152	0.088	0.101	-0.228	-0.190	**-0.702**
Zn	**0.034**	**0.035**	**0.053**	**0.038**	**0.040**	**0.052**	**0.055**	**0.042**		**0.577**	**0.537**	0.324	-0.009	0.131	0.109	**-0.523**	0.033	**-0.659**
V	**0.052**	**0.070**	**0.059**	**0.084**	**0.081**	**0.070**	**0.070**	**0.086**	**0.078**		**0.600**	0.319	**0.678**	**0.581**	0.257	-0.061	-0.312	**-0.555**
S	0.168	0.155	0.202	0.210	0.137	0.153	0.187	0.196	0.262	0.262		0.220	0.264	0.077	0.057	-0.405	-0.303	-0.403
CO2	0.592	0.571	0.669	0.668	0.564	0.619	0.649	0.587	0.615	0.616	0.829		0.068	-0.040	-0.127	-0.158	-0.186	-0.290
Sc	0.335	0.364	0.302	0.409	0.453	0.504	0.377	0.449	0.482	0.237	0.537	0.537		**0.673**	0.354	0.150	-0.315	-0.187
Ga	0.296	0.323	0.245	0.358	0.392	0.446	0.365	0.406	0.352	0.198	0.601	0.999	0.214		0.469	-0.173	-0.074	-0.110
Al2O3	0.126	0.165	0.123	0.196	0.170	0.208	0.175	0.181	0.138	**0.057**	0.390	0.756	0.316	0.224		0.087	-0.301	-0.468
TiO2	0.446	0.502	0.469	0.514	0.484	0.518	0.445	0.438	0.472	0.248	0.782	0.991	0.477	0.560	0.217		-0.312	0.158
Zr	0.671	0.676	0.725	0.670	0.615	0.602	0.669	0.667	0.515	0.530	1.011	1.286	1.010	0.741	0.515	0.780		-0.041
Y	0.565	0.613	0.563	0.646	0.528	0.536	0.644	0.609	0.515	0.338	0.783	1.091	0.658	0.531	0.310	0.331	0.828	
La	0.420	0.469	0.395	0.461	0.423	0.435	0.493	0.495	0.376	0.246	0.710	0.989	0.579	0.449	0.224	0.343	0.601	**0.089**
Ce	0.710	0.775	0.668	0.768	0.727	0.749	0.783	0.813	0.687	0.457	0.996	1.210	0.708	0.625	0.438	0.479	0.876	0.134
Pr	0.448	0.522	0.406	0.486	0.494	0.519	0.572	0.536	0.345	0.312	0.898	0.845	0.837	0.588	0.194	0.465	0.700	0.148
Nd	0.444	0.492	0.437	0.517	0.424	0.441	0.535	0.501	0.388	0.282	0.697	0.915	0.671	0.549	0.233	0.348	0.828	**0.066**
Gd	0.328	0.372	0.323	0.357	0.316	0.322	0.391	0.361	0.256	0.215	0.662	0.871	0.688	0.503	0.158	0.292	0.574	0.176
Tb	0.369	0.410	0.376	0.433	0.369	0.397	0.441	0.381	0.307	0.215	0.633	0.828	0.601	0.484	0.183	0.228	0.632	0.105
Dy	0.425	0.464	0.432	0.477	0.383	0.380	0.487	0.446	0.336	0.255	0.714	1.024	0.682	0.481	0.208	0.299	0.410	**0.055**
Ho	0.321	0.362	0.320	0.399	0.321	0.348	0.411	0.355	0.266	0.186	0.634	0.768	0.555	0.400	0.136	0.254	0.560	0.101
Er	0.372	0.411	0.378	0.439	0.341	0.348	0.445	0.406	0.313	0.209	0.618	0.921	0.567	0.430	0.173	0.276	0.461	**0.036**
Tm	0.284	0.291	0.272	0.321	0.232	0.237	0.345	0.288	0.205	0.165	0.475	0.863	0.558	0.390	0.137	0.319	0.489	0.115
Yb	0.322	0.360	0.325	0.375	0.313	0.324	0.380	0.345	0.262	0.161	0.637	0.883	0.496	0.365	0.130	0.198	0.436	**0.074**
Lu	0.263	0.284	0.302	0.309	0.230	0.240	0.306	0.244	0.194	0.147	0.538	0.690	0.558	0.453	0.128	0.213	0.366	0.193
Cu	0.449	0.477	0.470	0.512	0.521	0.600	0.515	0.454	0.405	0.441	0.745	0.993	0.920	0.709	0.348	0.740	0.814	0.937
Nb	1.084	1.170	1.056	1.146	1.108	1.120	1.147	1.181	1.017	0.712	1.516	1.873	0.920	0.834	0.539	0.541	0.577	0.374
Ba	0.287	0.313	0.329	0.358	0.282	0.287	0.362	0.271	0.233	0.305	0.560	0.826	0.641	0.649	0.247	0.498	0.444	0.635
CaO	3.334	3.421	3.480	3.485	3.237	3.247	3.228	3.231	3.317	2.884	3.559	4.401	3.323	3.351	2.607	2.026	2.137	2.675
Sr	1.025	1.101	1.076	1.213	0.995	1.035	1.088	1.091	1.146	0.775	1.152	1.589	0.949	1.094	0.593	0.388	0.985	0.818

gioclase was the aluminous phase and not spinel, Ca would behave like a compatible element.

Two other things are also surprising about the observed correlation structure:
- the REE are not affiliated with other incompatible elements, such as Zr or Ti (except perhaps to Nb, according to the log-ratio correlation coefficients), and
- the compatible (+ Si) group contains also S and H_2O.

The decoupling between the REE and Zr + Ti has already been observed above in the discussion of the spidergrams. This strengthens the idea that REE on the one side, Zr and Ti on the other side, are governed by different kinds of geochemical processes.

The connection between H_2O, S and the compatible elements suggested by the log-ratio variances deserves closer attention. Fig. 5.4 gives some element-element plots. According to these H_2O increases with SiO_2 and decreases with MgO (the outlier specimen is again Hi 1-96). This Can be related to the serpentinization reaction. Low-grade serpentinite can be either one of the parageneses brucite-chrysotile or talc-chrysotile, depending on the protolith's Si/Mg ratio (Bucher & Frey 1994). As the olivine, contained in the majority of the specimens, is of secondary origin (Bearth 1967, Ganguin 1988; it was formed by a prograde reaction from brucite, Bucher & Frey 1994), most specimens primarily must have had the paragenesis brucite - chrysotile. This is also evident from the low number of specimens contain-

Tab. 5.5 continued

	La	Ce	Pr	Nd	Gd	Tb	Dy	Ho	Er	Tm	Yb	Lu	Cu	Nb	Ba	CaO	Sr
SiO2	-0.570	-0.553	-0.487	-0.636	-0.467	-0.536	-0.756	-0.595	-0.773	-0.408	-0.785	-0.560	0.206	-0.890	0.207	-0.526	-0.433
MgO	-0.584	-0.579	-0.579	-0.640	-0.481	-0.533	-0.723	-0.596	-0.742	-0.360	-0.763	-0.466	0.188	-0.907	0.193	-0.521	-0.465
H2O(t)	-0.407	-0.404	-0.288	-0.535	-0.363	-0.489	-0.696	-0.503	-0.700	-0.361	-0.696	-0.697	0.180	-0.773	0.118	-0.610	-0.475
Cr	-0.453	-0.485	-0.376	-0.609	-0.374	-0.502	-0.645	-0.623	-0.721	-0.452	-0.678	-0.452	0.153	-0.774	0.126	-0.519	-0.569
Co	-0.564	-0.582	-0.624	-0.547	-0.396	-0.519	-0.564	-0.575	-0.602	-0.206	-0.713	-0.329	0.057	-0.922	0.226	-0.435	-0.374
Ni	-0.496	-0.545	-0.596	-0.497	-0.309	-0.514	-0.436	-0.570	-0.501	-0.115	-0.610	-0.248	-0.062	-0.845	0.251	-0.398	-0.378
Fe2O3(t)	-0.591	-0.542	-0.657	-0.705	-0.483	-0.573	-0.727	-0.728	-0.796	-0.537	-0.758	-0.491	0.135	-0.807	0.100	-0.345	-0.407
MnO	-0.665	-0.656	-0.613	-0.663	-0.431	-0.419	-0.651	-0.557	-0.713	-0.327	-0.673	-0.223	0.235	-0.922	0.306	-0.371	-0.448
Zn	-0.483	-0.563	-0.200	-0.510	-0.214	-0.356	-0.469	-0.408	-0.586	-0.164	-0.551	-0.214	0.287	-0.832	0.340	-0.553	-0.705
V	-0.457	-0.303	-0.595	-0.664	-0.649	-0.484	-0.760	-0.659	-0.745	-0.643	-0.669	-0.707	-0.025	-0.606	-0.189	-0.387	-0.283
S	-0.459	-0.400	-0.715	-0.414	-0.539	-0.389	-0.581	-0.588	-0.487	-0.155	-0.741	-0.484	0.047	-0.760	0.061	-0.308	-0.145
CO2	-0.280	-0.191	0.010	-0.150	-0.190	-0.070	-0.421	-0.032	-0.336	-0.299	-0.374	0.103	0.096	-0.596	0.104	-0.425	-0.201
Sc	-0.184	-0.006	-0.614	-0.378	-0.636	-0.331	-0.524	-0.372	-0.364	-0.464	-0.280	-0.597	-0.188	-0.072	-0.093	-0.192	0.056
Ga	-0.078	0.002	-0.306	-0.317	-0.263	-0.247	-0.156	-0.217	-0.181	-0.108	-0.555	-0.018	-0.072	-0.276	-0.280		-0.199
Al2O3	-0.383	-0.271	0.078	-0.414	-0.261	-0.318	-0.495	-0.283	-0.530	-0.458	-0.444	-0.603	0.419	-0.025	0.086	0.030	0.280
TiO2	-0.031	0.117	-0.269	-0.036	-0.009	0.263	0.027	0.061	0.006	-0.287	0.259	0.152	-0.229	0.244	-0.177	0.534	0.634
Zr	-0.121	-0.167	-0.229	-0.166	-0.185	-0.273	0.256	-0.229	0.078	-0.066	0.091	0.320	0.009	0.364	0.314	0.419	0.063
Y	0.746	0.781	0.599	0.817	0.422	0.687	0.855	0.706	0.958	0.652	0.851	0.263	-0.572	0.513	-0.498	0.078	0.029
La		0.957	0.824	0.900	0.694	0.527	0.636	0.678	0.735	0.380	0.640	-0.090	-0.256	0.646	-0.293	-0.199	-0.249
Ce	0.066		0.695	0.861	0.534	0.519	0.562	0.620	0.705	0.245	0.606	-0.157	-0.342	0.604	-0.723	-0.130	-0.169
Pr	0.056	0.178		0.931	0.946	0.730	0.599	0.979	0.712	0.577	0.733	0.489	0.480	0.577	0.227	-0.440	-0.512
Nd	0.028	0.106	0.023		0.701	0.703	0.709	0.841	0.842	0.561	0.671	0.141	0.509		-0.192	-0.160	-0.165
Gd	0.073	0.246	0.025	0.073		0.503	0.600	0.657	0.536	0.481	0.552	0.253	0.021	0.468	0.294	-0.191	-0.207
Tb	0.117	0.252	0.081	0.075	0.101		0.573	0.832	0.659	0.612	0.681	0.439	-0.066	0.196	-0.023	-0.142	-0.175
Dy	0.091	0.235	0.117	0.074	0.082	0.094		0.666	0.940	0.747	0.912	0.478	-0.476	0.592	-0.071	0.166	-0.040
Ho	0.074	0.220	0.025	0.043	0.058	0.034	0.064		0.756		0.750	0.347	-0.041	0.384		-0.227	-0.192
Er	0.063	0.189	0.084	0.042	0.081	0.066	0.014	0.038		0.764	0.890	0.383	-0.484	0.599	-0.135	0.059	-0.038
Tm	0.128	0.333	0.114	0.097	0.082	0.070	0.049	0.049	0.035		0.651	0.453	-0.236	0.273	-0.130	-0.206	-0.221
Yb	0.081	0.231	0.084	0.077	0.070	0.059	0.023	0.033	0.017	0.041		0.486		0.627	0.010	0.101	-0.038
Lu	0.196	0.427	0.134	0.163	0.106	0.092	0.088	0.078	0.080	0.058	0.052		-0.041	0.306	0.575	0.279	0.131
Cu	0.672	1.015	0.329	0.650	0.499	0.551	0.728	0.501	0.673	0.549	0.614	0.470		-0.059	0.137	-0.059	-0.171
Nb	0.297	0.342	0.332	0.369	0.389	0.515	0.330	0.432	0.338	0.463	0.344	0.451	0.964		-0.232	0.437	0.412
Ba	0.466	0.974	0.309	0.436	0.235	0.343	0.361	0.267	0.345	0.196	0.281	0.156	0.556	0.880		-0.331	-0.183
CaO	2.963	3.178	3.346	2.922	2.872	2.849	2.523	2.859	2.617	2.813	2.568	2.444	3.126	2.095	3.324		0.896
Sr	0.926	1.144	1.152	0.881	0.837	0.844	0.774	0.798	0.741	0.789	0.712	0.642	1.232	0.674	1.009	0.920	

ing talc. Now the more olivine-rich protoliths will form more brucite during serpentinization. The resulting rocks would have MgO rising with H_2O and SiO_2 decreasing. If the brucite is changed back into olivine during prograde metamophism, the observed relationships are produced.

Considering the joining of S to the ferromagnesian (+ Si) elements, it is to be noted that in the case under study the log-ratio method is impaired by the very low spread of the S contents. It is clear that with low spread of raw contents also the ratios will show only limited variation.

The relationship between SiO_2 and MgO, although showing good correlation according to the log-ratio variances, differs from the model studied above insofar as the element-element plot discloses an antipathetic relationship between these oxides. This, however, is also not necessarily very significant, as the sum of both oxides accounts for ca. 80 % of the total mass, and thus the unit-sum constraint enforces such a relationship. The good correlation according to the log-ratios is thus rather a proof for the superiority of the log-ratio method over crude statistical analysis.

Ce is not very well correlated to the other REE and to Y, which can be taken as a sign of its higher mobility. This difference to the other REEs is retained even if those specimens having a negative Ce-anomaly in the REE diagram are dropped from the statistical analysis. The fact that thus mobility of Ce can be demonstrated even in those specimens lacking a distinct Ce anomaly serves to lend additional credibility to the reality of the negative Ce anomaly in those other specimens.

To sum up, the log-ratio variance matrix is on the whole similar to the model, showing two large groups of elements: the compatible elements and the REE. Deviating from the model is the fact that Zr and Ti are not correlated to the REE and that Ca is not contained in either group.

5.6   Results of the log-ratio method: principal component analysis

Besides the log-ratio variance matrix a principal component analysis was performed on the serpentinite data. Upon scrutiny of the scattergrams it turned out that the specimens As 19-96 and Hi 1-96 are probably outliers (as could be already suspected from the discussion of the spidergrams and of the general chemistry; with Hi 1-96 this was already explained above by its provenance) and thus the analysis was repeated with these specimens omitted. The results presented here are those of this second analysis; the differences however are only minor.

Table 5.6 gives the factor loadings of the first 6 factors. Fig 5.7 contains some of the scattergrams appertaining. According to the scattergrams the pca can be judged successful; the point clouds are reasonably elliptic, i.e. the distribution is reasonably close to the logistic normal distribution.

The first component, responsible for ca. 44 % of the total variance, is dominated by a very high positive loading on CaO. Subordinate positive loadings are on TiO_2, Zr, Sr and on Nb. The negative loadings are distributed rather evenly between the volatiles, the ferromagnesian elements and SiO_2 and are consequently of much less size. This latter group is largely identical to the group of compatible elements already sorted out above by their good correlations, which is to be expected from this method: as these elements are well correlated, no major component of variance can exist between members of this group.

The second component, accounting for ca. 23 % of total variance, is characterized by the opposition between Y and the REE (particularly the LREE), joined by Nb, on the one hand and another group of elements, comprising the compatible and major phase elements SiO_2, $Fe_2O_3(t)$, MnO, MgO, Cr, Co, Ni, plus CaO plus Sr plus S. The loadings of this latter group are of similar size except the one on CaO, which is much larger.

Tab. 5.6: Pca-factor loadings for the serpentinites. It is suggested to color the positive loadings > 0.1 in red, the negative loadings < -0.1 in blue in order to obtain a clearer picture

	43.92 %	23.26 %	8.85 %	6.87 %	5.70 %	4.54 %
SiO2	-0.1420	0.1317	0.0297	-0.0458	0.0008	-0.0630
TiO2	0.1172	0.0355	0.1635	0.2039	0.0895	-0.1052
Al2O3	-0.0044	0.0376	0.0439	-0.0457	0.1684	-0.0571
Fe2O3(t)	-0.1354	0.1752	0.0294	-0.0719	-0.0484	-0.0652
MnO	-0.1356	0.1616	-0.0586	-0.0062	-0.0623	-0.0843
MgO	-0.1577	0.1486	0.0137	-0.0460	-0.0360	-0.0327
CaO	0.7132	0.4210	-0.0696	0.0234	-0.0621	-0.0905
H2O(t).	-0.1511	0.0989	0.0842	-0.1009	0.0415	-0.0858
S	-0.1649	0.2162	0.1130	0.0074	-0.2318	-0.2585
CO2	-0.2098	0.0722	-0.1148	0.7183	0.1617	0.4542
Sc	-0.0704	0.0960	0.5111	-0.1051	0.2251	0.2582
V	-0.0610	0.0674	0.1075	-0.0266	0.0294	0.0222
Cr	-0.1613	0.1303	-0.0003	-0.1204	-0.0500	-0.0650
Co	-0.1295	0.1366	-0.0342	0.0084	-0.1561	-0.0542
Ni	-0.1277	0.1331	-0.0522	-0.0139	-0.2396	-0.0473
Zn	-0.1320	0.0826	-0.1202	-0.0959	-0.0565	-0.0211
Ga	-0.0737	0.0199	0.3277	-0.2716	0.2119	0.2576
Sr	0.3166	0.2227	0.1646	0.2697	0.1049	-0.0115
Y	0.0842	-0.2201	0.1255	0.0755	-0.2317	-0.0077
Zr	0.1164	0.0579	-0.3437	-0.3418	-0.1536	0.5663
La	0.0283	-0.2198	0.0705	-0.0311	-0.0389	-0.0541
Ce	0.0563	-0.3184	0.2534	0.0351	-0.0473	-0.0601
Pr	-0.0110	-0.2809	-0.1931	0.0350	0.2068	-0.1244
Nd	0.0345	-0.2224	-0.0038	0.0864	-0.0707	-0.1031
Gd	0.0145	-0.1516	-0.1133	0.0227	-0.0032	-0.1356
Tb	0.0196	-0.1527	-0.0277	0.1249	-0.0243	-0.1184
Dy	0.0733	-0.1547	-0.0523	-0.0328	-0.2099	0.0266
Ho	0.0147	-0.1544	-0.0396	0.0865	0.0248	-0.0322
Er	0.0540	-0.1515	0.0033	0.0138	-0.1515	0.0122
Tm	0.0050	-0.1007	-0.0635	-0.0236	-0.1345	-0.0536
Yb	0.0491	-0.1203	0.0020	-0.0042	-0.0777	0.0459
Lu	0.0376	-0.0285	-0.1497	0.0799	-0.0370	0.0752
Nb	0.2538	-0.2484	0.0486	-0.2386	0.2219	0.1372
Ba	-0.0719	0.0167	-0.3079	-0.0565	0.0009	0.1743
Cu	-0.0488	0.0629	-0.3473	-0.1127	0.6356	-0.2993

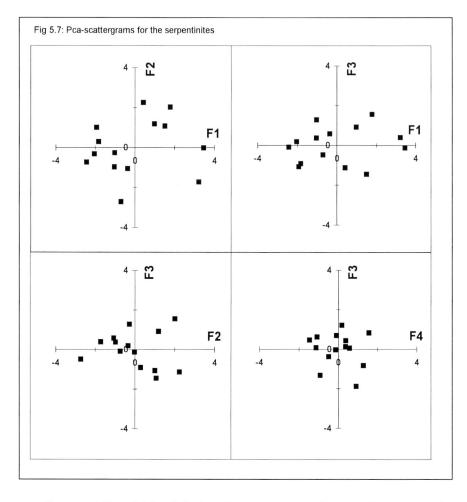

Fig 5.7: Pca-scattergrams for the serpentinites

Component 3, explaining 9 % of total variance, is a combination of negative loadings on Zr, Ba, Cu and positive loadings on Sc, Ga and perhaps Ce. Some of these elements did not appear in the previous components. The fact that Sc and Ga show the same sign on their loadings seems geochemically significant and can be related to their both entering preferentially chlorite, (cf. Burton & Culkin 1972, Frondel 1970), and also to these elements' behavior being independent from the element groups figuring in component 1 and 2.

Component 4, accounting for ca. 7 % total variance, is viewed as a single-element factor of CO_2, despite minor loadings on TiO_2, Sc, Ga, Zr and Nb. These loadings are probably caused by the orthogonality constraint of the procedure.

Component 5 is a single-element factor significantly loaded only on Cu. The factor has no parallel in the model, but this can not be expected, as the model did not contain any siderophile elements. It might have been expected however that Cu has some relationship to other elements, for example to S. Yet this is not the case.

It should be pointed out that Al_2O_3 is not loaded significantly on any of the components.

On the whole, only the second component behaves as predicted in the model presented above. The behavior of CO_2 and S is probably not very significant, taking into consideration the very low contents of these two elements. The joining of H_2O to the compatible elements, particularly to MgO, was already discussed above. Certain points however do merit special attention:

– V is not significantly loaded on any of the components; moreover, it is not linked to TiO_2 in its behavior;
– the incompatible elements Zr, Ti and partly also Nb behave differently from the REE.

As already stated above in the paragraphs on the general framework, there are two major groups of processes expected to control the chemical composition of the serpentinites: magmatic processes, namely melt-residue or melt-cumulate relationships, and metasomatic processes, in particular the serpentinization. As the behavior of Zr, TiO_2 and Nb is different from the behavior of the REE + Y, an explanation for this has to be found in one of these different kinds of processes. Magmatically, this could have been accomplished by some kind of buffering. Alternatively, Ti, Nb and Zr contents could be governed by magmatic processes, whereas the REE contents are overprinted by metasomatism. To see whether this latter hypothesis is compatible with the results of the pca presented above some discussion of the serpentinization process is necessary.

5.7 Serpentinization and the log-ratio results

From stable isotope studies water/rock ratios around 0.2 have been calculated for serpentinization processes (see e.g. Burkhard & O'Neill 1988). This is not so much larger than the minimum of ca. 11 % needed on mineral chemistry grounds. Yet whether serpentinization is an isochemical process should not be taken for granted.

Janetzky & Seyfried (1986) performed hydrothermal experiments with mixtures of olivine, enstatite, diopside and spinel at 200-300 °C and 500 bar. They used standard seawater at water/rock ratios of 10 and 30. The resulting minerals were serpentine-minerals + anhydrite ± Mg-hydroxidsulfatehydrate ± magnetite ± brucite ± tremolite-actinolite. When their data are analyzed by Grant's (1986) isocon method (fig 5.8), it is found that a straight line passes through the data points of MgO and FeO. The slope of this line is almost exactly as is it would be if neither FeO nor MgO are added or removed, but only diluted by addition of water. SiO_2 is slightly enriched, but falls almost on the same line. All other elements are removed, particularly Na_2O and CaO; their mobility is also attested by the spread between the three experimental runs. Calcium is lost even though it can be accommodated in the amphibole. The authors did not try a mass balance, but the SiO_2 gain seems not explicable. That the gain is only apparent and all other elements have been lost does not seem probable in view of the small spread of the MgO and FeO values.

Besides the element losses mentioned, variable oxidation of Fe is observed. There

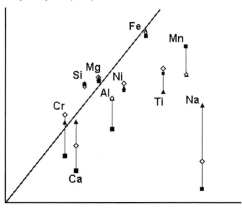

Fig 5.8:
Isocon diagrams after Grant (1986) for serpentinization experiments of Janetzky & Seyfried (1986)

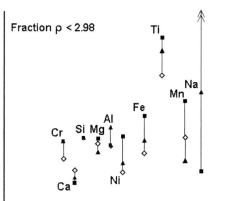

Fraction ρ < 2.98

▲ ◇ ■ 3 different experimental runs

are no Fe oxidation state data for the starting materials; however, $Fe_2O_3/(Fe_2O_3+FeO)$ is much higher in the alteration products (0.2 to 0.7) than seems likely for unaltered peridotites. According to the authors Fe is more strongly oxidized under alkaline conditions than at low pH. Oxidation is effected through O_2 and through reduction of SO_4^{--}, both contained in the seawater (Seyfried & Dibble 1980).

Under natural conditions, anhydrite and the $Mg-Si-SO_4$-compound would probably be lost, resulting in an even larger loss of CaO and partly of MgO. According to Seyfried & Dibble (1980) Mg-hydroxysulfate is precipitated from seawater at elevated temperatures and redissolved at T below 250 °C. Janetzky and Seyfried also analyzed the fraction of the alteration products specifically lighter than 2.98 g/cm^3, supposed to consist mainly of serpentine group minerals. Here also SiO_2 and MgO display the smallest spread, underlining their small mobility. Cr and TiO_2 are enriched, Na_2O is in one specimen extremely enriched. FeO is depleted, as it is not taken up easily by serpentine minerals and forms magnetite instead.

According to Seyfried & Dibble (1980), seawater is strongly undersaturated in Mg at 300 °C, whereas it is over-saturated at lower temperatures. This implies that gain or loss of Mg is dependent on the temperature.

Field studies came to diverging results as to the chemical changes associated with serpentinization. Komor et al. (1985a) in a careful investigation of the serpentinization of a wherlite-dunite layered ultramafic body came to the conclusion that within the marges of error of the method no metasomatism can be proved. The 1σ values for their method are given as 3 % (SiO_2, MgO), 1 % (CaO) and 0.6 % (Al_2O_3, FeO). They have no data on the oxidation state of Fe. From their data as given in Komor et al (1985b), an average MgO/SiO_2 ratio of the dunites of 1,13 ± 0,06 (N = 8) can be calculated. As the standard deviation is quite small, this also lends support to the assumption that MgO and SiO_2 are not mobilized. The much larger σ for the former wehrlites is surely primary.

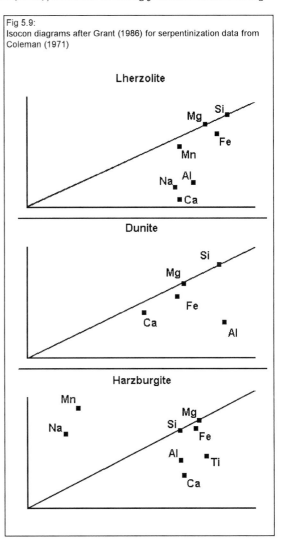

Fig 5.9:
Isocon diagrams after Grant (1986) for serpentinization data from Coleman (1971)

Lherzolite

Dunite

Harzburgite

Coleman (1971) investigated the serpentinization of three different protoliths, a lherzolite, a dunite and a harzburgite, using data from Nicolas (1969) and Coleman & Keith (1971). Applying again Grant's (1986) isocon diagram (fig. 5.9), it is found for two of the three investigated cases (the third case – a chrysotile vein in a harzburgite – is somewhat untypical) that MgO/SiO_2 ratios have remained constant during serpentinization. If it is assumed that MgO contents have remained constant, as seems sensible due to the presence of modal brucite, the constancy of SiO_2 is implied and it is further found that serpentinization was accompanied by a volume increase. It then also follows from the isocon diagram, that Fe and Mn have been slightly, Na, Ca, Al strongly removed. The dunite shows less Ca loss than the lherzolite.

Nicolas (1969) interpreted his data in much the same way. MgO is added or removed in the serpentine-mineral forming reaction (not necessarily to the whole rock) depending on whether there is more olivine than pyroxene in the precursor rock or the other way round. He proposes the theory that in serpentinization the framework of the SiO_4 tetrahedra is transferred without much destruction from the precursor minerals, whereas all other elements are only accommodated as they fit into the newly forming crystal structure. Elements left over are either removed or will form minerals of their own.

According to Coleman (1971) and Nicolas (1969), Fe is strongly oxidized during serpentinization, $Fe_2O_3/(Fe_2O_3+FeO)$ being changed from 0.12 resp. 0.42 to 0.47 resp. 0.67).

To sum up, the views are rather diverging, but there is a certain agreement that Ca and Mg behave antipathetically, just as they would do during melt extraction.

How can these findings be applied to the serpentinites studied here? Firstly, it is found that many specimens contain forsterite, which is formed in serpentinites during prograde metamorphism (see above), when the paragenesis chrysotile/lizardite + brucite changes to antigorite + forsterite. So removal of primary MgO can not have been very pronounced. Secondly, the MgO/SiO_2 ratio of the studied serpentinites is remarkably constant, as already pointed out above in the discussion of the log-ratio variance matrix. It lies between 0.98 and 1.18 for all but two specimens, compatible with the value of 1.13 reported in Nicolas (1969). Even those specimens which were not completely serpentinized, as is attested by their relictic pyroxenes (AS 1-96, AS 9-96, Hi 21-96), have the same MgO/SiO_2 ratio. The only exception is specimen Hi 1-96, already found above to have been metasomatized by the enclosing calcschists. Thus MgO and SiO_2 contents may well be close to their primary values.

CaO contents are, for most specimens, much lower than is usual for a lherzolite. A harzburgitic protolith, on the other hand, is not possible in view of the Al contents and of the relics of cpx. Maaløe & Aoki (1977) report average CaO contents of 1.92

% for a spinel lherzolite, 0.80 % for a garnet lherzolite, and 2.42 % for primitive mantle. The average of the specimens studied here (excepting Hi 1-96) is 0.38 %; none of the specimens even reaches the value of 1.92 %. So it seems likely that Ca was lost during serpentinization.

So the first factor of the pca might well be interpreted as due to metasomatism. Ca, the element most strongly lost, is opposed to almost all other elements, which have remained constant. The REE have perhaps been slightly reduced, leading to intermediate factor loadings. Of course, magmatic processes and metasomatic processes, operating in similar directions, might also very well be both contributing to component 1.

5.8 Log-ratio analysis and the oxidation state of Fe

Before pursuing these questions of interpretation any further, the relationship between the oxidation state of Fe and the geochemical composition of the specimens as a whole will be studied. With the log-ratio method, FeO and Fe_2O_3 cannot be treated as geochemical components (elements, oxides) in the way the other elements are treated. They are not components in the canonical sense, because the one can be transformed into the other. Fe^{II} and Fe^{III} can never be independent of each other in the way that e.g. Al_2O_3 can be independent of MgO. It is, however, possible to relate the Fe^{II}/Fe^{III} ratio to the geochemical composition of the specimens as a whole. As a simple way to do this it is proposed here to try a multidimensional least squares regression of $\ln(Fe_2/Fe_3)$ to the significant factors of the pca. This means that the values of $\ln(Fe_2/Fe_3)$ are approximated by a hyperplane in principal component space.

Logarithmized ratios were used rather than the ratios themselves because this is more in keeping with the log-ratio method. Also, logarithmizing the ratios results in more symmetric relationships between ratios larger than 1 and those < 1. Besides, using raw ratios does not lead to significantly different results.

The result of the least squares regression is the following:

$\ln(Fe_2/Fe_3) = -0.75 + 0.29\ F1 - 0.13\ F2 - 0.18\ F3 + 0.26\ F4 + 0.22\ F5 + res.Error.$

Considering the different "length" of the principal components (the major axes of the 1σ-ellipsoid in component space are proportional to the square root of the percentages of variance explained) it becomes clear that the only significant dependence of the oxidation state is on component 1. The same is borne out by the F-statistics of

Table 5.7: Regression parameters

m5	m4	m3	m2	m1	b
0.217	0.258	-0.180	-0.129	0.289	-0.747

σ5	σ4	σ3	σ2	σ1	σb
0.139	0.152	0.144	0.094	0.070	0.125

t5	t4	t3	t2	t1	critical value at 5 % two sided
1.563	1.696	-1.245	-1.372	4.109	2.31

r^2	error of y
0.732	0.468

F	degrees of freedom	critical value at 5 %
4.381	8	3.33

Sum of squares

regr.	res.
4.789	1.749

the parameters (table 5.7). That is, iron gets more oxidized with decreasing Ca-contents and with increasing ferromagnesian elements.

The measure of definiteness of the 5-factor regression is 0.73, which is significant according to the F-statistic. The same holds for the r^2 of the 1-component regression of 0.41.

The relatively low r^2 shows three things:

1. oxidation state is not very well predicted from the chemical composition of the specimens.
2. yet some dependence seems to exist.
3. prediction from F1 is not much worse than prediction from F1 .. F5

In fig. 5.10 the coordinates in F1 are plotted against $\ln(Fe^{II}/Fe^{III})$.

As for the success of the regression as a whole, it has to be taken into account that a) the FeO-determinations are less precise than the other measurements, and b) iron is easily variably oxidized by many late stage processes that do not leave a distinct trace on chemical composition. Thus on the whole the regression analysis can be considered to be successful.

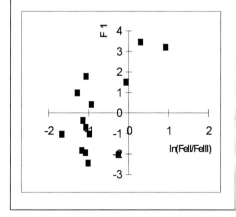

Fig. 5.10: Plot of 1st component of pca against $\ln(Fe^{II}/Fe^{III})$. Explanation see text.

How can this information be integrated into the interpretation of the pca and of the chemistry of the specimens in general? If component 1 (F1 henceforth) was the expression of some melt-residue relationship, the direction of decreasing Ca contents would be the direction of increasing melt extraction. The same would hold for a melt-cumulate relationship. How will Fe^{II}/Fe^{III} behave in a melting peridotite? To discuss this, the phase diagram SiO_2 - FeO - Fe_2O_3 (fig. 5.11) will be taken as an analogue of a spinel peridotite. A system fayalite-magnetite will produce a melt enriched in Fe_2O_3 under conditions of constant bulk composition (no oxygen added to or removed from the system) as long as the magnetite content is low. So a restite's Fe^{II}/Fe^{III} ratio should increase with increasing melt extraction. This not only follows from phase relationships, but it is also a common experience with magmatic series, that the more differentiated rocks have lower Fe^{II}/Fe^{III}. This is at variance with an interpretation of F1 as caused by melt extraction. It is, on the contrary, very well compatible with an interpretation of F1 as caused by serpentinization, as it was found above that this process is concomitant with iron oxidation.

As a check to this, the relationships between petrography, Fe^{II}/Fe^{III} ratio and pca are considered. Firstly a strong link is found between the presence of relictic pyroxene (a measure of completeness of serpentinization) and the Fe^{II}/Fe^{III} ratio. Of the five specimens with a $\ln(Fe^{II}/Fe^{III})$ significantly larger than 1, three are found to contain relictic pyroxene. This confirms that Fe^{II}/Fe^{III} can be used as a measure of completeness of serpentinization. Secondly, there is also a strong link between the presence of px and high scores on F1 in the pca. Of 6 specimens with a positive score on F1, 4 do contain px.

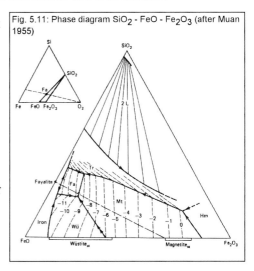

Fig. 5.11: Phase diagram SiO_2 - FeO - Fe_2O_3 (after Muan 1955)

There are only two further specimens known to contain relictic px, but these were left out of the pca due to incomplete analytical data. To sum up, a strong link between F1 and serpentinization seems to exist.

As it has been stated already above that both serpentinization and magmatic processes tend to work in the same direction, it is likely that F1 contains the effect of both processes. In this perspective, the loadings on Ti, Nb and Zr are probably caused by melt extraction, as are the negative loadings on the ferromagnesians, whereas the loadings on Ca and Sr are caused by the combined effect of melt extraction and serpentinization.

This interpretation implies that those specimens with positive scores on F1 are closest to primary composition, both from a magmatic and from an alteration point of view. It is interesting to note that the "immobile" trace element contents of these specimens (AS 1-96, AS 3-96, AS 9-96, Hi 21-96, Hi 27-3-96, Hi 32-9-96, to which, on account of its px and high FeO and CaO contents, can be added specimen Hi 15-96, left out in the pca) are spanning the whole range of the element contents encountered. This means that low REE contents can not be solely due to leaching during serpentinization.

5.9   Preliminary Summary

The first point to be settled is the question, whether the protolith of the serpentinites was a restite or a cumulate. The contents of incompatible trace elements are generally in the range of 0.001 to 1 times chondritic, and thus very low. A cumulate, developing from some melt having incompatible trace element contents high above mantle concentrations, would be supposed to have contents not lower than the mantle, in particular as it would contain some trapped interstitial melt. So those serpentinites with very low incompatible trace element contents can surely not be cumulates.

Furthermore, a cumulate would be supposed to have a more extreme modal composition, leading to a peculiar trace element signature. Now most of the more enriched serpentinites have the same trace element patterns as the depleted ones. The only candidate for a cumulate might be specimen AS 19-96 with its relatively high REE-contents, REE-pattern falling from Sm to Lu, negative Eu-anomaly and high contents of some other incompatible elements, particularly P. Zr, however, is very low. This specimen comes from the large Lichenbretter serpentinite body. It is difficult to envisage how a cumulate could be found within such a large body of restites.

A major problem in interpreting the serpentinite geochemistry is that the two major geochemical processes, mantle partial melting and restite formation, and serpentinization and related metasomatism, are both producing similar effects. A restite will be depleted in Ca and Al, and those elements will also be lost upon serpentinization. Also the differential depletion in REE is mimicked by the effects of REE mobilization, which can conceivably have taken place during serpentinization (Ottonello et al. 1979), and which would have affected the LREE more strongly than the HREE.

Indeed it is believed that the REE contents have been affected by both processes. The patterns are probably shaped mainly by the restite formation, as it is not the case that the specimens not thoroughly serpentinized are generally having higher REE contents. So the general shape of the REE pattern is thought to be caused by magmatic processes. It is also thought that some mobilization of REE has taken place. This is argued for by the negative Ce anomaly and by the intermediate factor loadings on the REE in the first (i.e. serpentinization) component of the pca.

From this it follows that the restite contained clinopyroxene, as the REE curves are generally rising from LREE to HREE. The relative fertility of the restite is also attested by its high Al contents.

An other primary mineral contained in the protolith must have been spinel, because Cr was retained in the restites. This is interesting, as it fixes the depth of melt segregation to the range between 70 km and 25 km.

Efforts to find out the primary modal composition, perhaps by norm calculations, are however futile because the major element chemistry has been altered completely, particularly regarding Mg, Ca, Na and K. This is deplorable, because according to Boyd (1989), oceanic peridotites can be distinguished from cratonic peridotites in a plot of modal olivine against Mg#.

The peculiar positive Ti and Nb anomalies in the spidergrams can be most easily be explained by buffering in some minor exotic mineral, contained in the restite. It is tentatively proposed here that Ti-clinohumite might be a possible candidate. It has to be conceded however that the question of Ti-chu's occurrence in the mantle is beeing discussed controversially (see above chapter 5.3, p. 50/51 and the discussion in Weiss (1997: 3f). It seems not likely that the REE have been mobilized so strongly that these anomalies remain as testimonies of the formerly higher contents of incompatible trace elements. The lack of correlation between Ti and Nb on the one hand, the REE on the other, is explained by the buffering as well. Further discussion of melt-residue relationships is deferred until later, when also the rodingites have been discussed.

The negative Zr anomaly remains unexplained. On the other hand, it is perhaps not so large, and mostly only apparent due to positive anomalies of the neighboring elements. Indeed, when Zr is compared to the elements Sm and Tb (the elements most close to Zr in the spidergram not having a positive anomaly), it is found that $Zr_N/sqrt(Sm_N*Tb_N)$ is < 0.5 for 6 specimens, > 1.5 for 3 specimens, and in between these limits, i.e. around 1, for 8 specimens. Still this may be regarded as a significant result, taking into account the possibility of REE leaching, tending to diminish any prior negative Zr anomalies.

A possible explanation might be that it is inherited from the source. This leads to the question whether these rocks were derived from oceanic or from subcontinental mantle. Now not very much is known about the subcontinental mantle. McDonough (1990) gives some figures, taken from data on spinel-peridotitic xenoliths from kimberlites. Now this can be hardly regarded as a representative sample, but nevertheless these data make it clear that there is substantial variation in the continental lithosphere. The average contents given by McDonough are for most elements at about 1 to 2 times chondritic concentrations, for Th and Nb decidedly higher. This now offers an alternative explanation for the positive Nb anomalies seen in the serpentinites. At any rate, although the data certainly do not make it possible to decide whether the serpentinites were originally derived as restites from subcontinental or from oceanic lithosphere, the former alternative remains a possibility.

6.   Geochemistry of the metagabbros

6.1.   General description of the metagabbro geochemistry

Most major and trace elements show concentrations typical for gabbros (cf. for example the averages in Wedepohl 1969, table 7.4. p. 238). According to major oxides, in particular to TiO_2 contents and to MgO-Fe_2O_3(tot) ratios, two gabbro types can be distinguished: normal gabbros and Fe-Ti gabbros (fig 6.1). As shown by Ganguin (1988), this chemical difference results also in petrographically distinct rock types, although with strongly retrogressed rocks the distinction becomes difficult. The dividing line between the gabbro types has been drawn differently in the literature. Some authors draw it on petrological grounds (Pfeifer et al 1989, Ganguin 1988); Fe-Ti gabbros then are dark black-green rocks rich in rutile, titanite or ilmenite and up to 80 % omphacite (when not retrogressed). According to this criterion only 4 of the studied samples belong to this group (Hi 32-2-96, Hi 32-4-96, Hi 32-5-96, Hi 32-6-96). Here a preliminary dividing line is drawn according to the clustering in the TiO_2 – MgO – Fe_2O_3(tot) diagram (fig. 6.1). This line crosscuts Ganguin's (1988) intermediate group. Later another division will be drawn founded on the totality of the chemical data.

The two gabbro types will be discussed separately, where necessary. For most elements however, there is no significant difference between normal gabbros and Fe-Ti-gabbros. The number of normal gabbro specimens is 16; for two specimens only the trace elements are known. There are 9 Fe-Ti gabbro specimens.

Minimal, maximal and geometric mean element/oxide contents are summarized in table 6.1.

SiO_2 contents vary between 41 % and 54 %, the geometric mean of all specimens being 47,4 %. Al_2O_3 varies between 13 % and 23 %, with the geometric mean at 16,7 %. The next most important oxide is CaO, lying between 6.8 % and 19 %, with a geometric mean of 10.92 %. MgO contents are between 5.4 % and 17.7 %, with all but one specimen lying below 12 %. The maximum for the Fe-Ti gabbros is at 8.5 %. The geometric mean of all specimens is 8.04 %, being somewhat lower for the Fe-Ti gabbros than for the normal gabbros; there is however much overlap between these groups. For the latter group Fe_2O_3(tot) contents are lying between 3 % and 12 %, geometric mean 5.55 %, whereas the Fe-Ti gabbros are between 6.6 % and 16.2 %, geometric mean 11.51 %. Thus although the geometric mean values are significantly different, there is some overlap between the two groups. The ratio Fe_2O_3(tot) / MgO however very nicely discriminates between the two groups: normal gabbros: 0.4 % – 0.6 % – 0.8 %, Fe-Ti gabbros: 1.1 % – 1.8 % – 2.9 %.

Table 6.1: minimal, geometric mean and maximal element/oxide contents of metagabbros

| | normal gabbros | | | Fe-Ti gabbros | | | all gabbros |
	min	max	geom. mean	min	max	geom. mean	geom. mean
SiO_2	44.0	53.6	48.0	41.7	50.4	46.4	47.4
TiO_2	0.134	0.856	0.344	1.155	8.010	3.005	0.804
Al_2O_3	13.0	23.3	17.6	13.1	19.6	15.4	16.7
Fe_2O_3(t)	2.99	11.99	5.55	6.55	16.24	11.51	7.38
MnO	0.069	0.230	0.100	0.115	0.331	0.190	0.129
MgO	5.57	17.68	8.33	5.39	8.52	6.38	8.04
CaO	6.84	18.89	11.35	6.89	13.25	10.28	10.92
Na_2O	1.12	4.16	2.88	2.70	6.28	3.60	3.14
K_2O	b.d.	0.32	0.08	b.d.	0.41	0.10	0.09
P_2O_5	0.015	0.070	0.026	0.027	0.481	0.156	0.052
Fe/Mg	0.40	0.83	0.60	1.05	2.93	1.80	4.68
$\ln(Fe^2/Fe^3)$	-0.45	1.42	0.83 (mean)	0.23	1.59	1.00 (mean)	0.89 (mean)
Mg#	74	85	79	44	59	56	69
H_2O tot.	1.56	5.20	2.94	0.99	4.04	1.81	2.47
S	0.0	0.1	0.03	0.0	0.4	0.05	0.04
CO_2	0.03	0.37	0.06	-0.02	3.39	0.06	0.06
Li	0.5	13.6	3.40	0.9	29.3	7.53	4.70
Sc	6	47	25.7	23	52	37.8	29.9
V	28	201	90.8	139	848	363.5	156.2
Cr	54	2599	335.3	24	186	86.2	177.7
Co	20	80	35.9	17	58	37.0	36.3
Ni	78	604	185.8	21	145	59.9	119.3
Zn	b.d.	80	30.4	23	70	35.2	32.4
Ga	10	18	12.5	14	20	15.6	13.6
Rb	0.6	7.5	1.24	0.1	5.2	1.63	1.45
Sr	130	694	272.1	14	550	149.1	219.1
Y	1.9	11.0	5.37	3.6	34.0	21.96	9.72
Zr	13	65	26.9	103	302	140.0	51.3
Nb	0.1	4.2	0.70	0.7	19.4	6.60	1.63
Sn	0.3	3.9	0.77	0.6	10.2	2.23	1.21
Ba	1.8	25.4	5.64	0.6	65.0	8.13	6.47
La	0.5	3.6	1.36	1.0	17.0	6.87	2.39
Ce	1.1	7.3	3.05	3.5	47.5	21.05	5.80
Pr	0.2	1.2	0.60	0.5	7.0	2.86	1.03
Nd	0.9	6.0	2.75	2.4	27.7	13.89	4.60
Sm	0.3	1.9	0.88	0.7	6.9	3.72	1.45
Eu	0.3	0.8	0.51	0.2	3.6	1.29	0.69
Gd	0.3	2.1	0.92	0.7	8.1	4.34	1.62
Tb	0.0	0.4	0.16	0.1	1.1	0.67	0.26
Dy	0.4	2.6	1.10	0.7	6.8	4.39	1.84
Ho	0.1	0.5	0.23	0.1	1.5	0.94	0.39
Er	0.2	1.1	0.52	0.3	3.6	2.27	0.92
Tm	0.0	0.2	0.09	0.1	0.6	0.38	0.16
Yb	0.2	1.7	0.53	0.3	3.3	2.14	0.90
Lu	0.0	0.9	0.10	0.0	0.5	0.35	0.16

The oxidation state of Fe, as measured by $\ln(FeO/Fe_2O_3)$ (for a justification of this measure see chapter 5.8 above) is quite variable, lying between -0.5 and 1.4 for the normal gabbros (only two specimens are < 0; average: 0.83) and between 0.2 and 1.6 for the Fe-Ti gabbros (average: 1.0).

All other oxides are minor. For the normal gabbros, TiO_2 contents are between 0.13 % and 0.85 %, geometric mean 0.344 %, and between 1.15 % and 8.01 %, geometric mean 3.005 % for the Fe-Ti gabbros. This results in the nice discrimination between the two groups in the $Fe_2O_3(tot)/MgO$ - TiO_2 diagram (fig. 6.1), proposed by Miyashiro (1973) to illustrate magmatic differentiation.

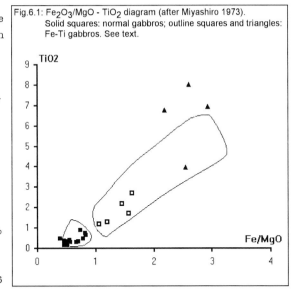

Fig.6.1: Fe_2O_3/MgO - TiO_2 diagram (after Miyashiro 1973). Solid squares: normal gabbros; outline squares and triangles: Fe-Ti gabbros. See text.

Some difference between the normal gabbros and the Fe-Ti gabbros is also shown by MnO (0.07 % – 1.00 % – 0.23 % and 0.11 % – 0.19 % – 0.33 % respectively) and P_2O_5 (0.015 % – 0.026 % – 0.07 % and 0.027 % – 0.156 % – 0.481 % respectively).

The P_2O_5 geometric mean of all specimens is 0.052. This seems very low when compared to the average in Wedepohl (1969), p. 238 of 0.24 %. The discrepancy is at least partly due to the difference between geometric mean and average. The average of the specimens studied here is 0.103 %.

Na_2O is between 1.12 % and 6.28 %, geometric mean 3.14 %. K_2O values are between b.d to 0.41 %, geometric mean 0.09 %, much lower than usual for gabbros (0.56%) and thus most probably influenced by alteration.

The classical differentiation index Mg# (calculated with FeO = 85 % of FeO(tot), as the oxidation state of Fe is certainly not primary) ranges from 74 to 85 for the normal gabbros (geometric mean 79) and from 44 to 69 for the Fe-Ti gabbros (geometric mean 56). Plots of some oxides and trace elements against Mg# are shown in fig. 6.2. Most major oxides, in particular CaO and to a lesser degree Al_2O_3, show much scatter. Al_2O_3 seems to generally decrease with falling Mg#. With compatible (Cr, Ni) and incompatible trace elements and minor oxides (TiO_2, P_2O_5, Zr, Y) there is much less scatter; typical differentiation trends can be discerned.

The AFM diagram (fig. 6.3), another classical way to depict major and minor oxide information with a view to differentiation, shows again a clear separation between normal gabbros and Fe-Ti gabbros. There is much spread towards the A apex, as the alkali elements were seemingly subject to alteration. Thus no differentiation trend emerges. Also this graph was constructed for volcanic rocks, not for cumulates. The 4 Fe-Ti gabbro specimens Hi 32-2-96, Hi 32-4-96, Hi 32-5-96,

Fig 6.2: Element against Mg# plots.
Solid squares: normal gabbros; outline squares and triangles: Fe-Ti gabbros.

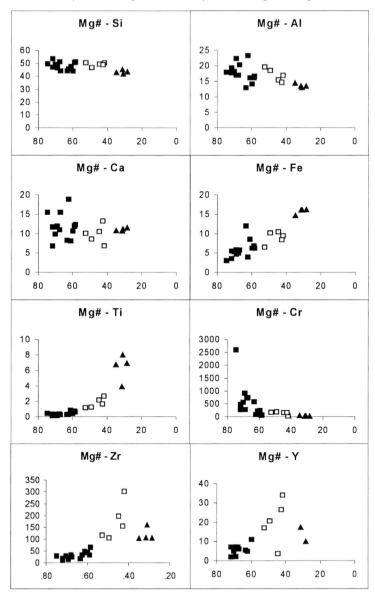

Hi 32-6-96 are closest to the F apex and are lying on a olivine control line.

The CIPW norm minerals have been calculated and are given in tab. 6.2. In the calculation, S and CO_2 contents have been neglected and FeO and Fe_2O_3 have been recalculated from total Fe_2O_3 (with Fe_2O_3 set at 15 % of total Fe), as the oxidation state of Fe is certainly not primary. A plot of the norm minerals pl, cpx and ol shows the gabbros to be mainly leukocratic to mesocratic olivine-gabbros. The normal and Fe-Ti gabbros can not be distinguished on this plot. The extremely Mg-rich gabbros however tend to be closest to the olivine apex.

In table 6.1, maximal, minimal and geometric mean contents are also given for the trace elements. The ferromagnesian elements Cr and Ni have higher geometric mean contents in the normal gabbros than in the Fe-Ti gabbros, although the ranges overlap. For V and for most incompatible elements (LILE, HFSE, REE) the opposite is the case. With the Zr contents, a clear distinction between the two gabbro types is seen.

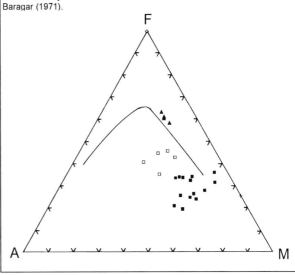

Fig 6.3: AFM triangle.
Solid squares: normal gabbros; outline squares and triangles: Fe-Ti gabbros. Boundary between calc-alkaline and tholeiitic rocks after Irvine & Baragar (1971).

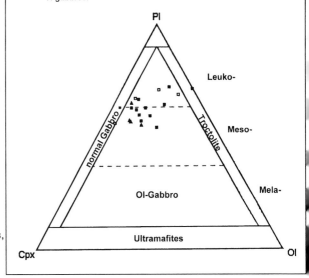

Fig 6.4: Classification of metagabbros according to CIPW norm minerals. Solid squares: normal gabbros; outline squares and triangles: Fe-Ti gabbros.

Element-element plots are shown and discussed below in chapter 6.4.

The oxide and element contents reported compare very well with other metagabbro analyses from the ZSF in the literature (Bearth 1967, Beccaluva et al. 1984, Dal Piaz & Ernst 1978, Ganguin 1988, Pfeifer et al 1989, Widmer 1996), although some papers report more extreme values than those found in this study. For the ferrogabbros and intermediate gabbros, Ganguin (1988) reports some very low Al_2O_3 (9.13 %), Zr (49 ppm) and Cr (b.d.) values, as well as high Sr (900 ppm), Sc (70 ppm), Co (70 ppm) and Ni (200 ppm) values. Beccaluva et al. (1984) report higher TiO_2 values for normal gabbros than those analyzed here and found elsewhere in the literature.

Tab 6.2 CIPW norm minerals of the metagabbros

	CM 2-98	Hi 4-96	Hi 16-96	Hi 18-96	Hi 19-96	Hi 24-96	Hi 30-1-96	Hi 30-2-96	Hi 30-3-96	Hi 30-6-96	Hi 32-1-96	Hi 32-8-96
ap	0.04	0.04	0.04	0.04	0.14	0.05	0.15	0.20	0.05	0.07	0.07	0.07
ilm	0.7	0.7	0.3	0.3	1.2	0.7	0.6	1.9	0.5	1.0	1.4	1.4
or	1.9	0.6		0.5	0.5	0.5	0.3	0.7	0.4	0.5	0.1	0.1
ab	35.1	22.5	33.6	19.0	30.5	31.7	1.1	33.4	5.0	25.5	29.3	29.3
an	29.3	38.8	35.9	49.4	29.4	28.2	55.9	34.0	51.7	28.3	27.8	27.8
cor			0.5									
mag	0.8	1.4	1.3	1.0	1.4	1.1	0.9	2.2	1.3	1.6	1.5	1.5
di: wo	12.3	11.0		4.5	13.7	11.7	15.9	5.5	11.6	13.8	13.6	13.6
en	9.1	7.9		3.2	9.1	8.4	11.0	3.8	8.2	9.2	9.1	9.1
fs	2.0	2.1		0.9	3.5	2.2	3.7	1.3	2.3	3.5	3.5	3.5
hy: en		5.8							2.8			
fs		1.5							1.0			
ol	8.8	7.6	27.8	17.8	10.0	13.0	2.9	13.2	16.3	13.8	12.1	12.1
ne	0.0		0.7	3.4	0.5	2.4	7.5		2.6	2.8	1.4	1.4

	Hi 32-10-96	Hi 32-11-96	Hi 27-1-96	Hi 27-2-96	Hi 30-4-96	Hi 30-5-96	Hi 32-12-96	Hi 32-2-96	Hi 32-4-96	Hi 32-5-96	Hi 32-6-96
ap	0.0	0.1	0.4	1.2	0.4	0.7	0.6	0.3	0.1	1.0	0.1
ilm	0.4	0.7	2.2	5.1	2.4	4.4	3.3	13.3	7.6	15.4	13.8
or		0.5	1.0	1.6	2.5	0.4	0.5	0.2		0.1	
ab	28.7	16.0	34.4	41.8	27.7	34.9	27.1	22.8	25.3	21.7	24.6
an	32.5	35.9	33.5	17.4	35.3	26.9	23.4	24.3	23.4	22.6	29.5
cor											
mag	1.2	3.3	1.4	2.1	2.3	2.5	1.9	3.6	3.6	3.6	3.4
di: wo	7.4	6.6	6.8	6.0	3.4	11.6	18.1	13.7	12.8	13.0	11.8
en	5.4	4.5	4.4	3.6	2.1	7.0	10.6	7.6	6.5	8.1	7.5
fs	1.3	1.6	1.9	2.0	1.1	4.0	6.7	5.5	6.0	4.2	3.6
hy: en		7.1				0.5					0.1
fs		2.5				0.3					0.0
ol	19.6	21.3	11.7	12.6	22.1	6.8	4.0	7.9	13.6	8.3	5.6
ne	3.5		2.1	6.6	0.7		3.8	0.9	1.0	2.1	

6.2. REE diagrams and spidergrams

Part of the trace element information is summarized in the form of the well-known REE diagrams and spidergrams. These diagrams have been calculated on a water-free basis. The normalization factors are to be found in Tab. 5.1 (in chapter 5.2). In all diagrams the normal metagabbros and the Fe-Ti metagabbros have been marked differently.

The REE diagram (fig. 6.5) displays subparallel curves with a generally falling aspect, but rising slightly from La to Pr (some specimens: from La to Sm). Ce shows some scatter and so seems to have been analytically difficult. Specimen 32-10-96 deviates in two ways: it does not show the rise in the LREE, and its falling tendency is more pronounced than with the other specimens. This specimen belongs to the few ones containing Mg-chloritoid and talc.

The Fe-Ti gabbros tend to have higher REE contents than the normal gabbros. These latter all have a positive Eu anomaly, which is largest for those specimens with the lowest REE content. The Fe-Ti gabbros have no or a small negative Eu anomaly.

The **Mantle-normalized spidergram** (after Thompson 1982, fig. 6.6) shows the following traits: the curves are roughly subparallel, rising from Ba to La and dropping from there to Yb. There is a very conspicuous positive Sr anomaly; only some Fe-Ti gabbros display a negative Sr anomaly. The Fe-Ti gabbros have generally higher trace element contents. The K and P values of the normal gabbros are close to or below the l.o.d. Nevertheless it can be stated that the normal gabbros (and the Fe-Ti gabbro Hi 32-4-96) have a negative P anomaly. Also Nb shows a negative anomaly with the normal gabbros. With some, but not all of the Fe-Ti gabbros, Ti displays large positive anomalies. The small negative Ti anomalies seen with some other specimens are probably not significant. Of the mobile elements, Ba is particularly low.

The **MORB-normalized spidergram** (fig. 6.7, after Pearce 1983) shows one peculiar additional trait: whereas the range of Sc contents is rather restricted, the spread of the Cr values is second only to Ba. According to Sc_N/Cr_N ratios, three distinct groups of specimens can be distinguished: specimens with rising curves (Sc_N/Cr_N around 0.2; specimens CM 2-96, Hi 4-96, Hi 16-96, Hi 18-96, Hi 24-96, Hi 30-3-96, Hi 32-8-96- Hi 32-10-96, Hi 32-11-96), flat curves (Sc_N/Cr_N around 1.2; specimens Hi 30-1-96, Hi 27-1-96, Hi 30-2-96, Hi 30-4-96, Hi 30-5-96, Hi 30-6-96, Hi 32-12-96) and falling curves (Sc_N/Cr_N around 8; specimens Hi 19-96, Hi 27-2-96, Hi 32-1-96, Hi 32-2-96, Hi 32-4-96, Hi 32-5-96, Hi 32-6-96). There exists the following connection between this grouping and the normal versus Fe-Ti character of the gabbros: all of the gabbros with rising Sc-Cr are normal gabbros. Of the 7 specimens

Fig. 6.5: REE diagrams of metagabbros. Normal gabbros: solid lines; Fe-Ti gabbros: broken lines. Bold line: limit of detection.

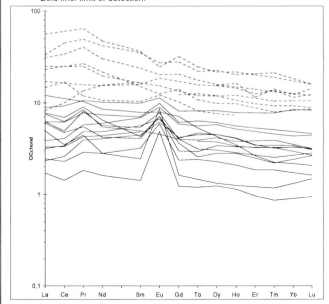

Fig. 6.6: Mantle-normalized spidergram of metagabbros (after Thompson 1982) Normal gabbros: solid lines; Fe-Ti gabbros: broken lines. Bold line: limit of detection. Values below the l.o.d. have been set to 0.5 l.o.d.

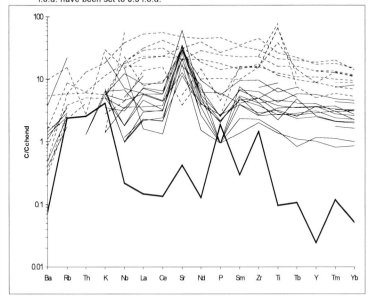

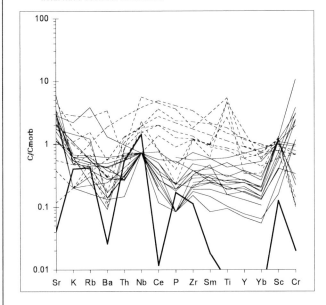

Fig. 6.7: MORB-normalized spidergram of metagabbros (after Pearce 1983). Normal gabbros: solid lines; Fe-Ti gabbros: broken lines. Bold line: limit of detection. Values below the l.o.d. have been set to 0.5 l.o.d.

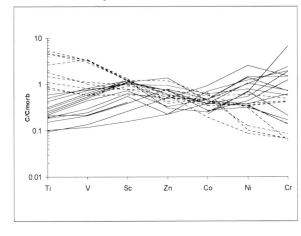

Fig. 6.8: Ferromagnesian elements, normalized to MORB (according to GERM). Normal gabbros: solid lines; Fe-Ti gabbros: broken lines.

with flat Sc-Cr, 4 are Fe-Ti gabbros. None one of them has a positive Ti anomaly. Of the 7 gabbros with falling Sc-Cr, 5 are Fe-Ti gabbros, and 4 of them have a positive Ti anomaly.

In order to better understand this grouping, a diagram containing also other ferromagnesian elements has been prepared (fig 6.8). It is seen that Ti and V behave similarly, V showing less spread than Ti. The elements Sc, Co and Zn show generally little spread. The behavior of Ni is opposite to Ti and V: Specimens high in Ti and V, particularly the Fe-Ti gabbros, are low in Ni and vice versa. In this respect Cr behaves similar to Ni, but even more extremely. But Ni and Cr do also show diverging behavior: in the normal gabbros, some curves are rising from Ni to Cr, others are falling. The crossing over of the curves of normal and Fe-Ti gabbros happens at Sc, Co and Zn. Yet Fe-Ti gabbros tend to be higher in Sc than normal gabbros. When Ti_N/Cr_N ratios are compared to Sc_N/Cr_N ratios the same three groups emerge. Only some specimens occupy more intermediate positions (Hi 30-6-96 between "rising" and "flat", Hi 19-96 and Hi 32-1- 96 are in "flat" instead of "falling").

These findings lead to the following preliminary conclusions: The generally falling curves of the REE diagrams point to derivation from a comparatively fertile mantle source containing clinopyroxene or orthopyroxene. Garnet as a residual phase can be excluded, as it would result in much steeper curves. Olivine in the mantle could be responsible for the slight rising from La to Pr or Sm, as some sets of distribution coefficients for olivine would produce such a pattern (a study of the GERM dataset produced 3 such sets of D's as opposed to one set with the opposite behavior; Frey 1969, Mahood & Hildreth 1983, Fujimaki & Tatsumoto 1984, McKenzie & O'Nions 1991; also the set of D's proposed by Rollinson 1993 has a slight bulge; cf. also Prinzhofer & Allègre 1985)

The curves' parallel character could point to olivine as fractionating or cumulate phase, as this mineral influences the REE contents only very weakly. The high Mg# of the normal gabbros point to olivine as a cumulus phase.

The pronounced positive Eu anomaly of the normal gabbros is the signature of plagioclase cumulates.

The additional information from the mantle-normalized spidergram is very well fitting into this picture: the less mobile incompatible trace elements behave very similar to the REE as discussed above. As long as no Ti-bearing phases enter the crystallization (i.e. with the normal gabbros), Ti belongs to these incompatible elements. The huge positive Sr anomaly of the normal gabbros again can be ascribed to plagioclase as cumulus phase. However, no clear correlation can be observed between the Eu and Sr anomalies. The negative P and Nb anomalies could point to some residual phase in the mantle residue, such as Ti-clinohumite (see the discussion in the chapter on serpentinites above). The elements Ba, Rb, Th and K are generally showing curves falling from K to Ba. This is contrary to expectations, as the degree of mantle incompatibility was thought increase in this direction by the authors of the diagram (Thompson et al 1984). A study of published D values however reveals more or less equal partition coefficients for all these elements in olivine, clinopyroxene, orthopyroxene and plagioclase (Hart & Dunn 1993, Hart & Brooks 1974, Dunn & Sen 1994, Philpotts & Schnetzler 1970; there are no D's for these elements for spinel). Furthermore, the normalization coefficients are perhaps somewhat misleading. Many modern estimates of mantle compositions show curves falling from K to Ba in this diagram (bulk silicate earth: McDonough & Sun 1995; depleted mantle: Jacobsen (1997), personal communication according to GERM). Thus a mantle-derived melt would have a rising tendency with Ba_N being about 15 times lower than K_N. Producing a cumulate from this melt, in turn, might or might not steepen the curves. Lastly, as the disturbed parallelism of the curves shows, these elements are certainly overprinted by later mobilization during the metamorphic history. This could again, depending on the exact properties the mobilizing fluid, have resulted in a general steepening of the curves.

The peculiar grouping displayed in the MORB-normalized spidergram and in the ferromagnesian diagram could be explained by the onset of three different phases crystallizing. Whenever a new phase enters fractionation, a sudden change in the trace element partitioning will occur, resulting in different element ratios. Not only the ratios will display a discontinuous distribution, but also the element contents, due to overstepping of equilibrium conditions. The phases involved in the case at hand might be olivine (excluding Sc and more or less neutral to Cr) plus spinel (incompatible to Sc, compatible to Cr), clinopyroxene (more or less neutral to Sc and compatible to Cr) and some Ti mineral (compatible to both elements).

The overall picture is thus quite complex, with four or five different cumulus or fractionating phases being involved (olivine, spinel, clinopyroxene, plagioclase, Ti-bearing phase). These questions will be pursued further below in the chapter on magmatic modeling.

6.3. Results of the log-ratio method 1. Model

As above, when the serpentinites were treated, a model is presented with the purpose of helping to understand the results of applying the log-ratio method to the actual metagabbro data. Again, the model is highly simplified. In a first step, melts are produced from a 4-phase lherzolite by partial melting. In a second step, these melts are mixed with different proportions of plagioclase being in equilibrium with the respective melts. The melting model, as above, produces constant amounts of the major elements, since the proportions of melting phases are kept constant (no phase is completely molten out). Only the ratio of FeO to MgO is changed with the purpose to produce more Fe-rich melts at smaller degrees of partial melting (fayalite being more fusible than forsterite). As the model is not very sensitive to this parameter, this was done by a simple ad-hoc formula.

The log-ratio method is applied to this model as a kind of calibration for the empirical findings. Two models have been considered, one dominated by different degrees of partial melting, the other by different proportions of mixing. Table 6.3 gives the parameters of the model in terms of source composition, melting modalities and distribution coefficients. The chemical composition of the model lherzolite is greatly simplified; only 4 major elements (oxides) are taken into account, viz. SiO_2, MgO, CaO and Al_2O_3. The compositions of the 4 phases olivine, orthopyroxene, clinopyroxene and spinel are idealized accordingly. In addition to these 4 oxides 7 trace elements, labeled Cr, Y, Zr, P, Ni, Ba, Sr are introduced into the model as representative more or less incompatible elements and as trace elements characterizing specific cumulus and fractionating minerals.

Applying the familiar formulas for batch melting with these parameters, 3 partial melts were produced. Model A uses the melt proportions 4, 6 and 8 %, model B 4, 8, 12 %. For these 3 melts equilibrium plagioclase compositions (with simplified major oxide composition) were calculated and mixed with the melts in the proportions 0, 8, 17, 26, 35 % (model A) and 0, 5, 10, 15, 20 % (model B). These percentages were chosen to cover the range of CIPW-normative plagioclase in the analyzed specimens. Thus two models were produced, each with a total of 15 "specimens".

Tab 6.3: Gabbro model parameters

phase proportions in the source (modified after Kinzler 1997)

ol	opx	cpx	spl
55	24.5	18	2.5

Phase proportions in melt (after Johnson et al. 1990)

ol	opx	cpx	spl
10	20	68	2

Distribution coefficients and source composition (after Bédard 2000)

	ol	opx	cpx	plag	spl	source composition
Cr	1.25	1.9	3.8	0.02	77	2692
Y	0.07	0.2	0.412	0.026	0.002	4.4
Zr	0.02	0.021	0.26	0.01	0.015	21
P	0.008	0.02	0.13	0.075	0.0006	164
Ni	10	3.5	2	0.04	10	2140
Ba	0.01	0.0026	0.00068	0.9	0.0005	17
Sr	0.008	0.006	0.1283	1.3	0.0006	8

Applying the log-ratio method to these datasets leads to the results given in table 6.4 and 6.5. The scattergrams for these models are given in fig 6.9.

As for the correlation structure of the elements, both models show some discrepancies between log-ratio variances and log-ratio correlation coefficients. The major element chemistry is in both cases dominated by the effects of mixing, resulting in good correlations even between Fe and Mg, the elements most strongly influenced by melting in the models. Some of the mixing relationships result in strong positive log-ratio correlation coefficients, although log-ratio variances are quite high (e.g. Si-Al). The reason for this is that the endpoints of the mixing line have quite different element ratios. For other element pairs the opposite occurs, particularly in the mixing-dominated model A, where for example Mg and Y have a low log-ratio variance, caused by both elements being excluded from plagioclase, although the log-ratio correlation coefficient is low in response to the diverging directions of melting and mixing.

Unequivocally correlated in the mixing dominated model A are the following 3 groups of elements:

Tab 6.4: log-ratio results for model gabbro A (mixing-dominated)

log-ratio variances (lower left) and log-ratio correlation coefficients (upper right)

	SiO$_2$	Al$_2$O$_3$	CaO	FeO	MgO	Cr	Y	Zr	P	Ni	Ba	Sr
SiO$_2$		0.186	**0.963**	0.155	0.371	0.366	-0.190	-0.504	-0.698	0.282	-0.854	-0.375
Al$_2$O$_3$	0.780		-0.074	-0.909	-0.771	-0.774	-0.987	-0.942	-0.818	-0.827	-0.537	0.599
CaO	0.000	0.815		0.416	0.608	0.603	0.080	-0.264	0.531	-0.752	-0.589	
FeO	0.017	1.011	0.013		**0.966**	0.967	0.936	**0.741**	0.522	**0.986**	0.145	-0.869
MgO	0.020	1.027	0.014	0.002		**1.000**	0.815	0.544	0.286	**0.995**	-0.115	-0.949
Cr	0.020	1.029	0.014	0.002	0.000		**0.818**	0.549	0.291	**0.996**	-0.109	-0.949
Y	0.029	1.066	0.023	0.003	0.008	0.008		**0.929**	**0.787**	**0.868**	0.478	-0.689
Zr	0.040	1.075	0.034	0.010	0.020	0.020	0.003		**0.959**	0.624	**0.768**	-0.388
P	0.057	1.094	0.051	0.024	0.039	0.039	0.012	0.003		0.378	**0.917**	-0.122
Ni	0.022	1.043	0.016	0.001	0.000	0.000	0.006	0.017	0.034		-0.018	-0.933
Ba	0.084	1.064	0.080	0.057	0.080	0.080	0.040	0.021	0.009	0.073		0.278
Sr	0.047	0.641	0.052	0.083	0.104	0.104	0.085	0.075	0.071	0.103	0.059	

Principal component analysis:

	F1 89.77%	F2 9.57%	F3 0.65%
SiO$_2$	0.0147	-0.2057	0.4308
Al$_2$O$_3$	0.9218	-0.0291	-0.2562
CaO	-0.0056	-0.2120	0.3654
FeO	-0.1137	-0.1536	-0.0625
MgO	-0.1195	-0.2942	-0.0811
Cr	-0.1205	-0.2933	-0.0859
Y	-0.1429	-0.0273	-0.2030
Zr	-0.1456	0.1470	-0.2239
P	-0.1504	0.3199	-0.2551
Ni	-0.1286	-0.2583	-0.1246
Ba	-0.1220	0.5971	-0.1408
Sr	0.1123	0.4095	0.6369

- Si and Ca
- Fe, Mg, Cr, Ni
- Zr, P, Y, Ba.

For the melting-dominated model B there are only two groups:

- Si, Ca, Mg, Cr, Ni
- Y, Zr, P, Ba, Sr.

These groupings reflect the fractionating and mixing phases. When melting is the dominant process, a division between compatible and incompatible elements is created. In this respect Fe occupies a intermediate position and is consequently not contained in any of the two groups. When mixing gets important, the group of compatible elements splits into two groups, one ferromagnesian (Fe, Mg, Cr, Ni) and one plagioclase group (Si, Ca; note also the relationship Al-Sr), reflecting the endpoints of the mixing relationship.

Considering now the pca's belonging to the models, it found that the first components of both models are in essence identical to first components that would arise in

Tab 6.5: log-ratio results for model gabbro B (melting-dominated)

log-ratio variances (lower left) and log-ratio correlation coefficients (upper right)

	SiO_2	Al_2O_3	CaO	FeO	MgO	Cr	Y	Zr	P	Ni	Ba	Sr
SiO_2		0.345	**0.994**	0.644	**0.868**	**0.893**	-0.319	-0.885	-0.966	**0.825**	-0.994	-0.890
Al_2O_3	0.178		0.244	-0.486	-0.155	-0.099	-0.990	-0.738	-0.571	-0.234	-0.413	0.092
CaO	0.000	0.189		**0.722**	**0.916**	**0.936**	-0.216	-0.832	-0.935	**0.881**	-0.981	-0.933
FeO	0.008	0.244	0.006		**0.936**	**0.913**	0.518	-0.221	-0.436	**0.961**	-0.594	-0.912
MgO	0.007	0.247	0.005	0.007		**0.998**	0.184	-0.548	-0.722	**0.997**	-0.834	-0.997
Cr	0.009	0.250	0.007	0.012	0.001		0.126	-0.595	-0.759	**0.990**	-0.862	-0.998
Y	0.024	0.271	0.022	0.006	0.026	0.035		**0.720**	0.543	0.263	0.376	-0.127
Zr	0.056	0.303	0.054	0.029	0.064	0.078	0.008		**0.974**	-0.478	**0.913**	0.595
P	0.100	0.348	0.099	0.064	0.114	0.133	0.031	0.007		-0.663	**0.760**	0.760
Ni	0.006	0.250	0.005	0.004	0.000	0.002	0.020	0.054	0.101		-0.787	-0.989
Ba	0.188	0.422	0.187	0.142	0.212	0.237	0.090	0.044	0.016	0.194		**0.865**
Sr	0.105	0.231	0.108	0.089	0.142	0.161	0.058	0.031	0.022	0.130	0.029	

Principal component analysis:

	F1	F2	F3
	60.49 %	39.34 %	0.10 %
SiO_2	0.1885	0.0818	0.4296
Al_2O_3	0.5059	-0.7571	-0.2913
CaO	0.1811	0.1068	0.3582
FeO	0.0535	0.1713	-0.0881
MgO	0.1852	0.2601	-0.0656
Cr	0.2267	0.2844	-0.0231
Y	-0.0792	0.1268	-0.2446
Zr	-0.2264	0.0402	-0.2623
P	-0.3632	-0.0377	-0.2590
Ni	0.1500	0.2485	-0.1044
Ba	-0.5475	-0.1848	-0.0648
Sr	-0.2746	-0.3404	0.6155

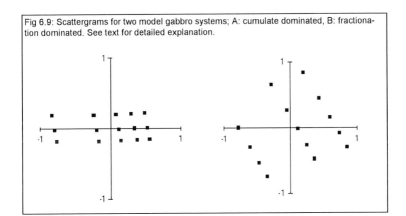

Fig 6.9: Scattergrams for two model gabbro systems; A: cumulate dominated, B: fractionation dominated. See text for detailed explanation.

pure mixing respectively melting systems, i.e. systems, where only one of either processes, mixing or melting, is operating.

Component 1 of the melt-dominated model B reflects the opposition between compatible and incompatible elements, whereas with component 1 of the mixing-dominated model A, the division is between elements contained in plagioclase (with the most heavy loading on Al) and the rest of the world. With the second components it is however interesting to note that they do not simply reflect the second model process. To name just one of the differences, factor 2 of the melt-dominated model B has no significant loadings on Ca and Si in the plagioclase group. Thus the second components are not as easily interpreted as the first ones. This problem has to do with the orthogonality constraint built into the method and with the fact, easily seen in fig. 6.9, that the components are oblique to the processes. Whenever one process has a dominating influence, it will be reflected rather clearly in the first component. This does not hold for subordinate processes and for the following components.

6.4.   Results of the log-ratio method 2: correlation structure

When considering the log-ratio statistics presented below, it has to be taken into account that for some elements the data are quite incomplete. There are values below the l.o.d, particularly for K, and many values are lacking altogether due to the calibration problems discussed above in chapter 3, particularly with Rb and Li. Thus for the LILE the results rely on a smaller base than for the other elements. Also, the particular way the algorithm deals with null cases leads to some distortions like correlation coefficients >1. This has to be kept in mind. The results were however checked by studying subsets of the specimens and of the elements analyzed and the results were seen to be broadly robust.

Table 6.6 gives the matrix of log-ratio variances and of log-ratio correlation coefficients for the metagabbros. It is seen that the elements and oxides analyzed are grouped according to their being correlated into two large and one small group. The first group, encompassing Si, Al, Ca, Na, Mg, Fe, Mn, H$_2$O, Ga, Zn and Co, obviously consists of elements behaving compatibly during fractionation. The second group, containing Ti, V, the REE (except Eu) and Y, Sn, Sc, Zr, P and Nb is the complementary group of incompatible elements. Ba, K, Li and Rb with less strong correlations are forming the small group.

Closer scrutiny reveals that the large groups can be subdivided into clusters of elements more intimately related. These relationships can be pictured as in fig. 6.10. The close relationship of Si, Al, Ca and Ga points to the importance of plagioclase in the magmatic processes. Si and Mg are also very close, as could be explained by olivine or orthopyroxene being involved. Ni, Co and Cr are only weakly correlated among each other, and Cr in particular does not show any affiliation to the

Tab. 6.6: log-ratio variances (lower left half) and log-ratio correlation coefficients (upper right half) for the metagabbros. Log-ratio correlation coefficients significant at the 5 % level and log-ratios < 0.3 are in bold print.

	SiO$_2$	Al$_2$O$_3$	Ga	CaO	Sr	MgO	H$_2$O(t)	Fe$_2$O$_3$(t)	MnO	Na$_2$O	Zn	Co	Ni	Cr	TiO$_2$	V	Sc	P$_2$O$_5$	Zr	Nb	Sn
SiO2		**0.968**	**0.890**	**0.893**	**0.538**	**0.928**	**0.870**	0.223	0.301	**0.710**	**0.530**	**0.733**	**0.862**	**0.726**	**-0.697**	-0.407	0.257	**-0.646**	**-0.807**	**-0.831**	**-0.61**
Al2O3	**0.025**		**0.904**	**0.893**	**0.605**	**0.874**	**0.890**	0.155	0.210	**0.651**	**0.440**	**0.678**	**0.854**	**0.702**	**-0.711**	**-0.508**	0.099	**-0.557**	**-0.747**	**-0.762**	**-0.57**
Ga	**0.043**	**0.064**		**0.903**	0.347	**0.763**	**0.697**	0.360	0.404	**0.549**	**0.484**	**0.684**	**0.706**	**0.462**	**-0.436**	-0.196	0.224	**-0.467**	**-0.562**	**-0.625**	**-0.47**
CaO	**0.063**	**0.064**	**0.067**		0.468	**0.762**	**0.673**	0.175	0.279	**0.449**	0.404	**0.663**	**0.756**	**0.653**	**-0.520**	-0.237	0.353	**-0.563**	**-0.683**	**-0.648**	**-0.53**
Sr	0.728	0.647	0.893	0.798		**0.485**	**-0.457**	-0.371	0.358	0.213	0.136	**0.580**	**0.611**	**-0.869**	**-0.752**	-0.045	**-0.556**	**-0.628**	**-0.587**	-0.315	
MgO	**0.091**	**0.108**	**0.200**	**0.190**	0.808		**0.920**	0.396	0.454	**0.556**	**0.488**	**0.862**	**0.929**	**0.765**	**-0.693**	-0.427	0.227	**-0.664**	**-0.846**	**-0.784**	**-0.673**
H2O(t)	**0.212**	**0.158**	0.350	0.346	0.597	**0.101**		0.176	0.205	**0.546**	0.392	**0.709**	**0.921**	**0.769**	**-0.851**	**-0.696**	-0.049	**-0.564**	**-0.820**	**-0.781**	**-0.623**
Fe2O3(t)	0.301	0.409	**0.218**	0.405	1.590	0.404	0.697		**0.949**	-0.086	**0.265**	**0.710**	0.242	-0.147	0.286	0.354	0.152	-0.153	-0.081	-0.153	-0.376
MnO	**0.270**	0.382	**0.203**	0.355	1.514	0.371	0.676	**0.019**		-0.047	**0.269**	**0.742**	0.263	-0.047	0.251	0.394	0.370	-0.219	-0.178	-0.252	**-0.46**
Na2O	**0.119**	**0.182**	**0.165**	**0.285**	0.892	0.316	0.445	0.422	0.406		**0.568**	0.254	0.438	0.437	**-0.530**	-0.311	0.118	**-0.430**	**-0.478**	**-0.650**	-0.269
Zn	**0.228**	0.321	**0.230**	0.345	1.063	0.380	0.580	0.339	0.336	**0.210**		**0.432**	0.254	0.164	**-0.484**	-0.161	0.349	**-0.610**	**-0.849**	**-0.562**	**-0.644**
Co	**0.181**	**0.226**	**0.209**	**0.238**	1.233	**0.118**	0.318	0.194	0.177	0.453	0.382		**0.788**	**0.548**	-0.355	-0.134	0.261	**-0.559**	**-0.654**	**-0.607**	**-0.758**
Ni	0.728	0.636	0.946	0.762	1.066	0.408	0.325	1.399	1.378	1.192	1.239	0.685		**0.834**	**-0.778**	**-0.581**	0.060	**-0.651**	**-0.849**	**-0.723**	**-0.644**
Cr	1.695	1.614	2.111	1.694	1.592	1.352	1.221	2.920	2.782	2.110	2.537	1.837	0.785		**-0.803**	**-0.564**	0.187	**-0.643**	**-0.842**	**-0.746**	-0.384
TiO2	1.690	1.922	1.375	1.737	3.555	2.215	2.802	0.843	0.870	1.551	1.643	1.700	4.138	5.842		**0.875**	**0.551**	**0.423**	**0.733**	**0.564**	0.028
V	1.048	1.282	0.840	1.067	2.719	1.445	2.032	0.521	0.495	0.985	0.967	1.083	3.092	4.455	**0.214**		**0.551**	0.064	**0.446**	0.319	0.004
Sc	0.328	0.482	**0.308**	0.351	1.294	0.538	0.908	0.356	0.391	0.333	0.469	1.634	2.483	1.009	0.399			**-0.484**	-0.274	-0.311	-0.278
P2O5	1.370	1.484	1.146	1.500	2.622	1.867	2.058	0.974	1.019	1.212	1.485	1.652	3.453	4.913	0.914	1.170	1.318		**0.783**	**0.550**	**0.702**
Zr	0.914	1.056	0.710	1.027	2.019	1.382	1.661	0.527	0.572	0.752	0.828	1.149	2.912	4.327	0.431	0.500	0.690	**0.273**		**0.754**	**0.704**
Nb	2.046	2.223	1.735	2.107	3.299	2.599	2.982	1.384	1.470	1.880	1.778	2.240	4.341	6.055	0.857	1.137	1.615	0.810	0.511		**0.781**
Sn	1.135	1.283	0.960	1.263	2.000	1.630	1.580	0.944	0.999	0.910	1.073	1.606	3.127	3.952	1.105	1.089	0.960	0.368	**0.267**	0.423	
La	0.868	0.953	0.747	1.007	1.272	1.273	1.353	0.628	0.702	0.799	0.897	1.101	2.425	3.856	0.831	0.971	0.972	**0.248**	**0.210**	0.580	0.305
Ce	1.107	1.213	0.959	1.260	1.335	1.600	1.688	0.779	0.844	0.986	1.128	1.415	2.944	4.172	0.613	0.851	1.073	**0.122**	**0.120**	0.512	0.298
Pr	0.815	0.913	0.705	0.933	1.108	1.244	1.336	0.631	0.688	0.735	0.860	1.122	2.418	3.696	0.742	0.846	0.820	**0.228**	**0.139**	0.520	**0.210**
Nd	0.799	0.925	0.688	0.935	1.162	1.250	1.375	0.590	0.642	0.675	0.771	1.128	2.523	3.741	0.450	0.618	0.713	**0.209**	**0.072**	0.419	**0.191**
Sm	0.703	0.851	0.603	0.823	1.100	1.106	1.321	0.477	0.509	0.593	0.636	0.973	2.350	3.550	0.484	0.473	0.508	0.390	**0.083**	0.476	0.238
Eu	0.394	0.505	**0.290**	0.515	0.910	0.695	0.902	**0.223**	**0.259**	0.336	0.381	0.617	1.772	3.076	0.582	0.449	0.395	0.751	0.297	0.876	0.518
Gd	0.820	0.988	0.695	0.958	1.268	1.275	1.523	0.549	0.591	0.662	0.655	1.101	2.649	3.838	0.473	0.488	0.553	0.390	**0.068**	0.423	**0.253**
Tb	0.821	1.008	0.715	0.953	1.273	1.240	1.518	0.547	0.565	0.679	0.625	1.079	2.567	3.756	0.431	0.382	0.460	0.498	**0.116**	0.529	0.310
Dy	0.693	0.870	0.567	0.798	1.762	1.071	1.401	0.402	0.418	0.587	0.585	0.895	2.457	3.696	0.506	0.357	0.379	0.516	**0.106**	0.646	0.297
Ho	0.693	0.866	0.563	0.795	1.803	1.064	1.381	0.392	0.407	0.604	0.617	0.887	2.452	3.687	0.506	0.361	0.387	0.482	**0.111**	0.668	0.318
Y	0.758	0.953	0.639	0.863	1.899	1.198	1.508	0.480	0.498	0.658	0.736	1.018	2.798	3.412	0.584	0.450	0.375	0.454	**0.149**	0.658	0.306
Er	0.748	0.933	0.606	0.861	1.864	1.158	1.493	0.423	0.438	0.620	0.619	0.943	2.644	3.885	0.483	0.483	0.398	0.554	**0.101**	0.650	0.312
Tm	0.766	0.962	0.619	0.883	1.864	1.166	1.507	0.441	0.451	0.651	0.651	0.982	2.668	3.819	0.519	0.373	0.386	0.502	**0.120**	0.649	0.296
Yb	0.718	0.895	0.574	0.813	1.762	1.107	1.434	0.426	0.428	0.617	0.606	0.933	2.531	3.769	0.504	0.342	0.365	0.525	**0.112**	0.623	0.294
Lu	0.809	0.980	0.669	0.883	1.777	1.172	1.447	0.544	0.527	0.733	0.587	0.987	2.595	3.773	0.776	0.526	0.459	0.738	0.294	0.811	0.517
K2O	0.604	0.586	0.737	0.680	0.769	0.718	0.487	1.109	1.102	0.594	0.706	1.009	1.245	1.762	2.337	1.764	1.022	2.101	1.483	2.294	1.817
Ba	0.833	0.823	0.927	1.003	0.945	1.035	0.939	1.262	1.132	0.828	1.203	1.221	1.224	2.011	2.414	2.348	1.850	1.005	1.875	1.470	2.684	1.834
Rb	0.521	0.587	0.720	0.798	1.193	0.743	0.701	1.087	1.026	0.394	0.388	1.032	1.695	1.806	2.417	1.821	0.717	1.964	1.538	2.243	1.540
Li	0.867	0.816	0.909	0.999	1.331	1.056	0.792	1.174	1.180	0.964	1.261	1.066	1.760	2.578	2.548	2.219	1.415	1.557	1.419	2.077	1.775
S	1.684	1.860	1.533	1.782	4.176	1.707	2.340	1.080	1.069	1.735	1.575	1.301	2.896	4.948	1.733	1.221	1.444	3.004	2.189	3.418	3.247
CO2	2.190	2.310	2.195	2.225	3.893	2.278	2.631	2.409	2.447	2.162	2.568	2.309	2.843	3.867	3.966	3.364	2.800	2.806	2.978	4.221	3.125

compatible elements except the weak one to Ni. The correlation between Mg (and other compatible elements) and H_2O is at first sight puzzling, as it cannot be explained by magmatic processes. It seems that those rocks with high Mg contents are most strongly hydrated during the subsequent metamorphic alteration. This can be explained using the known mineralogical changes during metamorphism (cf. chapter 2 on petrology). The following qualitative argument, summed up in table 6.9, uses idealized mineral chemistry, but will hold also for real compositions. Mineral reactions are not completely balanced as the main concern is with MgO. Of the 4 major magmatic minerals, plagioclase has the smallest MgO content. Upon metamorphism, 1 f.u (formula unit) of anorthite is changed to 0.5 f.u. epidote, containing 0.5 OH groups, plus some minor kyanite or white mica (and little additional water) to account for surplus Al. The mineral with the next higher MgO content is clinopyroxene, idealized here as diopside. This changes to omphacite and further on to amphibole. 1 f.u. of diopside will form 0.2 f.u. tremolite (containing 0.4 OH groups), plus 0.3 f.u. epidote (to account for the Ca), containing 0.3 OH groups, totaling 0.7. orthopyroxene, idealized as 1 f.u. enstatite, will form 0.4 f.u. tremolite (some Ca

Tab. 6.6 continued

	La	Ce	Pr	Nd	Sm	Eu	Gd	Tb	Dy	Ho	Y	Er	Tm	Yb	Lu	K2O	Ba	Rb	Li	S	CO2
SiO2	-0.741	-0.898	-0.861	-0.996	-1.035	-0.342	-0.934	-0.954	-0.841	-0.837	-0.778	-0.801	-0.825	-0.835	-0.630	0.300	0.246	-0.109	0.309	0.244	0.570
Al2O3	-0.599	-0.762	-0.721	-0.893	-0.989	-0.335	-0.917	-0.974	-0.880	-0.868	-0.841	-0.846	-0.885	-0.867	-0.651	0.380	0.308	-0.350	0.448	0.058	0.454
Ga	-0.674	-0.821	-0.818	-0.957	-1.006	-0.191	-0.855	-0.929	-0.721	-0.705	-0.695	-0.653	-0.669	-0.669	-0.501	0.052	0.086	-0.718	0.275	0.034	0.387
CaO	-0.672	-0.814	-0.741	-0.893	-0.898	-0.344	-0.840	-0.846	-0.702	-0.692	-0.648	-0.684	-0.711	-0.675	-0.473	0.064	0.132	-0.646	0.324	-0.439	-0.035
Sr	0.035	0.057	0.143	0.052	0.075	0.320	-0.037	-0.047	-0.689	-0.736	-0.700	-0.693	-0.680	-0.645	-0.428	0.544	0.490	-0.511	0.163	-0.435	-0.119
MgO	-0.720	-0.902	-0.867	-1.023	-1.013	0.369	-0.967	-0.925	-0.808	-0.793	-0.835	-0.812	-0.808	-0.814	-0.588	0.327	0.200	-0.417	0.216	0.418	0.646
H2O	-0.486	-0.663	-0.611	-0.777	-0.904	-0.357	-0.899	-0.905	-0.901	-0.868	-0.871	-0.889	-0.891	-0.893	-0.600	0.609	0.362	-0.206	0.583	0.336	0.672
Fe2O3(t)	-0.315	0.114	-0.518	-0.560	-0.465	0.206	-0.365	-0.374	-0.130	-0.101	-0.187	-0.075	-0.109	-0.151	-0.142	-0.469	-0.318	-0.891	0.056	0.376	0.518
MnO	-0.477	-0.502	-0.660	-0.702	-0.567	0.068	-0.472	-0.422	0.179	0.145	-0.234	-0.115	-0.136	-0.160	-0.107	-0.462	-0.153	-0.903	-0.077	0.321	0.422
Na2O	-0.596	-0.680	-0.674	-0.682	-0.709	-0.127	-0.555	-0.610	-0.553	-0.594	-0.538	-0.487	-0.546	-0.570	-0.471	0.315	0.255	0.583	-0.086	0.503	0.514
Zn	-0.569	-0.701	-0.696	-0.651	-0.550	-0.037	-0.328	-0.278	-0.318	-0.390	-0.485	-0.280	-0.333	-0.321	-0.030	0.220	-0.102	0.898	-0.527	0.568	0.299
Co	-0.604	-0.805	-0.823	-0.980	-0.927	-0.341	-0.837	-0.814	-0.638	-0.621	-0.687	-0.695	-0.648	-0.658	-0.443	-0.006	-0.002	-0.735	0.264	0.226	0.510
Ni	-0.499	-0.713	-0.613	-0.802	-0.805	-0.283	-0.847	-0.785	-0.812	-0.805	-0.971	-0.869	-0.876	-0.837	-0.632	0.440	0.130	-0.438	0.224	0.298	0.513
Cr	-0.588	-0.622	-0.601	-0.716	-0.721	-0.454	-0.725	-0.678	-0.746	-0.737	-0.435	-0.778	-0.722	-0.759	-0.544	0.557	0.329	0.465	0.096	-0.062	0.160
TiO2	0.348	0.567	0.419	0.762	0.769	0.695	0.721	0.770	0.711	0.710	0.595	0.714	0.670	0.703	0.400	-0.565	-0.364	-0.005	-0.612	-0.307	-0.450
V	-0.130	0.114	-0.064	0.215	0.414	0.457	0.422	0.572	0.610	0.604	0.476	0.627	0.585	0.632	0.410	-0.498	-0.330	0.415	-0.526	-0.189	-0.412
Sc	-0.824	-0.726	-0.745	-0.653	-0.348	-0.215	-0.212	-0.016	0.075	0.057	0.182	0.110	0.145	0.139	0.137	-0.228	0.071	0.339	-0.470	-0.132	-0.462
P2O5	0.813	0.921	0.861	0.916	0.698	0.052	0.662	0.522	0.495	0.543	0.578	0.448	0.517	0.482	0.267	-0.623	-0.239	-1.255	-0.273	-0.010	-0.068
Zr	0.652	0.834	0.748	0.877	0.874	0.303	0.882	0.790	0.810	0.800	0.727	0.818	0.782	0.796	0.511	-0.653	-0.330	-0.863	-0.570	-0.130	-0.275
Nb	0.695	0.731	0.781	0.923	0.937	0.448	0.891	0.785	0.690	0.665	0.647	0.661	0.660	0.706	0.491	-0.382	-0.418	-0.185	0.078	-0.455	-0.672
Sn	0.656	0.686	0.785	0.828	0.786	0.231	0.726	0.642	0.668	0.632	0.647	0.639	0.663	0.669	0.387	-0.597	-0.359	-0.328	-0.643	-0.120	-0.496
La		0.970	0.921	0.845	0.622	-0.171	0.685	0.400	0.403	0.374	0.491	0.350	0.411	0.283	0.096	-0.010	-0.150	-0.292	0.041	-0.101	-0.350
Ce	0.027		0.971	0.944	0.799	-0.087	0.834	0.602	0.607	0.592	0.597	0.571	0.658	0.550	0.311	-0.111	0.050	-0.522	-0.026	-0.377	-0.636
Pr	0.045	0.038		0.950	0.794	-0.103	0.794	0.568	0.583	0.565	0.645	0.535	0.587	0.478	0.240	-0.034	-0.135	-0.205	0.054	-0.396	-0.825
Nd	0.085	0.065	0.023		0.938	0.184	0.927	0.779	0.779	0.743	0.763	0.740	0.731	0.678	0.352	-0.145	-0.280	-0.284	-0.377	-0.311	-0.693
Sm	0.184	0.157	0.086	0.023		0.065	0.979	0.913	0.923	0.821	0.866	0.897	0.861	0.833	0.449	-0.151	-0.262	0.029	-0.570	-0.328	-0.751
Eu	0.450	0.521	0.359	0.240	0.223		0.093	-0.020	0.188	0.139	0.281	0.260	0.242	0.163	-0.002	0.117	-0.303	-0.185	-0.037	0.379	0.338
Gd	0.167	0.123	0.093	0.031	0.015	0.287		0.930	0.907	0.844	0.778	0.878	0.825	0.814	0.450	-0.268	-0.346	-0.033	-0.567	-0.274	-0.653
Tb	0.311	0.258	0.194	0.092	0.038	0.316	0.031		0.960	0.904	0.848	0.930	0.874	0.898	0.548	-0.230	-0.320	0.079	-0.790	-0.014	-0.581
Dy	0.287	0.250	0.172	0.082	0.026	0.219	0.039	0.018		0.974	0.943	0.983	0.960	0.956	0.634	-0.569	-0.425	0.071	-0.735	-0.115	-0.606
Ho	0.301	0.257	0.179	0.095	0.041	0.232	0.064	0.039	0.009		0.956	0.966	0.959	0.954	0.648	-0.544	-0.382	-0.133	-0.544	-0.215	-0.582
Y	0.267	0.262	0.162	0.100	0.057	0.235	0.098	0.066	0.026	0.020		0.935	0.935	0.962	0.659	-0.430	-0.388	-0.079	0.419	-1.129	-1.308
Er	0.333	0.275	0.207	0.106	0.042	0.232	0.053	0.030	0.009	0.015	0.028		0.959	0.987	0.695	-0.545	-0.371	0.116	-0.502	-0.466	-0.667
Tm	0.305	0.226	0.186	0.111	0.056	0.241	0.076	0.054	0.018	0.018	0.029	0.017		0.974	0.673	-0.484	-0.339	0.229	-0.449	-0.282	-0.617
Yb	0.351	0.282	0.221	0.123	0.058	0.238	0.077	0.042	0.016	0.017	0.017	0.006	0.012		0.704	-0.505	-0.317	0.089	-0.611	-0.227	-0.582
Lu	0.536	0.477	0.403	0.322	0.254	0.391	0.286	0.234	0.182	0.175	0.180	0.159	0.171	0.151		-0.379	-0.252	0.297	-0.091	-0.105	-0.186
K2O	0.918	1.114	0.871	0.907	0.844	0.650	1.028	0.994	1.153	1.139	1.151	1.214	1.178	1.140	1.229		0.827	0.942	0.518	0.176	0.341
Ba	1.278	1.167	1.184	1.253	1.156	1.096	1.347	1.318	1.326	1.295	1.387	1.352	1.334	1.268	1.376	0.264		0.807	0.474	-0.403	-0.172
Rb	1.098	1.393	1.016	1.105	1.035	0.766	1.241	1.231	1.073	1.065	0.923	1.081	1.061	1.049	0.948	0.235	0.580		0.583	-0.156	-0.048
Li	0.968	1.086	0.928	1.074	1.200	1.162	1.325	1.495	1.347	1.275	1.022	1.313	1.305	1.348	1.235	0.632	0.832	0.711		-0.332	0.293
S	2.550	2.878	2.704	2.554	2.220	1.581	2.333	2.057	1.942	1.978	2.620	2.087	2.125	1.978	2.077	2.917	3.004	1.684	2.481		0.760
CO2	2.882	3.478	2.866	2.778	2.896	2.615	3.131	3.007	2.880	2.849	3.118	3.050	3.081	2.914	2.849	3.846	4.293	1.450	1.269	0.389	

being added from plagioclase decomposition), containing 0.8 OH groups. Olivine, idealized as 1 f.u. forsterite, is changed to a mixture of talc, chlorite, and chloritoid. These minerals contain, for 1 MgO from forsterite, 1.3, 2.6 and 4 OH groups respectively, resulting, regardless of the proportions of the mixture, in the highest water content of all former magmatic minerals. Thus the more magnesian the precursor magmatic mineral, the stronger is the hydration upon retrogression.

Among the group of incompatible elements there are two subclusters, one consisting of Ti, V and Sc, the other of Zr, P, Nb, Sn, REE (except Eu) + Y. The elements S, C and Eu are not correlated to any of the other elements.

When normal gabbros and Fe-Ti gabbros are treated separately, the picture is somewhat changed. Let us first consider the normal gabbros (tab. 6.7). Generally, all correlations are much stronger. In particular, Ti and V are more clearly incompatible, showing closer correlations to the REE (chiefly HREE), to Zr, P and Sr. According to log-ratio variances, Ti's relationship to Mn, Ca and the other compatible elements seems closer, but the log-ratio correlation coefficient testifies against

Tab. 6.7: log-ratio variances (lower left half) and log-ratio correlation coefficients (upper right half) for the normal metagabbros. Log-ratio correlation coefficients significant at the 5 % level and log-ratios < 0.1 are in bold print.

	SiO2	Al2O3	Ga	CaO	Sr	MgO	H2O(t)	Fe2O3(t)	MnO	Na2O	Zn	Co	Ni	Cr	TiO2	V	Sc	P2O5	Zr	Nb	Sn
SiO2		0.919	0.803	0.573	0.236	0.718	0.786	0.442	0.305	0.558	-0.145	0.521	0.694	0.276	-0.355	-0.471	-0.469	0.145	-0.311	**-0.670**	-0.466
Al2O3	0.031		0.903	0.779	0.454	0.542	0.738	0.301	0.124	0.366	-0.271	0.421	0.645	0.209	-0.412	**-0.622**	**-0.622**	0.313	-0.129	-0.462	-0.374
Ga	0.034	0.026		0.788	0.543	0.377	0.576	0.301	0.203	0.297	-0.135	0.274	0.442	-0.044	-0.148	-0.431	-0.476	**0.599**	0.174	-0.293	-0.057
CaO	0.073	0.055	0.043		**0.647**	0.116	0.316	-0.033	-0.088	-0.037	-0.521	0.095	0.332	0.148	0.002	-0.288	-0.298	0.475	0.132	-0.187	-0.102
Sr	0.225	0.199	0.158	0.128		-0.155	0.018	-0.210	-0.190	-0.060	-0.375	-0.050	-0.064	-0.031	0.144	-0.074	-0.006	0.288	0.331	0.025	0.066
MgO	0.095	0.151	0.182	0.258	0.469		0.903	0.836	0.730	0.184	-0.061	0.926	0.908	0.447	-0.517	-0.577	-0.618	0.313	-0.544	**-0.600**	**-0.690**
H2O(t)	0.153	0.141	0.204	0.297	0.516	**0.061**		0.721	0.558	0.168	-0.203	0.840	0.891	0.373	-0.705	-0.733	-0.662	0.033	-0.534	**-0.600**	**-0.690**
Fe2O3(t)	0.141	0.216	0.189	0.283	0.429	0.091	**0.059**		0.945	0.168	-0.065	0.881	0.687	-0.037	-0.278	-0.390	-0.387	0.212	-0.239	-0.225	-0.138
MnO	0.146	0.246	0.191	0.268	0.429	0.091	0.215	**0.018**		-0.152	0.110	0.788	0.529	-0.059	-0.027	-0.118	-0.119	0.200	-0.215	-0.229	-0.120
Na2O	0.148	0.227	0.226	0.330	0.461	0.328	0.435	0.410	0.411		0.418	-0.058	-0.307	-0.510	-0.111	-0.017	0.033	0.280	0.307	-0.057	0.169
Zn	0.322	0.463	0.365	0.494	0.617	0.441	0.642	0.410	0.332	0.259		0.850	0.529		-0.493	-0.520	-0.425	-0.070	-0.449	-0.353	**-0.551**
Co	0.194	0.242	0.274	0.337	0.508	**0.040**	**0.093**	**0.061**	0.372	0.481	0.858		0.531	-0.644	-0.706	-0.601	0.060	-0.483	-0.468		**-0.620**
Ni	0.448	0.411	0.536	0.593	0.931	0.211	0.167	0.372	0.481	0.858	1.132	0.216		-0.436	-0.178	-0.009	-0.551	-0.593	-0.511	-0.628	
Cr	1.277	1.331	1.503	1.367	1.629	1.119	1.197	1.581	1.571	1.541	2.177	1.201	1.025								
TiO2	0.226	0.350	0.237	0.214	0.285	0.445	0.674	0.357	0.259	0.523	0.354	0.375	1.122	1.825		**0.884**	**0.773**	0.216	0.479	0.181	0.523
V	0.393	0.582	0.448	0.417	0.477	0.620	0.915	0.548	0.410	0.443	0.462	0.743	1.380	1.623	0.098		0.028	-0.242	0.129	-0.031	0.131
Sc	0.409	0.601	0.478	0.436	0.460	0.611	0.901	0.565	0.425	0.497	0.452	0.717	1.434	1.729	1.862	1.302		0.629	0.640		
P2O5	0.211	0.221	0.120	0.155	0.286	0.398	0.472	0.278	0.260	0.481	0.297	0.478	0.804	2.109	0.231	0.477	0.521		**0.629**	0.610	**0.823**
Zr	0.199	0.260	0.156	0.171	0.218	0.424	0.579	0.324	0.286	0.330	0.233	0.500	0.997	1.894	0.106	0.248	0.293	0.111		**0.610**	**0.765**
Nb	1.182	1.276	1.088	1.040	1.025	1.426	1.738	1.167	1.127	1.405	1.106	1.422	2.184	3.293	0.828	1.010	1.092	0.799	0.581		
Sn	0.531	0.640	0.617	0.482	0.523	0.838	1.090	0.577	0.536	0.652	0.475	0.935	1.593	2.574	0.231	0.423	0.503	0.204	0.140	0.356	
La	0.329	0.337	0.327	0.285	0.388	0.467	0.548	0.400	0.430	0.616	0.509	0.486	0.771	1.914	0.465	0.736	0.742	0.267	0.122	0.642	0.205
Ce	0.280	0.311	0.221	0.185	0.219	0.487	0.595	0.369	0.328	0.596	0.458	0.520	0.945	2.106	0.264	0.462	0.480	0.202	0.162	0.606	0.254
Pr	0.253	0.285	0.252	0.179	0.281	0.447	0.563	0.379	0.370	0.518	0.428	0.487	0.862	1.806	0.264	0.470	0.481	0.202	**0.078**	0.548	0.176
Nd	0.234	0.304	0.242	0.188	0.287	0.443	0.616	0.361	0.331	0.419	0.315	0.505	0.953	1.819	0.152	0.305	0.328	0.178	**0.090**	0.559	0.201
Sm	0.254	0.364	0.292	0.233	0.326	0.453	0.682	0.380	0.327	0.396	0.343	0.522	1.033	1.717	0.102	0.197	0.226	0.262	0.130	0.855	0.299
Eu	**0.071**	0.137	0.140	0.305	0.233	0.338	0.195	0.182	0.149	0.225	0.351		0.744	1.757	0.188	0.347	0.400	0.176	**0.089**	0.468	0.184
Gd	0.321	0.438	0.354	0.406	0.361	0.549	0.804	0.449	0.399	0.446	0.326	0.614	1.184	1.980	0.137	0.240	0.250	0.294	0.294	0.468	0.184
Tb	0.398	0.562	0.455	0.406	0.452	0.607	0.908	0.514	0.429	0.484	0.339	0.687	1.302	1.862	0.113	0.134	0.139	0.387	0.146	0.674	0.232
Dy	0.280	0.429	0.328	0.291	0.383	0.468	0.727	0.378	0.298	0.396	0.301	0.551	1.131	1.773	**0.055**	**0.095**	0.120	0.286	0.103	0.686	0.232
Ho	0.270	0.415	0.314	0.271	0.415	0.446	0.687	0.355	0.276	0.423	0.349	0.529	1.092	1.729	**0.054**	**0.091**	0.121	0.273	0.117	0.744	0.292
Y	0.287	0.433	0.357	0.273	0.419	0.441	0.571	0.362	0.297	0.560	0.393	0.446	0.958	0.597	**0.037**	**0.050**	**0.081**	0.359	0.147	0.742	0.305
Er	0.253	0.396	0.316	0.262	0.352	0.427	0.653	0.377	0.286	0.384	0.317	0.474	1.026	1.356	**0.040**	**0.069**	**0.099**	0.332	0.107	0.674	0.242
Tm	0.317	0.483	0.349	0.321	0.405	0.520	0.778	0.399	0.294	0.471	0.384	0.610	1.269	1.958	**0.054**	**0.077**	**0.097**	0.290	0.132	0.674	0.242
Yb	0.337	0.491	0.353	0.331	0.389	0.554	0.813	0.434	0.334	0.456	0.377	0.642	1.298	1.783	**0.050**	**0.071**	0.111	0.289	0.103	0.658	0.223
Lu	0.697	0.858	0.684	0.668	0.668	0.918	1.111	0.712	0.595	0.820	0.528	0.927	1.783	2.398	0.362	0.353	0.398	0.605	0.353	0.868	0.639
K2O	0.396	0.441	0.518	0.503	0.694	0.510	0.454	0.770	0.772	0.486	0.735	0.719	0.795	1.085	0.803	0.849	0.790	0.880	0.801	1.494	1.414
Ba	0.492	0.551	0.536	0.471	0.482	0.743	0.798	0.875	0.773	0.621	0.992	0.899	1.365	1.325	0.565	0.563	0.575	0.872	0.732	1.723	1.184
Rb	0.714	0.955	1.055	1.043	1.207	1.083	1.097	1.360	1.290	0.405	0.169	1.471	1.836	1.134	0.726	0.530	0.592	1.634	1.120	1.704	1.253
Li	0.926	0.820	0.942	0.911	1.084	1.016	0.873	1.142	1.219	1.314	1.756	1.009	1.313	2.166	1.538	1.747	1.719	1.435	1.463	1.987	2.115
s	0.796	0.944	0.923	1.211	1.441	0.700	0.810	0.731	0.766	0.632	0.575	0.897	1.064	2.355	1.144	1.236	1.202	1.036	1.012	2.434	1.314
CO2	0.586	0.612	0.656	0.894	1.087	0.467	0.424	0.566	0.633	0.567	0.748	0.574	0.700	1.817	1.143	1.338	1.405	1.003	1.011	2.671	1.618

this. The same holds of the spurious affiliation of P to the compatible elements, caused by this element being very close to the l.o.d with the normal gabbros, and its accordingly very reduced spread. The elements Na and Eu (and weakly also Sr) correlate with the group Al, Si, Ga. These elements are evidently related to plagioclase. The Cr-Ni relationship is even weaker than with the complete sample.

Again, the picture is somewhat different for the Fe-Ti gabbros (tab. 6.8). Ti and V now align with the compatible elements, in particular with Fe and Mn. The same holds for Zr and Sc, presumably due to these elements' relationship to the element Ti and to Ti-minerals. Ni and Cr show an albeit weak correlation.

Accordingly the group of incompatible elements is reduced to the REE + Y, now containing also Eu (cf. missing Eu-anomaly for Fe-Ti gabbros), plus Nb and Sn. The relationships among the LILE K, Ba, Rb and Li remain unchanged. The elements S, C, Sr and P are not having any correlations to other elements.

Some of these inter-element correlations are plotted in fig. 6.11. With the help of these figures, some of the ambiguous cases discussed above, where e.g. the log-ratio variance is low (presumably good correlation), but the log-ratio correlation coeffici-

Tab. 6.7 continued

	La	Ce	Pr	Nd	Sm	Eu	Gd	Tb	Dy	Ho	Y	Er	Tm	Yb	Lu	K2O	Ba	Rb	Li	S	CO2
SiO2	-0.191	-0.380	-0.369	**-0.568**	**-0.715**	0.434	**-0.636**	**-0.735**	**-0.753**	**-0.653**	**-1.068**	**-0.740**	**-0.750**	**-0.758**	**-0.747**	0.205	0.136	0.043	-0.018	0.225	0.316
Al2O3	0.074	-0.085	-0.067	-0.333	**-0.607**	0.353	**-0.553**	**-0.750**	**-0.786**	**-0.699**	**-1.017**	**-0.779**	**-0.818**	**-0.766**	**-0.663**	0.206	0.127	0.093	0.146	0.124	0.219
Ga	-0.018	0.103	-0.119	-0.290	**-0.566**	0.470	-0.483	**-0.652**	**-0.646**	0.544	**-1.026**	**-0.727**	**-0.566**	-0.505	-0.460	-0.057	0.089	-0.273	-0.104	0.298	0.219
CaO	0.146	0.273	0.234	0.038	-0.198	0.163	-0.214	-0.422	-0.405	-0.283	-0.484	-0.370	-0.362	-0.382	0.002	0.251	-0.123	-0.019	0.155	0.271	0.120
Sr	0.128	0.413	0.202	0.083	-0.057	-0.079	0.008	-0.119	-0.200	-0.288	-0.446	-0.185	-0.172	-0.086	-0.070	-0.127	0.334	0.056	0.084	-0.445	0.120
MgO	-0.134	-0.453	-0.421	**-0.633**	**-0.677**	0.090	**-0.666**	**-0.635**	**-0.643**	-0.538	**-0.708**	**-0.599**	**-0.665**	**-0.704**	**-0.590**	0.141	-0.129	0.050	0.034	0.294	0.312
H2O	-0.032	-0.328	-0.321	**-0.655**	**-0.860**	0.129	**-0.842**	**-0.878**	**-0.896**	**-0.758**	**-0.588**	**-0.796**	**-0.871**	**-0.887**	**-0.565**	0.360	-0.018	0.211	0.319	0.062	0.267
Fe2O3(t)	-0.011	-0.153	0.266	-0.395	0.479	0.196	-0.429	-0.447	-0.391	-0.286	-0.472	-0.489	-0.340	-0.399	-0.257	-0.383	-0.394	-0.651	-0.345	**0.632**	-0.058
MnO	-0.174	-0.124	-0.361	-0.427	-0.415	0.140	-0.393	-0.314	-0.217	-0.106	-0.343	-0.250	-0.088	-0.183	-0.097	-0.481	-0.291	-0.563	-0.349	**0.662**	-0.098
Na2O	-0.398	**-0.647**	-0.519	-0.403	-0.323	**0.555**	-0.247	-0.210	-0.260	-0.332	**-1.006**	-0.297	-0.386	-0.291	-0.336	0.219	0.113	0.140	-0.150	0.171	0.312
Zn	-0.120	-0.215	-0.196	0.020	-0.081	0.286	0.129	0.181	0.101	-0.040	-0.307	-0.002	-0.078	-0.023	0.176	-0.175	-0.429	0.175	-0.405	0.385	0.094
Co	0.010	-0.267	-0.249	-0.477	-0.537	-0.074	-0.524	-0.537	**-0.553**	-0.465	-0.344	-0.400	**-0.590**	**-0.615**	**-0.590**	-0.084	-0.218	-0.233	0.007	0.496	0.212
Ni	0.152	-0.210	-0.117	-0.401	**-0.569**	-0.032	**-0.608**	**-0.648**	**-0.697**	**-0.602**	-0.462	**-0.571**	**-0.820**	**-0.810**	**-0.686**	0.261	-0.198	0.055	0.136	0.161	0.414
Cr	-0.281	**-0.642**	-0.357	-0.496	-0.355	-0.519	-0.517	-0.299	-0.391	-0.321	**1.290**	0.153	-0.542	-0.467	-0.412	0.466	0.326	0.346	0.164	-0.119	0.349
TiO2	-0.411	0.059	-0.104	0.247	0.493	-0.063	0.466	0.638	0.744	0.749	0.814	0.801	0.774	0.805	0.365	**-0.699**	0.051	-0.643	-0.501	0.269	0.473
V	**-0.635**	-0.239	-0.338	0.038	0.414	-0.227	0.359	0.675	0.764	0.777	0.947	0.867	0.831	0.829	0.460	-0.373	0.214	-0.607	-0.638	0.459	-0.401
Sc	**-0.600**	-0.247	-0.323	0.004	0.351	-0.379	0.357	0.678	0.705	0.696	0.859	0.786	0.770	0.732	0.397	-0.247	0.210	-0.161	-0.540	0.361	-0.238
P2O5	0.347	**0.634**	0.359	0.362	0.043	0.332	0.101	-0.055	-0.002	0.058	-0.396	-0.248	0.065	0.106	-0.030	**-0.552**	-0.357	-0.649	-0.102	-0.373	-0.059
Zr	0.144	0.426	0.277	0.578	0.509	0.189	0.642	0.503	0.478	0.417	0.157	0.409	0.408	0.572	0.377	**-0.797**	-0.379	-0.884	-0.407	-0.149	-0.281
Nb	0.475	0.565	0.635	0.674	0.655	0.082	0.728	0.432	0.418	0.319	0.320	0.448	0.432	0.454	0.315	-0.232	-0.332	-0.050	-0.038	-0.350	0.409
Sn	0.463	0.632	0.513	0.725	0.655	0.358	0.677	0.524	0.565	0.405	0.349	0.583	0.541	0.589	0.144	**-0.914**	-0.453	-0.067	-0.232	-0.641	-0.148
La	0.019	0.969	0.694	0.363	0.140	0.432	-0.003	-0.072	-0.192	-0.477	-0.477	-0.135	-0.239	-0.167	0.094	-0.347	0.098	0.216	-0.291	0.118	
Ce	0.044	0.017	0.942	0.811	0.480	0.079	0.556	0.129	0.121	0.048	-0.141	-0.178	0.262	0.148	0.180	-0.150	0.062	-0.125	0.140	-0.394	-0.348
Pr				0.942	0.811	0.193	0.871	0.597	0.015	0.645	0.241	0.207	0.123	-0.103	-0.134	0.152	0.044	0.009	0.015	-0.237	-0.156
Nd	0.118	0.051	0.031		0.870	-0.031	0.916	0.638	0.602	0.487	0.214	0.330	0.421	0.425	0.225	-0.306	-0.413	-0.045	-0.012	-0.460	-0.094
Sm	0.212	0.128	0.091	0.024		-0.180	0.976	0.862	0.850	0.724	0.516	0.690	0.644	0.677	0.315	-0.351	-0.351	-0.227	-0.002	-0.236	-0.238
Eu	0.258	0.200	0.193	0.163	0.185		-0.052	-0.381	-0.270	-0.282	**-0.960**	-0.380	-0.097	-0.244	-0.189	-0.311	-0.148	0.364	-0.160	0.311	-0.353
Gd	0.214	0.130	0.098	0.025	0.012	0.221		0.869	0.828	0.651	0.460	0.681	0.586	0.636	0.636	0.452	-0.509	-0.309	-0.114	-0.412	-0.287
Tb	0.412	0.291	0.240	0.110	0.052	0.336	0.045		0.936	0.884	0.942	0.883	0.908	0.520	**-0.558**	-0.228	-0.026	-0.319	-0.028	-0.294	
Dy	0.351	0.222	0.185	0.078	0.029	0.215	0.045	0.023		0.930	0.894	0.885	0.900	0.540	-0.488	-0.149	0.010	-0.243	-0.087	-0.284	
Ho	0.393	0.244	0.208	0.102	0.055	0.222	0.089	0.058	0.014		0.924	0.941	1.017	0.419	0.139	-0.316	0.225	0.257	-0.652	-0.378	
Y	0.416	0.261	0.231	0.135	0.083	0.288	0.126	0.070	0.023	0.015		0.915	1.004	0.621	**-0.788**	-0.223	0.010	-0.222	-0.126	-0.355	
Er	0.447	0.277	0.245	0.120	0.055	0.212	0.079	0.047	0.012	0.022	0.013		0.915	1.004	0.621	-0.456	0.003	0.021	-0.247	-0.149	-0.399
Tm	0.399	0.205	0.218	0.128	0.080	0.212	0.113	0.085	0.028	0.028	0.017	0.022		0.952	0.569	-0.533	-0.028	0.059	-0.300	-0.069	-0.307
Yb	0.451	0.246	0.258	0.135	0.078	0.254	0.103	0.060	0.025	0.027	0.005	0.005	0.013		0.624	-0.357	-0.077	0.269	0.000	-0.055	-0.128
Lu	0.733	0.467	0.532	0.411	0.378	0.535	0.417	0.343	0.296	0.288	0.314	0.260	0.275	0.249		0.779	0.693	0.396	-0.110	-0.237	
K2O	0.569	0.622	0.523	0.608	0.624	0.568	0.818	0.861	0.730	0.709	0.430	0.781	0.734	0.792	1.094		0.779	0.693	0.396	-0.110	-0.273
Ba	0.932	0.594	0.726	0.743	0.684	0.598	0.874	0.847	0.683	0.653	0.683	0.657	0.605	0.633	0.953	0.197		0.695	0.523	-0.110	-0.273
Rb	1.056	1.078	0.859	0.839	0.690	0.755	0.777	0.745	0.681	0.789	0.729	0.647	0.613	0.701	0.523	0.080	0.217		0.790	-0.482	-0.191
Li	1.204	1.187	1.109	1.339	1.455	1.104	1.596	1.888	1.598	1.485	0.859	1.405	1.470	1.620	1.541	0.756	0.834	0.368		-0.364	-0.191
S	1.160	1.263	1.238	1.110	1.118	0.733	1.182	1.042	1.017	1.083	1.516	1.187	1.160	1.142	1.376	1.042	1.838	3.590	3.311		0.088
CO2	1.288	1.350	1.295	1.230	1.257	0.684	1.350	1.394	1.215	1.201	1.503	1.195	1.285	1.293	1.385	0.792	1.458	3.403	2.385	3.279	

89

ent is close to 0 (presumably bad correlation), can be resolved. This concerns element pairs like P-Si, Ti-Si, Zr-Sc. The diagram Zr-Ti discloses another fact, hidden by the statistics: the clear correlation among the normal gabbros shown by the log-ratio variance is confirmed. Some of the Fe-Ti gabbros are lining up with this trend, but four specimens (Hi 32-2-96, Hi 32-4-96, Hi 32-5-96, Hi 32-6-96) do not. We will comment on this finding below.

Few of the presumably good correlations are as obvious in the diagrams as the Mg-Ni relationship, resembling closely the textbook olivine fractionation line (cf. the outline in the figure, representing data from the Mid-Atlantic Ridge from Schilling et al. 1983). Others, like Ga-Al and the very similar Ga-Si, look quite inconspicuous. With the crude correlation coefficient, this element pair has insignificant values of around -0.5 for the Fe-Ti gabbros, 0.5 for the normal gabbros. The good correlation disclosed by the log-ratio method is due to the close clustering of the data points. This just as much deserves an explanation as a linear array like Mg-Ni. If Ga was governed by a process independent of Si and Al, a larger (or, at any rate, different) spread might be expected. As Ga is known to concentrate in plagioclase, due to its

Tab. 6.8: log-ratio variances (lower left half) and log-ratio correlation coefficients (upper right half) for the Fe-Ti metagabbros. Log-ratio correlation coefficients significant at the 5 % level and log-ratios < 0.1 are in bold print.

	SiO2	Al2O3	Ga	CaO	Sr	MgO	H2O(t)	Fe2O3(t)	MnO	Na2O	Zn	Co	Ni	Cr	TiO2	V	Sc	P2O5	Zr	Nb	La
SiO2		**0.863**	**0.847**	**0.871**	**-0.691**	**0.726**	0.029	0.605	**0.673**	0.415	0.441	**0.797**	0.305	**0.139**	0.423	0.517	0.670	**-0.428**	**-0.100**	**-0.480**	**-0.53**
Al2O3	**0.011**		0.590	0.605	-0.375	**0.754**	0.474	0.325	0.452	0.434	0.332	0.639	0.446	0.354	0.121	0.176	0.386	-0.408	-0.313	-0.468	-0.501
Ga	**0.022**	0.049		**0.860**	**-0.953**	**0.704**	-0.290	**0.850**	**0.824**	0.145	0.580	**0.863**	0.179	-0.226	**0.697**	**0.819**	**0.852**	-0.476	-0.109	-0.149	-0.637
CaO	**0.051**	0.086	0.037		-0.748	**0.682**	-0.277	**0.714**	**0.740**	0.032	0.419	**0.852**	0.358	0.146	0.582	0.658	**0.779**	-0.359	-0.167	-0.256	-0.654
Sr	1.317	1.220	1.612	1.715		-0.619	0.508	**-0.897**	**-0.855**	-0.160	-0.255	**-0.714**	0.181	0.415	**-0.867**	**-0.862**	**-0.835**	-0.286	-0.373	-0.495	**0.85**
MgO	**0.025**	0.023	0.038	0.073	1.361		0.267	**0.687**	**0.742**	-0.148	0.428	**0.904**	0.551	0.225	0.418	0.494	**0.738**	-0.625	-0.465	-0.359	**-0.94**
H2O(t)	0.191	0.125	0.295	0.378	0.774	0.184		-0.473	0.378	0.155	0.099	-0.068	0.572	0.643	**-0.692**	-0.644	-0.422	-0.223	-0.477	-0.368	-0.013
Fe2O3(t)	0.158	0.209	0.084	0.114	2.127	0.132	0.574		**0.973**	-0.188	0.376	**0.841**	-0.027	-0.439	**0.912**	**0.953**	**0.988**	-0.470	-0.067	-0.108	**-0.76**
MnO	0.171	0.212	0.109	0.120	2.179	0.143	0.582	**0.015**		-0.122	0.298	**0.872**	-0.023	-0.348	**0.878**	**0.885**	**0.972**	-0.386	-0.087	-0.195	**-0.71**
Na2O	**0.043**	0.046	0.095	0.169	1.126	0.104	0.175	0.305	0.328		-0.042	-0.157	-0.308	-0.069	-0.161	-0.141	-0.188	0.203	0.510	-0.210	0.491
Zn	0.106	0.123	**0.090**	0.154	1.341	0.112	0.263	0.231	0.285	0.175		0.537	0.554	-0.013	0.074	0.305	0.394	**-0.801**	-0.554	-0.349	-0.524
Co	0.179	0.205	0.123	**0.095**	2.132	0.131	0.499	**0.090**	**0.075**	0.381	0.223		0.405	0.044	**0.594**	**0.687**	**0.882**	-0.460	-0.383	-0.374	**-0.76**
Ni	0.441	0.400	0.490	0.437	1.253	0.362	0.327	0.734	0.766	0.608	0.337	0.480		**0.759**	-0.315	-0.177	0.066	-0.593	**-0.768**	-0.275	-0.641
Cr	0.789	0.713	0.981	0.840	1.073	0.780	0.500	1.409	1.386	0.864	0.939	1.084	0.340		-0.609	-0.584	-0.322	0.006	-0.394	-0.275	-0.132
TiO2	0.671	0.772	0.517	0.533	3.341	0.659	1.423	0.234	0.243	0.869	0.858	0.502	1.646	2.534		**0.966**	**0.875**	-0.309	0.149	0.119	-0.566
V	0.715	0.833	0.523	0.545	3.497	0.701	1.498	0.241	0.279	0.948	0.784	0.460	1.572	2.626	**0.057**		**0.922**	-0.417	0.085	0.049	**-0.685**
Sc	**0.114**	0.161	**0.062**	**0.075**	1.943	**0.094**	0.495	**0.007**	**0.019**	0.258	0.195	**0.072**	0.632	1.237	**0.295**	0.306		**-0.511**	-0.104	-0.214	**-0.82**
P2O5	0.892	0.907	1.012	1.083	2.238	1.017	1.036	1.338	1.326	0.694	1.346	1.465	1.911	1.514	1.960	2.237	1.287		0.576	0.193	**0.99**
Zr	0.117	0.149	0.161	0.241	1.301	0.182	0.338	0.321	0.362	**0.058**	0.496	0.848	1.064	0.776	0.899	0.285	0.531			0.251	0.573
Nb	0.490	0.506	0.470	0.597	1.967	0.500	0.686	0.643	0.735	0.439	0.631	0.905	1.063	1.443	1.001	1.155	0.648	0.879	0.343		0.241
La	0.263	0.274	0.357	0.476	0.499	0.372	0.317	0.669	0.704	0.116	0.431	0.794	0.986	1.045	1.315	1.491	0.620	0.213	0.104	0.392	
Ce	0.279	0.310	0.341	0.501	0.427	0.392	0.376	0.645	0.736	0.133	0.340	0.808	1.089	1.064	1.366	1.368	0.612	0.141	**0.090**	0.439	0.011
Pr	0.298	0.309	0.414	0.509	0.284	0.394	0.316	0.743	0.800	0.150	0.490	0.906	0.909	0.888	1.382	1.599	0.677	0.262	**0.094**	0.262	0.05
Nd	0.216	0.236	0.312	0.400	0.314	0.286	0.303	0.571	0.652	0.110	0.411	0.820	0.822	1.014	0.983	1.267	0.510	0.171	**0.052**	0.115	0.02
Sm	0.140	0.172	0.194	0.278	0.384	0.192	0.276	0.384	0.455	**0.086**	0.266	0.578	0.669	0.961	0.850	1.011	0.341	0.587	**0.030**	0.158	0.15
Eu	0.226	0.267	0.220	0.253	0.716	0.217	0.495	0.236	0.304	0.311	0.298	0.463	0.637	1.222	0.440	0.586	0.213	1.420	0.286	0.257	0.673
Gd	0.116	0.156	0.153	0.249	0.603	0.185	0.291	0.345	0.418	**0.060**	0.214	0.528	0.728	1.050	0.811	0.934	0.311	0.567	**0.023**	0.191	0.116
Tb	0.108	0.150	0.130	0.221	0.511	0.158	0.292	0.293	0.372	**0.080**	0.167	0.478	0.645	1.043	0.739	0.850	0.262	0.707	**0.048**	0.157	0.181
Dy	0.101	0.142	0.110	0.187	1.338	0.130	0.318	0.248	0.321	**0.096**	0.183	0.394	0.621	0.989	0.740	0.790	0.216	0.759	**0.052**	0.278	0.201
Ho	0.101	0.138	0.109	0.195	1.379	0.129	0.309	0.252	0.318	**0.088**	0.191	0.395	0.644	1.002	0.748	0.807	0.223	0.696	**0.046**	0.255	0.168
Y	0.148	0.174	0.189	0.280	1.384	0.168	0.315	0.372	0.445	0.118	0.298	0.541	0.655	0.953	0.868	1.017	0.323	0.304	**0.059**	0.235	0.05
Er	**0.078**	0.114	**0.086**	0.169	1.309	0.113	0.293	0.230	0.297	**0.068**	0.153	0.373	0.611	1.005	0.716	0.763	0.199	0.790	**0.048**	0.287	0.197
Tm	0.129	0.172	0.134	0.223	1.373	0.151	0.338	0.272	0.350	0.119	0.205	0.421	0.651	1.020	0.780	0.824	0.242	0.735	**0.062**	0.285	0.196
Yb	**0.097**	0.131	0.102	0.184	1.353	0.115	0.299	0.234	0.301	**0.096**	0.185	0.375	0.595	0.986	0.730	0.787	0.206	0.776	**0.064**	0.250	0.208
Lu	**0.094**	0.123	**0.095**	0.179	1.413	0.110	0.284	0.243	0.308	**0.092**	0.169	0.363	0.579	0.954	0.756	0.807	0.216	0.724	**0.069**	0.232	0.184
K2O	1.017	0.860	1.222	1.484	0.493	0.984	0.390	1.693	1.738	0.855	0.770	1.641	1.002	1.217	3.047	3.027	1.578	1.583	1.213	1.801	0.615
Ba	1.377	1.209	1.623	1.810	1.159	1.318	0.867	1.855	1.723	1.235	1.602	1.831	1.887	1.828	2.805	3.135	1.766	1.979	1.455	2.223	1.266
Rb	0.427	0.317	0.582	0.634	0.962	0.394	0.134	0.941	0.909	0.391	0.523	0.828	0.447	0.522	1.871	2.081	0.839	0.807	0.617	0.831	0.470

ionic radius being only slightly larger than Al's (Burton & Culkin 1972), this can be explained by the role played by plagioclase in the magmatic development of the gabbros. The similarity of the geochemical behavior of Al and Ga extends also to spinel. Ga has quite high distribution coefficients around 3 between spinel and melt (Horn et al. 1994).

The good correlation of Ti and V comes as no surprise, as these elements are having similar ionic radii. With the normal gabbros, they both behave largely incompatibly, the only minerals with significant contents being the pyroxenes. With the Fe-Ti gabbros, ilmenite as a potential cumulus phase could make both elements compatible (Correns 1978, Landergren 1974).

Thus, according to the correlation structure, considered together with the trace element chemistry, the following picture emerges: Part of the metagabbro chemistry is controlled by olivine. Not only was olivine present in the mantle source (this is not surprising and was already seen from the REE diagrams), but it probably played also a role as cumulus phase, considering the high Mg# and high normative olivine contents.

Therefore it has to be assumed that gabbroic magmas were derived by different degrees of partial melting from a source containing olivine (plus spinel) and also cpx, considering the falling REE curves. This magma further differentiated by form-

Tab. 6.8 continued

	Ce	Pr	Nd	Sm	Eu	Gd	Tb	Dy	Ho	Y	Er	Tm	Yb	Lu	K2O	Ba	Rb
SiO2	**-0.664**	**-0.744**	**-1.025**	**-1.013**	-0.259	**-0.761**	**-0.951**	-0.176	-0.231	**-0.568**	-0.046	**-0.267**	**-0.231**	-0.192	**-0.400**	-0.102	-0.021
Al2O3	**-0.750**	**-0.696**	**-1.034**	**-1.241**	-0.439	**-1.130**	**-1.429**	-0.516	-0.528	**-0.696**	-0.387	-0.573	-0.517	-0.429	0.077	0.284	0.441
Ga	-0.577	**-0.884**	**-1.101**	**-0.871**	0.030	-0.512	-0.467	0.119	0.095	-0.424	0.244	0.045	0.129	0.183	-0.816	-0.411	-0.368
CaO	**-0.760**	**-0.753**	**-0.924**	**-0.782**	0.127	-0.626	-0.587	0.002	-0.067	-0.348	0.047	-0.095	-0.024	0.007	**-0.741**	-0.447	-0.243
Sr	**0.953**	**1.122**	**1.402**	**1.778**	0.582	**1.527**	**1.791**	-0.566	**-0.690**	-0.597	-0.600	-0.528	-0.663	**-0.809**	**0.740**	0.505	0.357
MgO	**-1.077**	**-1.040**	**-1.242**	**-1.229**	-0.059	**-1.244**	**-1.247**	-0.244	-0.282	-0.483	-0.224	-0.254	-0.192	-0.147	-0.203	0.065	0.159
H2O	-0.211	0.000	-0.292	-0.533	-0.570	**-0.683**	**-0.925**	-0.568	-0.563	-0.487	-0.564	-0.502	-0.550	-0.464	**0.837**	0.649	**0.830**
Fe2O3(t)	**-0.710**	**-0.941**	**-0.989**	-0.854	0.396	-0.480	-0.285	0.158	0.131	-0.336	0.215	0.111	0.204	0.160	**-0.695**	-0.296	-0.539
MnO	**-0.805**	**-0.936**	**-1.087**	**-0.801**	0.284	-0.658	-0.561	-0.014	-0.022	-0.464	0.040	-0.059	0.041	0.008	-0.661	-0.135	-0.401
Na2O	0.363	0.282	0.049	-0.167	**-0.763**	0.155	-0.340	-0.048	-0.010	-0.180	0.139	-0.111	-0.154	-0.111	0.085	0.231	0.144
Zn	-0.212	**-0.717**	**-1.016**	**-0.738**	-0.040	-0.401	-0.135	-0.002	-0.074	-0.592	0.126	-0.029	0.068	0.040	0.312	-0.208	-0.011
Co	**-0.817**	**-1.012**	**-1.417**	**-1.131**	-0.002	**-0.947**	**-0.908**	-0.134	-0.160	-0.624	-0.111	-0.150	-0.095	-0.046	-0.476	-0.178	-0.193
Ni	**-0.842**	-0.489	**-0.674**	**-0.683**	0.005	**-0.846**	**-0.672**	-0.260	-0.359	-0.315	-0.304	-0.265	-0.221	-0.169	0.258	-0.064	0.491
Cr	-0.164	0.096	-0.275	-0.373	-0.380	**-0.681**	**-0.838**	-0.333	-0.389	-0.206	-0.452	-0.320	-0.371	-0.288	0.262	0.134	0.598
TiO2	-0.649	-0.652	-0.271	-0.121	**0.706**	-0.017	0.221	0.216	0.192	-0.074	0.279	0.137	0.238	0.168	**-0.880**	-0.364	-0.647
V	-0.503	**-0.801**	-0.648	-0.341	0.586	-0.133	0.108	0.284	0.243	-0.212	0.364	0.211	0.296	0.244	**-0.775**	-0.462	**-0.724**
Sc	**-0.820**	**-0.979**	**-1.026**	**-0.727**	0.378	-0.593	-0.417	0.119	0.066	-0.345	0.165	0.065	0.144	0.086	**-0.682**	-0.135	-0.491
P2O5	**1.112**	**0.919**	**1.333**	0.537	**-0.812**	0.619	0.150	0.042	0.201	**1.141**	-0.077	0.128	-0.022	0.121	-0.008	0.023	0.286
Zr	0.633	0.634	0.653	0.773	-0.274	**0.848**	0.589	0.593	0.636	0.581	0.605	0.564	0.474	0.428	-0.588	-0.133	-0.463
Nb	0.135	0.523	**0.953**	**1.033**	0.534	**0.906**	**1.234**	0.461	0.552	0.602	0.438	0.432	0.580	0.653	-0.554	-0.424	-0.125
La	**0.968**	0.825	**0.929**	0.225	**-1.192**	0.491	-0.063	0.021	0.194	**0.843**	-0.125	-0.068	0.079	0.535	0.210	0.145	
Ce		**0.868**	0.967	0.344	**-1.318**	0.692	0.202	0.244	0.400	**0.947**	0.192	0.386	0.119	0.280	0.400	0.249	0.027
Pr	0.040		**1.015**	0.703	**-0.741**	0.838	0.286	0.259	0.354	**0.827**	0.146	0.317	0.159	0.229	0.388	0.075	0.027
Nd	0.019	0.010		**0.898**	0.569	0.832	0.410	0.417	0.435	**0.831**	0.298	0.400	0.297	0.283	0.354	-0.456	-0.245
Sm	0.135	0.087	0.019		-0.150	0.775	0.733	0.782	0.763	**0.911**	0.633	0.787	0.633	0.557	0.264	-0.364	**-0.851**
Eu	0.705	0.540	0.107	0.212		-0.547	-0.089	0.142	-0.034	**0.848**	0.132	0.015	0.160	-0.055	0.244	-0.425	-0.239
Gd	0.086	0.098	0.046	0.015	0.263		0.816	0.835	0.873	0.722	0.785	0.860	0.857	0.676	0.225	-0.133	**-1.111**
Tb	0.149	0.144	0.065	0.016	0.186	0.011		1.060	1.031	0.750	0.991	1.047	0.967	0.882	0.371	**-0.806**	**-1.341**
Dy	0.159	0.161	0.076	0.020	0.181	0.016	0.003		0.969	0.803	0.925	0.972	0.961	0.952	**-0.741**	-0.593	-0.557
Ho	0.130	0.143	0.072	0.021	0.209	0.012	0.003	0.003			0.925	0.972	0.961	0.952	**-0.741**	-0.593	-0.557
Y	0.030	0.055	0.024	0.012	0.051	0.029	0.028	0.026	0.022		0.634	0.774	0.765	0.769	**-0.558**	**-0.808**	-0.153
Er	0.160	0.173	0.084	0.028	0.174	0.016	0.004	0.005	0.007	0.039		0.906	0.919	0.850	-0.652	-0.621	-0.627
Tm	0.140	0.159	0.088	0.027	0.220	0.020	0.010	0.004	0.005	0.030	0.014		0.939	0.908	-0.657	-0.593	-0.610
Yb	0.175	0.174	0.087	0.030	0.173	0.027	0.006	0.005	0.004	0.026	0.007	0.009		0.944	-0.660	-0.592	-0.485
Lu	0.149	0.162	0.088	0.036	0.208	0.025	0.011	0.010	0.010	0.005	0.025	0.013	0.013		**-0.712**	-0.647	-0.407
K2O	0.712	0.721	0.743	0.791	0.825	0.805	0.768	1.212	1.196	1.183	1.130	1.241	1.149	1.168		**0.831**	**0.771**
Ba	1.230	1.388	1.665	1.491	1.835	1.546	1.593	1.694	1.646	1.818	1.627	1.724	1.630	1.855	0.404		0.543
Rb	0.624	0.530	0.549	0.621	0.659	0.667	0.655	0.620	0.588	0.496	0.586	0.856	0.560	0.539	0.351	0.917	

Table 6.9 Mineral reactions responsible for correlation MgO - H_2O

an $=$ $CaO \cdot Al_2O_3 \cdot Si_2O_4 + 0.5\,(OH)$ $\rightarrow$ $0.5\,ep$ $=$ $0.5\,(CaO)_2 \cdot (Al_2O_3)_{1.5} \cdot (Si_3O_{11}) \cdot (OH) + Al + Si$

di $=$ $CaO \cdot MgO \cdot Si_2O_4 + 0.7\,(OH) + Al$ $\rightarrow$ $0.2\,tr$ $=$ $0.2\,(CaO)_2 \cdot (MgO)_5 \cdot (Si_8O_{22}) \cdot (OH)_2 +$

$+ 0.3\,ep$ $=$ $0.3\,(CaO)_2 \cdot (Al_2O_3)_{1.5} \cdot (Si_3O_{11}) \cdot (OH)$

en $=$ $(MgO)_2 \cdot Si_2O_4 + 0.8\,(OH) + 0.8\,CaO$ $\rightarrow$ $0.4\,tr$ $=$ $0.4\,(CaO)_2 \cdot (MgO)_5 \cdot (Si_8O_{22}) \cdot (OH)_2$

fo $=$ $(MgO)_2 \cdot SiO_2 + n\,(OH) + Al$ $\rightarrow$ $x * 0.67\,tlc$ $=$ $x * 0.67\,(MgO)_3 \cdot (Si_4O_{10}) \cdot (OH)_2 +$

$+ y * 0.67\,chl$ $=$ $y * 0.67\,(MgO)_3 \cdot (Al_2O_3) \cdot (SiO_2)_2 \cdot (OH)_4 +$

$+ z * cld$ $=$ $z * (MgO)_2 \cdot (Al_2O_3)_2 \cdot (SiO_2)_2 \cdot (OH)_4 +$

ing plagioclase cumulates (positive Eu- and Sr-anomalies) in the way of a layered intrusion. In the Allalin gabbro, these layers can be still observed despite the polymetamorphic history (Meyer 1983a). Also layers enriched in olivine or cpx were formed. The latter is attested by the fact that Cr and Ni do not behave coherently.

The Fe-Ti gabbros could have formed partly from

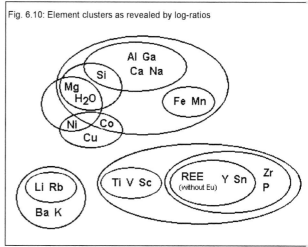

Fig. 6.10: Element clusters as revealed by log-ratios

magma depleted in Sr and Eu, squeezed off from the plagioclase cumulates (these are having small negative Eu- and Sr-anomalies). Some Fe-Ti gabbros however do not show any Eu- or Sr anomaly; perhaps these are derived from melts produced by very small degrees of partial melting, being consequently enriched in incompatible elements and crystallizing early Fe-Ti oxides. This remains at the moment somewhat unclear.

As for later alteration, there are no clear signals in the statistics.

Fig. 6.11: element-element plots for the metagabbros. Solid squares: normal gabbros; triangles and open squares: Fe-Ti gabbros

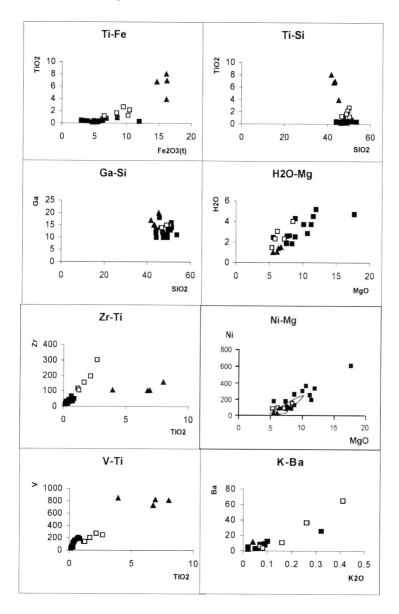

6.5. Results of the log-ratio method 3: principal component analysis

Besides the log-ratio variances, a principal component analysis was performed on the metagabbro data. Rather than to study the complete sample of all metagabbro specimens, the two subsets – normal gabbros and Fe-Ti gabbros – are considered separately. The factor loadings for both cases are contained in tables 6.10 and 6.11, scattergrams are given in figs. 6.12 and 6.13

The normal gabbro pca will be discussed first. On the whole it is rather similar to the melt-dominated model discussed above (chapter 6.3) and will be interpreted accordingly. Differently from the case of the serpentinites dealt with above, the two element groups found in the loadings of the first component do not correspond so well to the results of the log-ratio variance matrix. The first component, explaining about 37 % of sample variance, shows high positive loadings on Al, Mg, K, H_2O, S, CO_2, Cr, Co, Ni, and high negative loadings on Ti, Sc, V, Zr, Nb, Sn, Y + REE without Eu. The REE are more strongly loaded on the HREE than on the LREE. These are compatible vs. incompatible elements, but in contrast to the log-ratio correlation analysis, Si, Ca, Fe, Mn, Ga are missing in the list of compatible elements, K, Cr, Ni are included. This seems also conspicuous in relation to the model system, where Si and Ca are included in the compatible elements as reflected in component 1. These elements seem to occupy an intermediate position between compatibility and incompatibility.

The LREE display a gradient in their loadings, which is at variance with the interpretation of component 1 as reflecting compatibility and incompatibility: the LREE, although being usually more incompatible than the HREE due to their larger ionic radii, have smaller negative loadings. Maybe this gradient is more due to analytical scatter than to any real effects, considering that when all metagabbros are analyzed, the gradient is in the opposite direction. H_2O joins the compatible elements in virtue of its correlation with Mg, discussed already above in chapter 6.4

The second component, responsible for ca. 21 % of total sample variance, has positive loadings on the LILE K, Li, Rb, Ba, to which are added Sc, V, Cr. Negative loadings are on Fe, Mn, Mg, P, Co, Ni, Sn, Y, S, CO_2. This component unites two geochemical processes: the role played by the LILE and a differentiation between Ni plus Co (olivine) and Cr (spinel and clinopyroxene). The positive loading on Y, opposed to loadings close to zero on the REE, is often observed with pca's as a independent component of Y's behavior. On the whole, this component seems again to reflect magmatic processes. That P goes along with the olivine-compatible elements is not understood.

Component 3 explains about 15 % of sample variance. The significant negative loadings are on Al, Ca, Sr and also on Li, H_2O, Ni, Nb and LREE. Positive loadings

Tab. 6.10: factor loadings (pca) for normal gabbros (N = 13). (it is suggested to color the positive loadings > 0.1 in red, the negative loadings < -0.1 in blue in order to obtain a clearer picture)

	F1 36.76 %	F2 21.13 %	F3 15.12 %	F4 12.18 %	F5 6.81 %
SiO_2	0.0876	-0.0364	-0.0151	-0.0096	0.0654
TiO_2	-0.1107	0.0130	0.0673	0.0553	0.1983
Al_2O_3	0.1166	-0.0593	-0.1058	-0.0255	0.0860
Fe_2O_3 (t)	0.0846	-0.1817	-0.0275	0.0562	0.0697
MnO	0.0534	-0.1502	0.0044	0.0770	0.1387
MgO	0.1671	-0.1038	-0.0009	0.0803	0.0149
CaO	0.0363	-0.0188	-0.1464	0.0418	0.1439
Na_2O	0.0744	-0.0046	0.1830	-0.1361	0.0018
K_2O	0.1758	0.2303	0.0003	-0.1819	-0.2601
P_2O_5	-0.0589	-0.1968	-0.1155	0.0099	0.1090
H_2O (t)	0.2397	-0.0911	-0.0916	-0.0158	0.0324
S	0.1544	-0.3077	0.4439	-0.1516	-0.2303
CO2	0.3097	-0.2001	0.2269	-0.2299	0.0916
Li	0.2409	0.2524	-0.4027	-0.3838	0.1307
Sc	-0.1131	0.1166	0.1829	0.1136	0.1945
V	-0.1313	0.1046	0.1830	0.0721	0.2278
Cr	0.3385	0.3571	0.0802	0.5756	-0.2173
Co	0.1620	-0.1405	-0.0773	0.1358	0.0388
Ni	0.3110	-0.1742	-0.1316	0.2163	-0.1621
Zn	-0.0359	-0.0681	0.2429	-0.1969	-0.1579
Ga	0.0614	-0.0840	-0.0872	-0.0248	0.1478
Rb	0.0447	0.4739	0.2261	-0.2888	-0.2012
Sr	-0.0078	-0.0047	-0.1647	0.0171	0.2386
Y	-0.0596	0.1794	-0.0511	0.2623	-0.0391
Zr	-0.1041	-0.0883	-0.0180	0.0167	0.0556
Nb	-0.3125	-0.0185	-0.2908	-0.0809	-0.3255
Sn	-0.2327	-0.1274	-0.0278	0.0146	-0.0880
Ba	0.0760	0.3254	-0.0472	-0.1230	0.2722
La	-0.0431	-0.0808	-0.2052	-0.0525	-0.3337
Ce	-0.0982	-0.0527	-0.1930	-0.0612	-0.0505
Pr	-0.0739	-0.0145	-0.1636	-0.0295	-0.1801
Nd	-0.1068	-0.0260	-0.0665	0.0056	-0.1333
Sm	-0.1184	0.0160	-0.0017	0.0435	-0.1067
Eu	0.0012	-0.0780	0.0056	-0.1028	0.0143
Gd	-0.1564	-0.0145	-0.0032	0.0226	-0.1342
Tb	-0.1644	0.0146	0.1293	0.0937	-0.0777
Dy	-0.1293	0.0149	0.0885	0.0603	-0.0044
Ho	-0.1151	0.0182	0.0643	0.0745	0.0628
Er	-0.1015	0.0613	0.0632	0.1359	0.0376
Tm	-0.1432	0.0444	0.0593	0.0037	0.0888
Yb	-0.1534	0.0338	0.0833	0.0397	0.0936
Lu	-0.1652	0.0668	0.1003	-0.1291	0.1474

are on Na, S, Sc, V, Zn and Rb. This component is the plagioclase cumulate component known from the model system. The positively loaded elements do not seem to be related geochemically and are thought to have entered this component through the orthogonality constraint.

Component 4, responsible for still 12 % variance, is showing an opposition between LILE (Na, K, Li, Rb, Ba) plus S, CO_2, Zn against the group Sc, Cr, Co, Ni plus Y.

There is no component which could be clearly identified as being related to mobilization of elements during metamorphism. It is however interesting to note in this

Fig. 6.12: Pca scattergrams for normal metagabbros

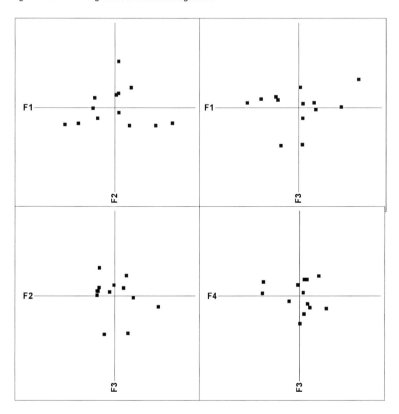

respect that the mobile LILE, although being incompatible, are only partly joined to the HFSE plus REE in the first component, but are mainly to be found in the second component.

Again, the picture is somewhat different for the Fe-Ti gabbros. Because the sample is much smaller here (only 9 specimens), the statistical results are less well founded than with the normal gabbros. Because with a smaller sample also the effects of the zero treatment are more detrimental, not the complete list of elements was used, but S, CO_2, Li and Sn were left out. It was found that changes in the list of elements used for the pca resulted in large differences between component structures, much larger than with the normal gabbros. The reason for this is also the small sample size. When all elements analyzed are used for the pca, component 1 of table 6.11 splits in two components, whereas component 2 of table 6.11 (with its

Tab. 6.11: factor loadings fo Fe-Ti gabbros (N = 9). It is suggested to color the positive loadings > 0.1 in red, the negative loadings < -0.1 in blue in order to obtain a clearer picture

	F1 50.53%	F2 23.47%	F3 13.30%	F4 12.30%	F5 4.87%
SiO_2	0.0418	-0.0677	0.0081	0.0511	-0.0769
TiO_2	0.3610	-0.0084	-0.2105	-0.0922	0.0214
Al_2O_3	0.0119	-0.1003	-0.0048	0.0727	-0.0177
Fe_2O_3 (t)	0.1916	-0.0994	-0.0885	-0.0020	-0.0404
MnO	0.1886	-0.1327	-0.1243	0.0533	-0.0797
MgO	0.0464	-0.1205	0.0101	0.0505	0.0152
CaO	0.1152	-0.0979	0.0955	0.0614	-0.1577
Na_2O	-0.0110	0.0127	-0.0422	0.0633	-0.0581
K_2O	-0.3635	-0.1806	-0.2055	-0.0753	0.2854
P_2O_5	-0.0974	0.4060	-0.0541	0.5097	-0.0323
H_2O (t)	-0.1214	-0.1524	0.0263	0.0708	0.0788
Sc	0.1666	-0.1127	-0.0547	-0.0011	-0.0764
V	0.3882	-0.0625	-0.1577	-0.1256	-0.0641
Cr	-0.1898	-0.2357	0.4916	0.2086	-0.2601
Co	0.1618	-0.2269	0.0208	0.1028	-0.1760
Ni	-0.0454	-0.3284	0.3920	-0.0238	0.0885
Zn	0.0364	-0.1373	0.0455	-0.0659	0.0370
Ga	0.0970	-0.0656	0.0069	0.0364	-0.0438
Rb	-0.1830	-0.1566	0.0630	0.2513	0.3728
Sr	-0.4125	0.0157	0.0268	-0.5269	-0.2934
Y	0.0166	0.1679	0.1003	0.0805	0.1372
Zr	0.0228	0.1294	-0.0508	0.0573	-0.0939
Nb	0.0641	0.2383	0.1176	-0.0086	0.4929
Ba	-0.3080	-0.2873	-0.6207	0.1730	-0.0016
La	-0.1172	0.1815	-0.0745	0.0615	-0.1208
Ce	-0.1114	0.2084	-0.0968	0.0383	-0.2660
Pr	-0.1312	0.2063	0.0300	-0.0242	-0.0990
Nd	-0.0766	0.2198	0.0378	-0.1194	0.0702
Sm	-0.0387	0.1207	0.0114	-0.1689	-0.0551
Eu	0.0544	-0.0427	0.0075	-0.3663	0.3561
Gd	-0.0226	0.1257	-0.0116	-0.1464	-0.1055
Tb	-0.0087	0.0927	0.0065	-0.1927	-0.0264
Dy	0.0476	0.0855	0.0583	-0.0160	0.0060
Ho	0.0443	0.0940	0.0407	0.0167	0.0242
Er	0.0453	0.0667	0.0317	-0.0221	0.0226
Tm	0.0463	0.1018	0.0602	-0.0056	-0.0047
Yb	0.0454	0.0686	0.0454	-0.0061	0.0720
Lu	0.0449	0.0742	0.0625	0.0301	0.0692

opposite loadings on Al and the REE) does not have a counterpart. Notwithstanding this instability, the pca with the reduced element list is discussed here, as it leads to a petrogenetically clearer picture.

Component 1, accounting for half of the sample variance, has positive loadings on Ti and related elements (Fe, Mn, Sc, V, Co) and on Ca, negative loadings on H_2O, LILE (K, Rb, Sr, Ba), Cr and LREE.

The second component is responsible for about 1/4 sample variance. It has high negative loadings on the ferromagnesian elements Fe, Mn, Mg, Sc, Cr, Co, Ni, Zn, on

Al, H$_2$O and on the LILE K, Rb and Ba. Significant positive loadings are on P, Zr, Nb, REE and Y. The HREE are loaded less strongly than the MREE and LREE.

Fig. 6.13: Pca scattergrams for Fe-Ti metagabbros

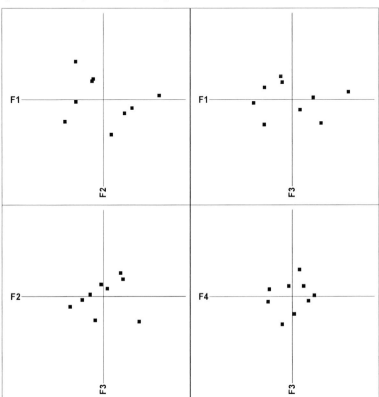

Component 3 and 4 are both responsible for about 12 % sample variance. Component 3 has significant negative loadings on Ti, Mn, K and Ba, positive loadings on Cr and Co. Component 4 has a high positive loading on P and high negative loadings on Eu and Sr.

The first two components can be interpreted as being related to compatibility vs. incompatibility. The first component reflects the Ti-mineral (or minerals) participating in the differentiation process. The positive loadings on Fe, Mn, V and Sc fit well into this interpretation. It is however somewhat surprising that Co is accepted, but Cr excluded.

Another possibility is that component 1 is not (or not solely) related to fractionation, but rather separates two distinct groups of Fe-Ti gabbros. The strongest hint that there are two such groups, probably having also different magmatic histories, was found above in chapter 6.4, fig. 6.11, plot of Ti against Zr, where the five specimens Hi 27-1-96, Hi 27-2-96, Hi 30-4-96, Hi 30-5-96, Hi 32-12-96 were found to continue the trend of the normal gabbros, whereas the 4 specimens Hi 32-2-96, Hi 32-4-96, Hi 32-5-96, Hi 32-6-96, are displaced to the Ti-rich side of this trend. Closer scrutiny reveals that these two groups are differing on several chemical characteristics. For the purposes of the present discussion, the first group will be called here "normal", the second group "extreme". These differences are displayed in table 6.13.

The many differences allow it to speak of two distinct kinds of Fe-Ti gabbros. The "normal" group more or less aligns with the normal gabbros, continuing their trends towards more enriched positions, whereas the "extreme" group is completely distinct. The former group should be rather called highly differentiated gabbros, whereas the latter group are the Fe-Ti gabbros sensu stricto. When, despite the very small sample size, separate pca's are performed on the two subsets, it is found that the highly differentiated gabbros have a first component quite similar to the normal gabbros, whereas the "extreme" Fe-Ti gabbros do not produce any interpretable picture. So no further efforts are presented here to interpret the pca in the light of fractionation processes.

Table 6.13: differences between two groups of Fe-Ti metagabbros

"normal"	"extreme"
negative coordinates with pca F1	positive coordinates with pca F1
$Zr/TiO_2 = 96.3 \pm 1.0$	$Zr/TiO_2 = 18.8 \pm 1.1$ (geometric mean, σ)
no Ti anomaly in spidergram	positive Ti anomaly in spidergram
	small Sr anomaly of either sign large negative Sr anomaly in
	spidergram (except specimen Hi-32-6-96: large +)
negative Eu anomaly	positive Eu anomaly
Sc_N/Cr_N around 1	Sc_N/Cr_N around 5 to 12
(except Hi 27-2-96 = 7)	
Sc/V = 0.12 - 0.25	Sc/V around 0.06
Fe_2O_3(tot) / MgO < 2	Fe_2O_3(tot) / MgO > 2 (cf. fig. 6.1)
	AFM triangle: intermediate position position close to F apex
	(around 60 % F), A low (12 - 13 %)
Mg # 59 to 69	Mg # 44 to 52
rich in epidote, no omphacite relics	low in epidote, many omphacite relics

Another puzzle is the petrographic difference between the two groups of Fe-Ti gabbros (s.l.). Those specimens with positive coordinates in component 1 abound in omphacite relics, whereas the specimens with negative coordinates are rich in epidote and lacking omphacite. Now the amount of epidote is mainly controlled by the degree of retrograde overprint. Although epidote is already stable and present at the eclogitic stadium, it does not become frequent before the retrograde stadia, growing chiefly in expense of garnet and phengite (Ganguin 1988, p. 105). The chemical differences between the two groups, as also reflected in the component loadings, are however not to be interpreted as a result of retrograde overprint. Rather the extreme Fe-Ti gabbros, due to their high omphacite content, reacted more competently during deformation (cf. Ganguin 1988, p. 49) and thus were not subjected to the same degree of retrograde overprint as the "normal" Fe-Ti gabbros.

6.6   Log-ratio analysis and the oxidation state of Fe

As with the serpentinites above, a regression of the state of oxidation of Fe, measured as $\log(FeO/Fe_2O_3)$, against the principal components was tried. With the Fe-Ti gabbros, no meaningful result could be expected due to this group being lumped together from two totally distinct rock types. With the normal gabbros, the regression is not significant at the 5 % level according to the F-statistics, although the mark is missed but narrowly. The regression produced depends most strongly on component 3. A Plot reveals that the correlation between component 3 and $\ln(Fe^{II}/Fe^{III})$ is produced by two specimens only (Hi 30-1-96 and Hi 30-3-96). There are no petrographical peculiarities of these specimens; their only chemical peculiarity is a high Ca content. On account of the F statistics and the dependence on only two specimens the regression is dismissed as insignificant.

6.7   Differentiation indices

From the paragraphs above it can be concluded that perhaps the pca coordinates can be used as a differentiation index. To test this, in fig. 6.14 some indices are plotted against each other and against the specimen coordinates in component 1. The results from the two separate pca's for the normal gabbros and for the Fe-Ti gabbros are plotted together. To get a clearer picture, the coordinates of the Fe-Ti gabbros have been multiplied by -1 (the signs of neither the factor loadings nor of the coordinates have any mathematical meaning), and both the coordinates of the Fe-Ti gabbros and of the normal gabbros have been shifted in opposite directions in order to link the two point clouds, produced from two separate pca's, into more or less continuous trends. Thus two different measures of differentiation (the two different pca's) are united in one plot. Of course it is also possible to produce a pca for all metagabbros together. It was found that the results are quite similar to those presented here.

The Mg number is the most time-honored of the differentiation indices. It is calculated here as $MgO/40.305/(MgO/40.305 + 0.85*Fe_2O_{3tot}/79.847)$, using total iron, as the oxidation state of Fe is surely not primary, and a correction factor of 0.85 to account for primary Fe^{III}. The correlation between Mg# and component 1 coordinates is very good. The slopes for the two gabbro types are slightly different, reflecting different fractionation paths. This difference in slope is also observed when a single pca for all specimens is used or when the two groups of Fe-Ti gabbros are analyzed separately.

The Mg# is seen to correlate also very well with TiO_2 content, although the slope is much less steep for the normal gabbros than for the Fe-Ti gabbros. A difference in slope exists also between the two groups of Fe-Ti gabbros.

A correlation between Mg# and the Y and Zr contents is only observed for the normal gabbros and the "normal" Fe-Ti gabbros (or highly differentiated Ti-rich normal gabbros). On the other hand, with the Fe-Ti gabbros, there is a correlation between Mg# and the two parameters normative plagioclase and Sr-anomaly (calculated as $2*Sr_N/(Ce_N+Nd_N)$). The Eu anomaly is perhaps weakly correlated with the Mg# of the normal gabbros and of the highly differentiated Ti-rich normal gabbros.

With component 1 coordinates, the picture is very similar. Component 1 coordinates are well correlated with TiO_2 contents, for the normal gabbros and the highly differentiated Ti-rich gabbros also with Zr and Y contents and the Eu anomaly (this latter correlation is more clear than with the Mg#), and for the Fe-Ti gabbros with Sr anomaly and normative plagioclase.

To sum up, the interpretation of the first components of the pca's as being related to magmatic differentiation is strengthened by the comparison with other differentiation indices. As the pca components are rendering a completely "intrinsic" measure, they seem to reflect differentiation more clearly than "extrinsic" measures like the Mg# (cf. the better correlation with TiO_2, Zr and Y). On the other hand, extrinsic measures allow the comparison of data from different suites of rock.

Fig. 6.14: Correlation of differentiation indices. Solid squares: normal gabbros; outline squares and tri-angles: Fe-Ti gabbros. See text for discussion.

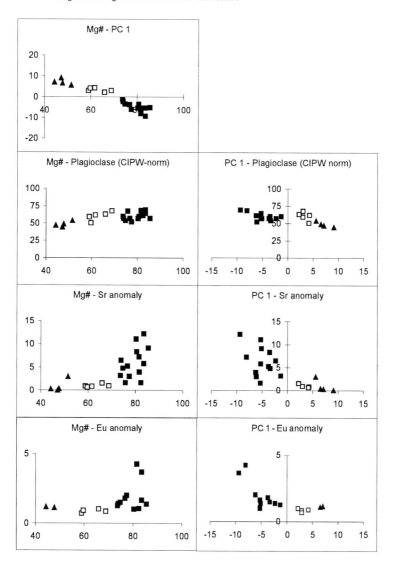

Fig. 6.14 continued

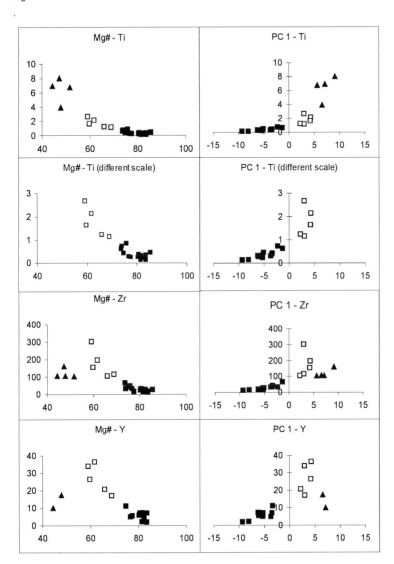

6.8 Chemical changes during retrograde metamorphism

As a tool to study chemical changes during metamorphism (or other processes potentially altering the chemical composition of rocks) in cases where there is no 1:1 relationship between protolith and the altered rock, Beach & Tarney (1978) proposed a statistical technique using correlation matrices. These matrices are calculated for two separate subsets of the rock suite studied, e.g. protoliths and metamorphic equivalents. Changes in the degree of correlation between element pairs then are interpreted in the light of the processes involved (e.g. metamorphic processes), which might be responsible for alteration. In their paper, Beach & Tarney used the crude correlation coefficient (and a derived quantity, the Fisher transform), as no other statistical tool was available at the time. Their technique can however be easily transferred to log-ratio variances. Himmelheber & Sheraton (in prep., see appendix 3) did this to the data set used by Beach & Tarney and could show that when using log-ratios a clearer and more easily interpretable picture emerges as compared to the crude correlation coefficients.

This method is tried here on the metagabbro specimens. Two subsamples were selected from the normal gabbros. The Fe-Ti gabbros were excluded, because this group does not contain strongly retrogressed specimens and in order not to distort the chemical relationships. The subsample of more or less unretrogressed normal gabbros contains the specimens Hi 4-96, Hi 19-96, Hi 32-8-96, Hi 32-11-96. Specimen Hi 32-10-96, although belonging petrographically to this group, was excluded due to its deviatory REE pattern. The subsample of strongly retrogressed specimens contains CM 2-98, Hi 16-96, Hi 18-96, Hi 24-96, Hi 30-1-96, Hi 30-3-96, Hi 30-6-96. These rocks contain albitic porphyroblasts and much hornblende, whereas the weakly to unretrogressed specimens are characterized by abundant omphacite relics. Different from the case of Lewisian gneisses discussed by Beach and Tarney (1978), the mineralogy of the retrogressed specimens is varying more strongly than the mineralogy of the specimens dominated by minerals from the HP-metamorphic stage.

Table 6.13 contains the log-ratio variance matrices of both samples and the log-ratio correlation coefficients.

In interpreting the differences between the two matrices, it has to be taken into account that probably a primary petrographic and chemical difference existed between the two subsamples. Primarily clinopyroxene-rich (now omphacite-rich) specimens may have reacted more competent to deformation and were thus potentially less prone to retrograde metamorphism as compared to more plagioclase-rich specimens. It has also to be kept in mind that both subsamples are extremely small.

The differences between the two subsamples can be also analyzed statistically. Following Beach & Tarney (1978), for the log-ratio correlation coefficients statistical testing is possible (putting aside the question of the legitimization of the log-ratio correlation coefficient itself) when instead of the correlation coefficients their Fisher-transforms are used. These are calculated as

$z = \ln((1+r)/(1-r))/2$.

For the difference between two log-ratio correlation coefficients then the sample error is $sqrt((n_1 + n_2 -6)/(n_1-3)/(n_2-3))$. To this the standard t-test can be applied. It turns out that the absolute value of the difference of two log-ratio correlation coefficients has to be larger than 1.726 to be significant at the 5 % level. There is no straightforward meaning attached to the sign of the difference of Fisher-transforms, because the signs of the transforms themselves have to be taken into account.

For variances, the significance of different values is tested by the F-test on their quotient. The larger variance has to be the numerator, the smaller one the denominator. In order to be able to know which of the two subsample has the larger and which the smaller variance, the following formula was used in the spread sheet: <Quotient> = IF(<Var1> > <Var2> ;<Var1>/<Var2>;-<Var2>/<Vari1>). This means that negative values represent correlations that are lower in the more strongly retrogressed subsample. It turns out that according to the F-test, positive values > 4.39 and negative values < -4.95 are significant at the 5 % level.

Both the Fisher transforms and the quotients of log-ratio variances as described are contained in table 6.14.

The following traits emerge from studying the tables:

Na looses its alignment with the incompatible elements upon retrograde overprint; its correlations with Ti, V, Sc, P, Zr, Nb, REE are much worse for the retrograde sample. This can be interpreted as due to the albitization observed petrographically. This effect is not seen in the raw chemical data. The change is significant according to the log-ratio variances.

K and Ba get correlated upon retrogression. This is presumably due to the greater importance played by sheet silicates in the retrogressed specimens. This change is only slightly below the 5 % significance limit.

Ti, V, Sc are having a more strongly incompatible character with the retrogressed specimens. There they seem also better correlated to the HREE, although this is not significant. There seems to be no way to relate this fact to the retrograde overprint, rather it seems to be related to the lack of pyroxenes in the respective protoliths. This lack is not adequately reflected in the CIPW norm data, probably due to a se-

Next pages:
Tab. 6.13: Log-ratio variances for little and for strongly retrogressed normal gabbros. Bold face: log-ratio correlation coefficients significant at the 5 % level and log-ratios < 0.1

Weakly retrogressed normal gabbros:

	SiO2	Al2O3	Ga	CaO	Sr	MgO	H2O(t)	Fe_2O_3(t)	MnO	Na2O	Zn	Cr	Co	Ni	TiO2	V	Sc	P2O5	Zr
SiO2		**0.930**	0.628	0.709	-0.230	0.638		0.267	0.271	0.163	-0.395	0.636	0.578	0.605	-0.106	0.066	0.300	**-0.916**	**-0.762**
Al2O3	**0.006**		**0.830**	**0.907**	-0.020	0.358		-0.094	-0.099	0.420	-0.644	0.704	0.260	0.392	0.005	0.272	0.482	**-0.982**	-0.551
Ga	**0.019**	**0.022**		**0.933**	0.412	-0.198		-0.506	-0.504	**0.850**	-0.563	0.348	-0.263	-0.183	0.493	0.746	**0.797**	**-0.809**	-0.011
CaO	**0.033**	**0.012**	**0.033**		0.358	-0.020		-0.489	-0.469	0.691	**-0.784**	0.653	-0.133	0.080	0.234	0.530	0.731	**-0.922**	-0.241
Sr	0.262	0.244	0.174	0.182		-0.708		-0.743	-0.637	**0.759**	-0.191	-0.128	-0.702	-0.630	**0.769**	**0.779**	**0.845**	-0.094	0.523
MgO	0.175	0.224	0.279	0.327	0.772			**0.824**	**0.824**	-0.644	0.029	0.483	**0.983**	**0.954**	-0.649	-0.672	-0.424	-0.360	**-0.954**
H2O(t)	0.248	0.292	0.373	0.408	0.973	**0.023**													
Fe2O3(t)	0.285	0.371	0.359	0.511	0.870	0.101	0.157		**0.987**	-0.744	0.583	-0.094	**0.905**	0.651	-0.442	-0.643	-0.606	0.122	-0.628
MnO	0.233	0.313	0.297	0.435	0.730	**0.089**	0.168	**0.010**		-0.717	0.585	-0.077	**0.910**	0.659	-0.385	-0.607	-0.519	0.097	-0.650
Na2O	0.114	**0.089**	**0.065**	**0.055**	**0.084**	0.567	0.694	0.672	0.581		-0.329	-0.057	-0.659	-0.640	**0.820**	**0.971**	**0.904**	-0.426	0.464
Zn	0.141	0.190	0.106	0.249	0.318	0.323	0.451	0.205	0.165	0.234		**-0.852**	0.206	-0.216	0.242	-0.098	-0.345	0.671	0.227
Cr	0.323	0.396	0.457	0.532	1.026	**0.032**	**0.052**	**0.077**	**0.081**	0.799	0.424		0.323	0.687	-0.531	-0.267	0.114	**-0.751**	-0.695
Co	0.797	0.856	1.009	0.999	1.724	0.276	0.220	0.569	0.573	1.479	1.164	0.253		**0.891**	-0.561	-0.644	-0.439	-0.259	**-0.899**
Ni	2.945	2.828	3.219	2.773	3.719	2.681	2.509	3.812	3.705	3.484	4.236	2.973	1.819		**-0.765**	-0.724	-0.397	-0.418	**-0.969**
TiO2	0.145	0.146	**0.085**	0.131	**0.081**	0.572	0.735	0.568	0.477	0.037	0.136	0.761	1.564	4.042		**0.932**	**0.778**	-0.030	0.593
V	0.139	0.121	**0.082**	**0.090**	**0.078**	0.606	0.751	0.667	0.574	**0.007**	0.210	0.826	1.579	3.764	**0.015**		**0.884**	-0.281	0.541
Sc	**0.071**	0.059	**0.041**	**0.036**	**0.069**	0.437	0.573	0.551	0.454	**0.019**	0.192	0.641	1.247	3.275	**0.041**	**0.027**		-0.559	0.197
P2O5	0.206	0.245	0.129	0.286	0.306	0.446	0.571	0.352	0.306	0.265	**0.052**	0.605	1.297	4.193	0.194	0.257	0.238		0.582
Zr	0.242	0.242	0.128	0.229	0.157	0.712	0.856	0.669	0.596	0.120	0.152	0.949	1.758	4.315	**0.092**	0.109	0.153	**0.087**	
Nb	1.077	1.151	0.872	1.208	0.939	1.532	1.737	1.088	1.061	0.953	0.501	1.690	2.935	7.112	0.718	0.867	1.004	0.396	0.437
Sn	0.997	1.039	0.733	0.984	0.370	1.675	2.003	1.221	1.145	0.710	0.462	1.922	3.324	5.797	0.531	0.652	0.699	0.314	0.290
La	0.277	0.336	0.215	0.403	0.444	0.393	0.518	0.283	0.243	0.432	**0.091**	0.516	1.121	4.147	0.337	0.427	0.362	**0.032**	0.218
Ce	0.259	0.304	0.174	0.346	0.321	0.485	0.625	0.383	0.329	0.327	**0.080**	0.642	1.316	4.241	0.241	0.317	0.282	**0.009**	0.120
Pr	0.123	0.152	**0.064**	0.185	0.245	0.367	0.481	0.324	0.275	0.182	**0.042**	0.531	1.179	3.871	0.138	0.182	0.156	**0.011**	**0.074**
Nd	0.135	0.151	**0.059**	0.165	0.173	0.473	0.602	0.434	0.373	0.116	**0.061**	0.662	1.383	4.011	**0.078**	0.110	0.114	**0.033**	**0.027**
Sm	0.120	0.127	**0.047**	0.126	0.117	0.497	0.634	0.487	0.414	**0.066**	**0.083**	0.698	1.425	3.923	**0.040**	**0.060**	**0.069**	**0.071**	**0.026**
Eu	**0.061**	0.084	**0.025**	0.113	0.194	0.305	0.422	0.288	0.232	0.112	**0.030**	0.455	1.109	3.723	**0.074**	0.108	**0.082**	**0.056**	**0.094**
Gd	0.188	0.200	**0.096**	0.195	0.113	0.606	0.768	0.549	0.469	**0.095**	**0.091**	0.805	1.619	4.284	**0.039**	**0.074**	0.101	**0.079**	**0.020**
Tb	0.178	0.180	**0.085**	0.163	0.083	0.620	0.777	0.607	0.520	**0.070**	0.126	0.841	1.611	4.078	**0.040**	**0.059**	**0.080**	**0.097**	**0.016**
Dy	0.113	0.121	0.044	0.124	0.135	0.483	0.614	0.467	0.399	**0.067**	**0.076**	0.678	1.413	3.944	**0.040**	**0.060**	**0.072**	**0.071**	**0.028**
Ho	**0.077**	**0.079**	**0.033**	**0.096**	0.240	0.399	0.482	0.413	0.372	**0.081**	0.100	0.588	1.263	3.695	**0.090**	**0.089**	**0.099**	0.108	**0.075**
Er	**0.080**	**0.079**	**0.049**	**0.076**	0.149	0.465	0.630	0.597	0.487	**0.034**	0.134	0.642	1.303	3.421	**0.015**	**0.024**	**0.023**	0.129	**0.007**
Tm	0.143	0.162	0.067	0.181	0.199	0.469	0.594	0.419	0.363	0.135	0.056	0.655	1.378	4.057	**0.094**	0.128	0.133	**0.025**	**0.030**
Yb	0.218	0.221	0.114	0.217	0.188	0.662	0.793	0.610	0.548	0.119	0.129	0.886	1.695	4.305	**0.091**	0.109	0.155	**0.078**	**0.004**
Lu	0.168	0.175	0.090	0.190	0.258	0.552	0.653	0.501	0.460	0.124	0.104	0.753	1.533	4.182	0.105	0.119	0.161	**0.081**	**0.033**
K2O	0.450	0.420	0.482	0.437	0.930	0.510	0.453	0.879	0.854	0.717	0.842	0.762	0.711	1.764	0.936	0.845	0.670	0.696	0.804
Ba	0.224	0.161	0.188	0.099	0.272	0.678	0.735	0.982	0.892	0.143	0.557	0.991	1.379	2.527	0.314	0.210	0.174	0.491	0.318
S	0.620	0.694	0.584	0.762	0.786	0.548	0.685	0.560	0.486	0.923	0.507	0.664	0.916	3.523	0.847	0.952	0.749	0.375	0.710
CO2	**0.046**	**0.054**	0.102	**0.090**	0.401	**0.095**	0.147	0.290	0.235	0.262	0.275	0.228	0.499	2.372	0.324	0.311	0.174	0.337	0.445

Strongly retrogressed normal gabbros:

	SiO2	Al2O3	Ga	CaO	Sr	MgO	H2O(t)	Fe_2O_3(t)	MnO	Na2O	Zn	Cr	Co	Ni	TiO2	V	Sc	P2O5	Zr
SiO2		**0.921**	**0.854**	0.476	0.139	**0.875**		0.732	0.640	0.718	-0.037	0.268	0.593	**0.867**	-0.648	-0.594	-0.570	0.269	-0.285
Al2O3	**0.039**		**0.958**	**0.771**	0.408	**0.774**		0.716	0.587	0.417	-0.358	0.101	0.594	**0.933**	-0.687	**-0.835**	**-0.783**	0.533	-0.101
Ga	**0.039**	**0.020**		**0.771**	0.333	0.681		0.706	0.584	0.399	-0.101	-0.066	0.509	**0.871**	-0.550	**-0.803**	**-0.772**	0.695	0.169
CaO	0.118	**0.083**	**0.060**		0.650	0.247		0.264	0.095	-0.171	-0.604	-0.221	0.218	0.652	-0.426	**-0.868**	**-0.779**	**0.760**	0.127
Sr	0.178	0.188	0.162	**0.075**		0.251		0.265	0.179	-0.549	**-0.908**	0.251	0.498	0.562	-0.363	-0.600	-0.390	0.302	0.048
MgO	**0.064**	0.100	0.128	0.269	0.257			**0.917**	**0.885**	0.596	-0.031	0.524	**0.889**	**0.896**	**-0.798**	-0.555	-0.508	0.012	-0.285
H2O(t)	0.112	0.085	0.142	0.297	0.348	**0.036**													
Fe2O3(t)	0.070	0.103	**0.085**	0.192	0.180	**0.041**	**0.088**		**0.984**	0.437	0.169	0.279	**0.912**	**0.871**	**-0.784**	-0.649	-0.615	0.206	0.055
MnO	0.071	0.135	**0.099**	0.181	0.150	**0.070**	0.151	**0.012**		0.445	0.323	0.309	**0.907**	**0.781**	-0.737	-0.528	-0.505	0.089	0.072
Na2O	0.202	0.358	0.343	0.577	0.701	0.263	0.355	0.326	0.311		0.536	0.223	0.213	0.318	-0.239	0.045	-0.065	-0.133	-0.310
Zn	0.608	0.912	0.678	0.884	0.976	0.737	1.029	0.538	0.436	0.407		-0.131	-0.137	-0.477	0.660	0.641	0.498	0.322	0.549
Cr	0.168	0.190	0.205	0.298	0.197	**0.055**	0.105	**0.050**	**0.075**	0.511	0.836		0.562	0.320	-0.447	0.046	0.113	-0.671	-0.592
Co	0.293	0.170	0.258	0.403	0.459	0.181	**0.086**	0.256	0.364	0.714	1.657	0.217		**0.840**	**-0.833**	-0.599	-0.514	-0.037	-0.131
Ni	0.552	0.713	0.758	0.807	0.556	0.426	0.636	0.560	0.525	0.754	1.197	0.401	0.830		**-0.769**	**-0.813**	-0.723	0.380	-0.086
TiO2	0.208	0.337	0.240	0.192	0.164	0.391	0.568	0.272	0.184	0.465	0.357	0.426	0.876	0.718		**0.774**	**0.803**	0.014	0.375
V	0.430	0.673	0.551	0.517	0.413	0.614	0.889	0.507	0.371	0.528	0.286	0.666	1.333	0.711	0.086		**0.967**	-0.591	-0.060
Sc	0.584	0.856	0.727	0.672	0.504	0.771	1.089	0.669	0.516	0.708	0.396	0.809	1.531	0.764	0.164	0.029		-0.541	-0.053
P2O5	0.263	0.210	0.126	0.103	0.247	0.471	0.483	0.310	0.301	0.712	0.504	0.518	0.585	1.320	0.267	0.639	0.799		0.644
Zr	0.223	0.294	0.176	0.159	0.156	0.371	0.503	0.200	0.139	0.553	0.352	0.357	0.745	0.873	0.061	0.241	0.357	0.145	
Nb	1.523	1.609	1.364	1.086	1.017	1.834	2.105	1.486	1.325	2.120	2.212	1.634	2.384	1.817	0.880	1.009	1.142	1.136	0.778
Sn	0.325	0.357	0.233	0.139	0.258	0.614	0.696	0.390	0.315	0.737	0.601	0.647	0.987	1.252	0.109	0.335	0.493	0.108	0.058
La	0.332	0.316	0.283	0.100	0.145	0.546	0.595	0.411	0.359	0.885	1.115	0.520	0.781	1.023	0.235	0.531	0.684	0.249	0.251
Ce	0.340	0.349	0.284	0.108	0.131	0.563	0.646	0.398	0.331	0.875	0.927	0.525	0.855	1.034	0.162	0.423	0.556	0.208	0.166
Pr	0.373	0.395	0.341	0.139	0.155	0.615	0.709	0.467	0.390	0.906	0.975	0.591	0.935	1.031	0.174	0.413	0.530	0.274	0.224
Nd	0.312	0.384	0.293	0.151	0.146	0.541	0.679	0.377	0.287	0.734	0.628	0.530	0.954	0.950	**0.060**	0.230	0.329	0.227	0.100
Sm	0.319	0.442	0.346	0.220	0.164	0.520	0.704	0.374	0.269	0.671	0.604	0.502	1.027	0.760	**0.037**	0.126	0.210	0.350	0.111
Eu	0.104	0.179	0.131	0.199	0.355	0.269	0.317	0.190	0.154	0.219	0.565	0.406	0.669	0.837	0.196	0.361	0.581	0.316	0.203
Gd	0.411	0.543	0.431	0.274	0.195	0.643	0.830	0.443	0.320	0.844	0.713	0.600	1.202	0.908	**0.035**	**0.097**	0.154	0.383	0.106
Tb	0.511	0.718	0.582	0.483	0.342	0.706	0.970	0.538	0.400	0.780	0.441	0.684	1.367	0.886	**0.084**	**0.053**	0.060	0.553	0.195
Dy	0.349	0.522	0.410	0.327	0.242	0.537	0.753	0.391	0.274	0.606	0.384	0.538	1.130	0.785	**0.029**	**0.051**	0.105	0.422	0.124
Ho	0.328	0.485	0.374	0.289	0.220	0.521	0.724	0.369	0.258	0.604	0.390	0.521	1.085	0.809	**0.021**	**0.067**	0.129	0.370	0.101
Er	0.367	0.553	0.432	0.366	0.274	0.537	0.763	0.383	0.264	0.589	0.374	0.527	1.163	0.756	**0.043**	**0.046**	0.114	0.467	0.124
Tm	0.404	0.587	0.487	0.360	0.283	0.630	0.837	0.465	0.330	0.703	0.561	0.631	1.292	0.834	0.043	**0.031**	**0.094**	0.520	0.176
Yb	0.373	0.553	0.428	0.351	0.256	0.554	0.784	0.399	0.278	0.625	0.409	0.541	1.163	0.761	0.035	**0.045**	0.102	0.444	0.115
Lu	1.200	1.459	1.247	1.152	1.014	1.374	1.637	1.036	0.862	1.388	0.880	1.246	2.253	1.795	0.603	0.527	0.637	1.135	0.608
K2O	0.328	0.514	0.594	0.605	0.645	0.407	0.505	0.595	0.552	0.271	0.793	0.690	0.884	0.655	0.486	0.510	0.658	1.063	0.821
Ba	0.560	0.715	0.779	0.678	0.639	0.683	0.843	0.797	0.717	0.711	1.207	0.793	1.193	0.441	0.627	0.639	0.768	1.225	0.924
S	1.114	1.326	1.304	1.839	2.011	1.023	1.014	1.041	1.072	0.586	0.942	1.349	1.576	2.315	1.546	1.535	1.651	1.758	1.545
CO2	1.140	1.179	1.233	1.724	1.836	0.851	0.733	0.902	1.027	0.935	1.421	0.987	0.990	2.005	1.894	2.172	2.482	1.753	1.700

Table 1

	Nb	Sn	La	Ce	Pr	Nd	Sm	Eu	Gd	Tb	Dy	Ho	Er	Tm	Yb	Lu	K2O	Ba	S
SiO2	-0.851		-0.747	-0.922	-0.922	-0.861	-0.645	-0.462	-0.675	-0.646	-0.592	-0.129		-0.876	-0.725	-0.536	0.247	0.270	
Al2O3	-0.876		-0.918	-0.996	-0.970	-0.756	-0.468	-0.602	-0.553	-0.444	-0.431	0.028		-0.797	-0.534	-0.394	0.356	0.598	
Ga	-0.547		-0.983	-0.864	-0.770	-0.279	0.103	-0.311	-0.007	0.127	0.132	0.369		-0.367	-0.013	0.035	0.029	0.740	
CaO	-0.783		-0.965	-0.911	-0.901	-0.522	-0.157	-0.600	-0.269	-0.090	-0.163	0.077		-0.608	-0.274	-0.286	0.303	0.823	
Sr	0.057		-0.289	-0.014	-0.082	0.368	0.668	0.160	0.653	0.770	0.575	-0.024		0.236	0.401	0.092	-0.413	0.379	
MgO	-0.545		0.028	-0.311	-0.405	-0.820	-0.930	-0.308	-0.863	-0.952	-0.893	-0.530		-0.750	-0.906	-0.715	0.322	-0.367	
H2O																			
Fe2O3(l)	0.005		0.402	0.111	0.082	-0.363	-0.589	0.251	-0.465	-0.673	-0.521	-0.315		-0.269	-0.545	-0.339	-0.121	-0.803	
MnO	-0.028		0.402	0.117	0.050	-0.383	-0.578	0.268	-0.439	-0.643	-0.528	-0.429		-0.304	-0.589	-0.432	-0.181	-0.823	
Na2O	-0.113		-0.742	-0.477	-0.380	0.220	0.594	0.048	0.505	0.626	0.591	0.476		0.106	0.429	0.340	-0.307	0.637	
Zn	0.743		0.626	0.604	0.660	0.503	0.309	0.920	0.445	0.204	0.365	0.121		0.552	0.290	0.345	-0.765	-0.909	
Cr	-0.932		-0.489	-0.641	-0.772	-0.859	-0.740	-0.951	-0.811	-0.644	-0.787	-0.481		-0.874	-0.744	-0.738	0.806	0.576	
Co	-0.417		0.109	-0.223	-0.305	-0.723	-0.849	-0.130	-0.758	-0.888	-0.802	-0.499		-0.650	-0.842	-0.648	0.156	-0.515	
Ni	-0.671		0.001	-0.327	-0.468	-0.889	-0.988	-0.544	-0.946	-0.968	-0.986	-0.651		-0.830	-0.956	-0.827	0.543	-0.172	
TiO2	0.301		-0.344	-0.081	0.001	0.512	0.795	0.588	0.800	0.792	0.799	0.409		0.410	0.566	0.444	-0.792	0.093	
V	0.060		-0.616	-0.341	-0.237	0.354	0.705	0.282	0.649	0.719	0.709	0.494		0.242	0.512	0.417	-0.525	0.437	
Sc	-0.321		-0.723	-0.522	-0.540	-0.015	0.400	-0.011	0.359	0.477	0.358	0.073		-0.149	0.118	-0.048	-0.338	0.525	
P2O5	0.923		0.898	0.972	0.998	0.786	0.473	0.618	0.543	0.421	0.463	0.129		0.844	0.595	0.518	-0.331	-0.590	
Zr	0.764		0.187	0.498	0.627	0.948	0.955	0.505	0.913	0.933	0.944	0.614		0.911	0.986	0.850	-0.425	0.124	
Nb			0.681	0.834	0.940	0.926	0.712	0.802	0.765	0.633	0.734	0.423		0.961	0.798	0.755	-0.570	-0.516	
Sn	0.036																		
La	0.470	0.394		0.940	0.867	0.448	0.082	0.432	0.191	0.056	0.051	-0.268		0.524	0.184	0.107	-0.151	-0.734	
Ce	0.395	0.270	0.017		0.957	0.705	0.403	0.539	0.488	0.386	0.359	-0.090		0.751	0.477	0.333	-0.286	-0.592	
Pr	0.512	0.416	0.059	0.031		0.818	0.513	0.625	0.574	0.457	0.508	0.195		0.874	0.644	0.575	-0.336	-0.554	
Nd	0.488	0.364	0.119	0.061	0.015		0.909	0.697	0.917	0.865	0.907	0.522		0.990	0.949	0.831	-0.551	-0.195	
Sm	0.566	0.400	0.178	0.105	0.038	0.008		0.629	0.984	0.986	0.988	0.553		0.846	0.931	0.771	-0.636	0.060	
Eu	0.638	0.537	0.123	0.090	0.021	0.023	0.027		0.736	0.541	0.672	0.300		0.700	0.544	0.529	-0.927	-0.896	
Gd	0.471	0.297	0.193	0.108	0.060	0.069	0.018	0.010		0.052	0.967	0.968		0.858	0.885	0.712	-0.739	-0.103	
Tb	0.545	0.319	0.219	0.126	0.069	0.023	0.010	0.052	0.006		0.967	0.948		0.787	0.884	0.671	-0.584	0.136	
Dy	0.562	0.429	0.181	0.110	0.058	0.039	0.037	0.038	0.074	0.072		0.027		0.534	0.718	0.902	-0.205	0.247	
Ho	0.678	0.669	0.226	0.170	0.069	0.023	0.008	0.061	0.006	0.013	0.013			0.660	0.855	0.842	-0.660	0.028	
Er	0.262	0.362	0.316	0.188	0.094	0.029	0.011	0.051	0.015	0.017	0.008	0.044							
Tm	0.463	0.362	0.107	0.054	0.011	0.001	0.014	0.026	0.025	0.013	0.040	0.036			0.928	0.846	-0.508	-0.263	
Yb	0.431	0.339	0.208	0.118	0.063	0.021	0.023	0.023	0.022	0.053	0.004	0.022				0.925	-0.432	0.079	
Lu	0.482	0.482	0.209	0.137	0.057	0.027	0.035	0.063	0.050	0.056	0.026	0.020	0.025	0.015			-0.364	0.030	
K2O	1.966	1.893	0.707	0.735	0.600	0.681	0.707	0.638	0.873	0.797	0.708	0.573	1.018	0.680	0.780	0.710		0.533	
Ba	1.475	1.054	0.664	0.559	0.375	0.323	0.271	0.330	0.359	0.285	0.276	0.231	0.333	0.343	0.321	0.316	0.357		
S	1.119	0.898	0.216	0.292	0.409	0.540	0.617	0.511	0.660	0.664	0.641	0.730	1.020	0.531	0.723	0.759	0.738	1.023	
CO2	1.434	1.309	0.356	0.376	0.239	0.288	0.277	0.174	0.375	0.354	0.274	0.224	0.216	0.297	0.421	0.362	0.330	0.288	0.53

Table 2

	Nb	Sn	La	Ce	Pr	Nd	Sm	Eu	Gd	Tb	Dy	Ho	Er	Tm	Yb	Lu	K2O	Ba	S
SiO2	-0.717		-0.187	-0.401	-0.427	-0.695	-0.951	0.594	-0.886	-0.800	-0.859	-0.878		-0.827	-0.901	-0.717	0.386	0.140	
Al2O3	-0.511		0.164	-0.029	-0.101	-0.411	-0.814	0.494	-0.765	-0.870	-0.922	-0.896		-0.890	-0.964	-0.738	0.119	0.011	
Ga	-0.364		0.101	-0.026	-0.149	-0.357	-0.791	0.540	-0.718	-0.812	-0.876	-0.825		-0.916	-0.880	-0.645	-0.165	-0.219	
CaO	-0.026		0.667	0.569	0.487	0.219	-0.274	0.232	-0.220	-0.654	-0.681	-0.592		-0.577	-0.725	-0.597	-0.273	-0.093	
Sr	0.054		0.482	0.442	0.394	0.163	-0.047	-0.480	0.057	-0.243	-0.363	-0.328		-0.347	-0.373	-0.418	0.448	-0.054	
MgO	-0.684		-0.346	-0.546	-0.598	-0.862	-0.969	0.298	-0.934	-0.712	-0.810	-0.861		-0.872	-0.803	-0.547	0.341	0.099	
H2O(l)																			
Fe2O3(l)	-0.523		-0.299	-0.424	-0.563	-0.740	-0.926	0.342	-0.752	-0.664	-0.768	-0.784		-0.812	-0.731	-0.305	-0.160	-0.246	
MnO	-0.508		-0.427	-0.526	-0.666	-0.789	-0.894	0.333	-0.661	-0.553	-0.664	-0.692		-0.685	-0.511	-0.167	-0.261	-0.076	
Na2O	-0.641		-0.629	-0.771	-0.784	-0.742	-0.842	0.663	-0.842	-0.403	-0.392	-0.455		-0.486	-0.401	-0.291	0.645	0.217	
Zn	-0.574		-0.801	-0.611	-0.623	-0.182	-0.188	0.114	-0.267	0.375	0.454	0.443		0.072	0.391	0.258	0.076	-0.209	
Cr	-0.194		-0.436	-0.580	-0.504	-0.708	-0.339	-0.204	-0.443	-0.201	-0.298	-0.414		-0.285	-0.214	-0.411	0.319	0.597	
Co	-0.428		-0.214	-0.358	-0.453	-0.700	-0.758	-0.014	-0.697	-0.576	-0.696	-0.737		-0.769	-0.649	-0.343	-0.093	-0.026	
Ni	-0.503		0.051	-0.131	-0.219	-0.530	-0.838	0.193	-0.850	-0.802	-0.896	-0.891		-0.999	-0.907	-0.681	0.136	-0.028	
TiO2	0.309		-0.256	-0.002	0.054	0.473	0.613	-0.200	0.925	0.871	0.834	0.858		0.867	0.819	0.437	-0.409	-0.343	
V	0.135		-0.588	-0.427	-0.305	0.042	0.486	-0.180	0.658	0.846	0.841	0.778		0.902	0.858	0.494	0.055	0.071	
Sc	0.085		-0.538	-0.374	-0.240	0.085	0.493	-0.404	0.667	0.889	0.847	0.786		0.838	0.842	0.383	-0.010	0.034	
P2O5	0.061		0.403	0.465	0.318	0.336	-0.207	0.186	-0.100	-0.318	-0.359	-0.234		-0.503	-0.390	-0.240	-0.742	-0.659	
Zr	0.481		-0.063	0.181	-0.023	0.296	0.108	0.029	0.404	0.229	0.149	0.247		0.013	0.252	0.361	-1.213	-0.927	
Nb			0.246	0.423	0.336	0.504	0.651	-0.122	0.655	0.281	0.295	0.349		0.401	0.464	0.351	-0.504	-0.166	
Sn	0.614																		
La	0.925	0.114		0.954	0.947	0.720	0.373	0.056	0.500	-0.236	-0.157	-0.072		0.244	-0.252	-0.040	-0.033	0.091	
Ce	0.784	0.077	0.016		0.968	0.875	0.552	-0.146	0.636	-0.001	0.064	0.167		0.311	-0.009	0.143	-0.252	-0.119	
Pr	0.849	0.120	0.018	0.010		0.883	0.617	-0.209	0.676	0.074	0.150	0.233		0.433	0.043	0.102	0.005	0.058	
Nd	0.757	0.071	0.083	0.034	0.039		0.815	-0.347	0.879	0.448	0.510	0.604		0.620	0.432	0.408	-0.303	-0.231	
Sm	0.704	0.139	0.156	0.095	0.095	0.026		-0.504	0.977	0.712	0.792	0.822		0.894	0.786	0.606	-0.085	0.038	
Eu	1.185	0.194	0.297	0.318	0.359	0.292	0.294		-0.362	-0.636	-0.501	-0.482		-0.195	-0.434	-0.299	0.039	-0.058	
Gd	0.639	0.142	0.144	0.089	0.088	0.025	0.011	0.342		0.833	0.903	0.955		0.873	0.919	0.467	-0.598	-0.428	
Tb	0.900	0.319	0.434	0.314	0.308	0.151	0.088	0.526	0.055		0.980	0.969		0.853	0.961	0.649	-0.318	-0.232	
Dy	0.869	0.204	0.289	0.201	0.200	0.077	0.030	0.030	0.020	0.024		0.990		0.977	0.967	0.744	-0.213	-0.152	
Ho	0.843	0.166	0.256	0.171	0.174	0.057	0.023	0.306	0.013	0.033	0.002			0.970	0.955	0.756	-0.309	-0.242	
Er	0.807	0.215	0.331	0.237	0.247	0.106	0.041	0.308	0.033	0.036	0.009	0.013							
Tm	0.803	0.212	0.216	0.170	0.152	0.074	0.026	0.305	0.028	0.050	0.007	0.011	0.009		0.965	0.560	-0.125	-0.094	
Yb	0.771	0.197	0.322	0.224	0.232	0.094	0.034	0.329	0.017	0.026	0.006	0.008	0.003	0.008		0.763	-0.313	-0.183	
Lu	1.078	0.807	0.897	0.744	0.794	0.585	0.511	1.028	0.548	0.417	0.426	0.435	0.366	0.496	0.407		-0.434	-0.306	
K2O	1.392	0.934	0.563	0.622	0.526	0.553	0.458	0.500	0.725	0.719	0.529	0.542	0.556	0.538	0.577	1.515		0.914	
Ba	1.702	1.071	0.635	0.714	0.638	0.685	0.562	0.692	0.829	0.837	0.659	0.676	0.663	0.674	0.683	1.582	0.087		
S	4.453	2.163	2.293	2.246	2.349	1.990	1.950	1.231	2.291	1.724	1.634	1.643	1.615	2.053	1.698	2.016	1.008	2.411	
CO2	4.239	2.161	2.192	2.224	2.369	2.182	2.191	1.319	2.406	2.348	2.078	2.055	2.023	2.286	2.118	2.441	1.070	2.466	0.304

quence of crystallization different from the one assumed by the CIPW norm calculation.

P, Zr and Nb loose some of their incompatible character upon retrogression. This is probably also related to the generally more felsic character of the retrogressed subsample. As their incompatible character is reflected in negative correlation coefficients, the significance of this can not well be tested on the variances; for the log-ratio correlation coefficients this change is seen to be significant for P.

Eu has better correlations to the other REE with the weakly retrogressed subsample. All these specimens are having only a small Eu-anomaly, probably due to low primary plagioclase contents (again this is not reflected in the CIPW norm mineral compositions). Also this difference is of primary origin.

To sum up, from the correlation matrices presented here two retrograde processes are clearly showing up: albitization with its concomitant element mobility and probably enrichment of Na, and the development of sheet silicates, going along with a redistribution of K and Ba.

The other observed changes are most probably due to differences between the protoliths of the two subsamples.

Tab. 6.14: Fisher transforms and the quotients of log-ratio variances for little and for strongly retrogressed Mg-gabbros. Bold face: significant at the 5 % level

	SiO2	Al2O3	Ga	CaO	Sr	MgO	H2(l).	Fe2O3(l)	MnO	Na2O	Zn	Cr	Co	Ni	TiO2	V	Sc	P2O5	Zr
SiO2		0.066	-0.531	0.368	-0.374	-0.598		-0.660	-0.480	-0.739	-0.381	0.477	-0.023	-0.620	0.665	0.749	0.957	**-1.838**	-0.707
Al2O3	**-6.321**		-0.731	0.486	-0.453	-0.655		-0.994	-0.772	0.004	-0.391	0.774	-0.418	-1.270	0.846	1.483	1.579	**-2.931**	-0.519
Ga	-2.066	1.071		0.656	0.092	-1.032		-1.436	-1.223	0.835	-0.536	0.429	-0.831	-1.523	1.158	**2.072**	**2.117**	**-1.980**	-0.182
CaO	-3.614	**-6.793**	-1.807		-0.402	-0.272		-0.805	-0.605	1.021	-0.355	1.004	-0.356	-0.698	0.693	**1.916**	**1.973**	**-2.599**	-0.373
Sr	1.472	1.300	1.075	2.411		-1.139		-1.228	-0.935	1.609	1.325	-0.385	-1.419	-1.377	1.398	**1.736**	1.649	-0.406	0.532
MgO	2.731	2.234	2.179	1.215	3.010			-0.401	-0.228	-1.452	0.060	-0.055	0.971	0.424	0.321	-0.188	0.108	-0.389	-1.577
H2O(l)	2.215	3.437	2.635	1.371	2.796	-1.610													
Fe2O3(l)	4.085	3.595	4.230	2.653	**4.830**	2.442	1.781		0.089	-1.429	0.496	-0.381	-0.035	-0.560	0.580	0.009	0.014	-0.086	-0.793
MnO	3.297	2.314	2.999	2.397	**4.869**	1.274	1.112	-1.146		-1.380	0.335	-0.397	0.017	-0.255	0.539	-0.116	-0.019	0.008	-0.848
Na2O	-1.772	-4.005	**-5.305**	**-10.54**	**-8.325**	2.156	1.956	2.063	1.865		-0.940	-0.284	-1.008	-1.087	1.399	**2.062**	1.562	-0.321	0.823
Zn	-4.296	**-4.802**	**-6.399**	-3.543	-3.064	-2.285	-2.282	-2.624	-2.651	-1.737		-1.130	0.346	0.301	-0.546	-0.858	-0.906	0.478	-0.386
Cr	1.923	2.087	2.227	1.786	**5.197**	-1.737	-2.000	1.542	1.091	1.564	-1.971		-0.301	0.511	-0.112	-0.320	0.002	-0.164	-0.178
Co	2.717	**5.029**	3.912	2.475	3.752	1.523	2.555	2.219	1.574	2.072	-1.424	1.167		0.205	0.563	-0.072	0.097	-0.228	-1.332
Ni	**5.339**	3.967	4.249	3.436	**6.695**	**6.295**	3.944	**6.801**	**7.058**	4.619	3.539	**7.417**	2.192		0.008	0.220	0.494	-0.846	-1.997
TiO2	-1.430	-2.307	-2.835	-1.469	-2.018	1.461	1.294	2.086	2.601	**-12.53**	-2.633	1.786	1.784	**5.633**		0.642	-0.067	-0.045	0.289
V	-3.091	**-5.550**	**-6.695**	**-5.769**	**-5.314**	-1.014	-1.185	1.315	1.547	**-76.89**	-1.362	1.241	1.185	**5.292**	**-5.609**		-0.648	0.391	0.665
Sc	**-8.196**	**-14.46**	**-17.59**	**-18.61**	**-7.256**	-1.763	-1.902	-1.214	-1.137	**-36.65**	-2.056	-1.263	-1.228	4.286	-3.999	-1.073		-0.028	0.253
P2O5	-1.281	1.167	1.018	2.774	1.239	-1.055	1.182	1.135	1.016	-2.683	**-9.617**	1.168	2.216	3.177	-1.380	-2.488	-3.382		-0.099
Zr	1.085	-1.214	-1.375	1.434	1.004	1.920	1.701	3.342	4.277	-4.619	-2.319	2.654	2.362	**4.943**	1.492	-2.208	-2.329	-1.661	
Nb	-1.414	-1.399	-1.564	1.112	-1.083	-1.197	-1.212	-1.366	-1.249	-2.224	-4.414	1.034	1.231	3.914	-1.224	-1.163	-1.137	-2.867	-1.780
Sn	3.071	2.914	3.141	**7.067**	1.431	2.729	2.876	3.128	3.632	-1.038	-1.302	2.972	3.369	4.629	4.873	1.945	1.418	2.897	**5.008**
La	-1.200	1.066	-1.314	4.047	3.064	-1.387	-1.150	-1.453	-1.479	-2.048	**-12.28**	-1.008	1.434	4.052	1.433	-1.245	-1.890	**-7.871**	-1.149
Ce	-1.312	-1.147	-1.630	3.199	2.451	-1.161	-1.032	-1.037	-1.006	-2.675	**-11.61**	1.223	1.539	4.103	1.487	-1.336	-1.967	**-32.13**	-1.380
Pr	-3.026	-2.604	**-5.31**	1.329	1.583	-1.678	-1.475	-1.440	-1.419	**-4.99**	**-23.18**	-1.114	1.261	3.756	-1.263	-2.274	-3.408	**-24.50**	-3.017
Nd	-2.304	-2.538	-4.933	1.090	1.182	-1.145	-1.128	1.151	1.298	**-6.31**	**-10.34**	1.249	1.449	4.222	1.298	-2.089	-2.891	**-6.972**	-3.729
Sm	-2.656	-3.481	**-7.39**	-1.743	-1.400	-1.047	-1.111	1.300	1.542	**-10.09**	**-7.28**	1.389	1.387	**5.163**	1.093	-2.111	-3.066	**-4.958**	-4.344
Eu	-1.700	-2.130	**-5.21**	-1.756	-1.827	1.135	1.331	1.520	1.604	-1.961	**-18.66**	1.121	1.656	4.447	-2.664	-3.337	**-7.068**	**-5.608**	-2.168
Gd	-2.184	-2.716	-4.496	-1.404	-1.727	-1.061	-1.081	1.239	1.465	**-8.89**	**-7.84**	1.341	1.346	4.719	1.112	-1.320	-1.523	-4.841	**-5.214**
Tb	-2.867	-3.990	**-6.88**	-2.970	-4.112	-1.139	-1.248	1.127	1.301	**-11.11**	-3.494	1.230	1.178	4.605	-2.104	1.106	1.332	**-5.688**	**-11.55**
Dy	-3.078	-4.308	**-9.31**	-2.629	-1.798	-1.111	-1.227	1.194	1.460	**-9.10**	**-5.07**	1.261	1.251	5.023	1.383	1.172	1.457	**-5.934**	-4.396
Ho	-4.283	**-6.17**	**-11.31**	-3.018	1.091	-1.304	-1.503	1.118	1.443	**-7.47**	-3.908	1.127	1.184	4.568	4.246	1.329	-1.302	-3.437	-1.033
Er	-4.584	**-7.05**	**-8.87**	-4.836	-1.842	-1.155	-1.212	1.556	1.847	**-17.26**	-2.787	1.218	1.120	4.528	-2.895	-1.922	**-4.876**	-3.619	**-16.98**
Tm	-2.821	-3.622	-7.263	-1.990	-1.421	-1.342	-1.409	-1.109	1.102	**-5.21**	**-9.95**	1.037	1.066	4.864	2.169	4.135	1.419	**-20.83**	
Yb	-1.716	-2.499	-3.773	-1.613	-1.363	1.195	1.011	1.529	1.971	**-5.24**	-3.146	1.537	1.457	5.654	2.616	2.399	1.515	**-5.694**	**-32.24**
Lu	**-7.13**	**-8.34**	**-13.90**	**-6.07**	-3.922	-2.489	-2.509	-2.068	-1.874	**-11.22**	**-8.42**	-1.655	-1.469	2.331	**-5.732**	-4.438	-3.966	**-13.93**	**-18.19**
K2O	1.372	-1.224	-1.231	-1.383	1.443	1.252	-1.115	1.477	1.547	2.645	1.063	1.104	-1.244	2.694	1.928	1.855	1.093	-2.111	-1.021
Ba	-2.499	-4.433	-4.152	**-6.849**	-2.349	-1.007	-1.148	1.231	1.245	**-4.99**	-2.167	1.248	1.156	**5.725**	-1.997	-3.037	-4.421	-2.492	-2.904
S	-1.797	-1.910	-2.232	-2.413	-2.558	-1.869	-1.481	-1.859	-2.205	1.575	-1.859	-2.030	-1.720	1.522	-1.824	-1.612	-2.204	-4.694	-2.178
CO2	**-24.98**	**-21.66**	**-12.04**	**-19.18**	-4.583	**-8.95**	**-4.99**	-3.110	-4.372	-3.566	**-5.16**	-4.331	-1.986	1.183	**-5.842**	**-6.981**	**-14.26**	**-5.209**	-3.824

6.9 Ocean floor metamorphism and related chemical changes

In the literature (e.g. Coleman 1977, Humphris & Thompson 1978), two major metasomatic processes related to hydrothermal alteration of oceanic basaltic crust are distinguished: chloritization, accompanied by exchange of Mg for Ca, and spilitization, accompanied by enrichment in Na and probably depletion in Ca. The first process is well documented by experimental studies on the alteration of basalts by seawater (Hajash 1975, Mottl & Holland 1978, Wedepohl 1988) and in modern MORB specimens, whereas the second process can be reproduced experimentally only when Na-rich, Mg-free solutions are used. Also spilites, frequently found in ophiolites, are rarely found in dredged or drilled samples from the ocean floor. Besides these effects on Ca, Na and Mg, hydrothermal alteration of mafic oceanic crust also leads to changes in the contents of K, Mn, Li, Rb, Sr, Ba and other elements. With these elements gains as well as losses may occur, depending on the temperature at which the alteration occurs and on other parameters like water/rock ratio.

All these processes and effects have been extensively studied on basaltic rocks. With oceanic gabbros they are less well documented, partly because most ODP holes did not reach deep enough. Generally it has to be assumed that in normal oceanic

Table 6.14 continued

	Nb	Sn	La	Ce	Pr	Nd	Sm	Eu	Gd	Tb	Dy	Ho	Er	Tm	Yb	Lu	K2O	Ba	S
SiO2	-0.358		-0.778	-1.179	-1.144	-0.441	1.073	-1.184	0.583	0.330	0.607	1.239		-0.178	0.558	0.302	-0.155	0.136	
Al2O3	-0.793		**-1.742**	**-3.103**	**-1.991**	-0.550	0.631	-1.238	0.384	0.857	1.141	1.477		0.328	1.398	0.526	0.253	0.680	
Ga	-0.233		**-2.475**	-1.285	-0.871	0.086	1.179	-0.926	0.895	1.261	1.490	1.559		1.180	1.365	0.803	0.195	1.172	
CaO	-1.028		**-2.816**	**-2.180**	**-2.008**	-0.802	0.123	-0.930	-0.052	0.691	0.665	0.758		-0.047	0.638	0.416	0.593	1.259	
Sr	0.002		-0.823	-0.488	-0.499	0.222	0.854	0.684	0.725	1.269	1.034	0.316		0.603	0.817	0.538	0.043	0.453	
MgO	0.226		0.389	0.291	0.261	0.146	0.420	-0.625	0.389	-0.961	-0.310	0.708		0.369	-0.396	-0.283	-0.022	-0.484	
H2O(t)																			
Fe2O3(t)	0.586		0.735	0.564	0.719	0.571	0.955	-0.100	0.474	-0.017	0.439	0.730		0.859	0.321	-0.039	0.040	-0.855	
MnO	0.532		0.882	0.702	0.854	0.666	0.783	-0.071	0.325	-0.141	0.212	0.393		0.524	0.034	-0.294	0.084	-0.884	
Na2O	0.646		-0.215	0.503	0.562	1.280	1.639	-0.750	**1.785**	1.162	1.094	1.009		0.637	0.883	0.654	-1.084	0.533	
Zn	1.611		**1.837**	1.410	1.523	0.737	0.510	1.474	0.751	-0.187	-0.107	-0.354		0.549	-0.114	0.096	-1.083	-1.311	
Cr	-1.477		-0.067	-0.099	-0.470	-0.406	-0.599	-1.640	-0.654	-0.560	-0.757	-0.085		-1.058	-0.743	-0.509	0.783	-0.032	
Co	0.014		0.328	0.147	0.173	-0.046	-0.262	-0.116	-0.128	-0.754	-0.246	0.396		0.241	-0.456	-0.414	0.250	-0.544	
Ni	-0.259		-0.050	-0.208	-0.283	-0.828	-1.339	-0.805	-0.533	-0.954	-1.019	0.652		**2.541**	-0.385	-0.346	0.471	-0.146	
TiO2	-0.009		-0.097	-0.080	-0.053	0.052	0.372	0.878	-0.522	-0.263	-0.107	-0.851		-0.886	-0.513	0.009	-0.644	0.451	
V	-0.076		-0.044	0.101	0.073	0.329	0.346	0.471	-0.012	-0.338	-0.339	-0.498		-1.238	-0.719	-0.097	-0.639	0.398	
Sc	-0.418		-0.313	-0.187	-0.359	-0.100	-0.116	0.418	-0.429	-0.897	-0.870	-0.987		-1.364	-1.109	-0.452	-0.342	0.548	
P2O5	1.545		1.033	1.628	**3.029**	0.711	0.724	0.533	0.708	0.779	0.877	0.368		**1.788**	1.097	0.611	0.611	0.113	
Zr	0.481		0.252	0.365	0.759	1.504	1.781	0.526	1.118	1.466	1.627	0.462		1.520	**2.226**	0.879		1.759	
Nb	0.580		0.749	1.391	1.076	0.115	1.227	0.223	0.458	0.633	0.087			1.528	0.591	0.613	-0.594	-0.404	
Sn	**-17.18**																		
La	-1.968	3.454		-0.133	-0.478	-0.426	-0.309	0.406	-0.356	0.296	0.210	-0.203		0.333	0.443	0.147	-0.120	-1.029	
Ce	-1.985	3.510	1.064		-0.159	-0.475	-0.195	0.750	-0.217	0.408	0.312	-0.258		0.653	0.528	0.202	-0.037	-0.561	
Pr	-1.859	3.474	3.272	3.193		-0.240	-0.154	0.945	-0.169	0.419	0.409	-0.039		0.888	0.722	0.553	-0.355	-0.682	
Nd	-1.553	**5.136**	1.434	1.827	-2.650		0.377	1.225	0.195	0.831	0.945	-0.120		**1.936**	1.359	0.759	-0.307	0.038	
Sm	-1.243	2.888	1.146	1.106	-2.478	-3.238		1.294	0.175	1.534	1.459	-0.539		-0.199	0.603	0.320	-0.666	0.022	
Eu	-1.859	2.774	-2.420	-3.532	**-17.06**	**-12.50**	**-10.96**		1.320	1.357	1.365	0.836		1.065	1.074	0.896	-1.677	0.802	
Gd	-1.357	2.093	1.339	1.204	-1.485	-1.355	-1.153	**-6.594**		0.846	0.566	-1.019		-0.059	-0.184	0.385	-0.259	0.354	
Tb	-1.850	-1.002	-1.978	-2.489	-4.449	**-6.569**	**-10.68**	**-8.578**	**-9.603**		-0.486	-1.595		-0.202	-0.563	0.038	-0.339	0.373	
Dy	-1.547	2.106	-1.591	-1.821	-5.317	**-9.589**	**-27.61**	**-14.03**	-1.603	-1.827		**-1.866**		-0.954	-0.259	0.270	-0.577	0.182	
Ho	-1.242	4.034	-1.131	-1.005	-2.981	-1.464	1.603	**-8.109**	**5.891**	2.170	**14.19**			-1.502	-0.978	0.494	0.112	0.498	
Er	-3.083	1.686	-1.049	-1.266	-2.624	-3.592	-3.694	**-6.053**	-2.142	-2.153	-1.103	3.527							
Tm	-1.734	1.708	-2.014	-3.164	**-13.58**	**-75.30**	-1.831	**-11.83**	-1.131	-1.546	1.859	3.775	3.859		-0.375	0.811	-0.434	-0.175	
Yb	-1.789	1.721	-1.548	-1.898	-3.692	-4.507	-1.640	-4.162	1.362	-1.109	3.840	**6.377**	1.073	2.626		0.619	-0.138	0.264	
Lu	-2.238	-1.673	-4.298	**-5.418**	**-13.73**	**-21.42**	**-14.73**	**-16.40**	**-10.95**	-7.509	-16.18	-21.37	-14.42	**-19.95**	**-26.29**		0.083	0.346	
K2O	1.413	2.027	1.255	1.182	1.140	1.232	1.542	1.276	1.203	1.109	1.339	1.057	1.832	1.263	1.353	-2.133		-0.955	
Ba	-1.154	-1.016	1.046	1.277	-1.702	-2.120	-2.073	-2.094	-2.312	-2.937	-2.388	-2.930	-1.992	-1.964	-2.129	**-5.005**	4.090		
S	-3.981	-2.409	**-10.64**	**-7.686**	**-5.746**	-3.687	-3.161	-2.408	-3.469	-2.597	-2.550	-2.253	-1.583	-3.862	-2.347	-2.656	-1.366	-2.357	
CO2	-2.956	-1.650	**-6.152**	**-5.917**	**-9.899**	-7.580	-7.912	-7.558	-6.410	-6.632	-7.583	-9.188	-9.345	-7.688	-5.033	-6.750	-3.240	**-8.577**	1.746

crust the effects of hydrothermal alteration are less pronounced for the gabbros than for the basalts because of the deeper position of the former. It is however not clear whether the ZSF has been normal oceanic crust (see above chapter 1.1).

Meyer (1983a) on petrological reasons contended that the Allalin gabbro has not been hydrated prior to HP metamorphism. Barnicoat & Cartwright (1997) have claimed the opposite on the basis of $\delta^{18}O$ data (gabbros with magmatic relics: $\emptyset$ = 5.7 ‰ , eclogitic metagabbros: $\emptyset$ = 4.8 ‰ $\delta^{18}O$). There must have been some water present during the HP metamorphism, but this might also have been supplied by dehydration reactions of other rocks (e.g. sediments) subducted along with the gabbros.

Widmer (1996; cf. also Widmer et al. 2000) made a detailed study on the effects of ocean floor metamorphism on the metabasalts of the ZSF. According to him the effects of the two processes of chloritization and spilitization resulted in two distinct lithotypes, whose differences were preserved during all subsequent metamorphic stages. With the HP rocks, these are the eclogites and the glaucophanites, with rocks retrogressed in the greenschist facies, the prasinites and ovardites. Barnicoat & Cartwright (1995) found that the two lithotypes are also bearing different $\delta^{18}O$ signatures, which were preserved at least during HP metamorphism.

A third process of strong Ca enrichment, associated to ocean floor metamorphism, leads to the formation of epidosites and has been also documented in metabasalts from the ZSF (Barnicoat & Bowtell 1995).

These findings for the metabasalts from the ZSF can not be easily transferred to the metagabbros. For one, it is not possible to distinguish Mg enrichment and Ca depletion caused by oceanic chloritization from trends caused by differentiation. Thus the two lithotypes of eclogites and glaucophanites are not diagnostic. Whether the second process of Na enrichment has been operating is equally difficult to ascertain. The geometric mean of the NaO contents of the specimens studied is 3.33 %, slightly higher than the average of 2.26 % for (continental) gabbros given by Wedepohl (1969) and the preferred value for mature oceanic gabbros of 2.97 % given by GERM. These values may however very likely be different from unaltered oceanic gabbros, of which there are as yet too few data. Therefore there is no way to compare the metagabbros from the ZSF to unaltered protoliths.

When the metagabbros and the metabasalts are plotted together in a MgO-CaO-Na$_2$O triangle (fig 6.15), it is found that the metagabbros are rather poor in NaO, as compared for example to the metabasalts. Thus there is no proof for any effect of ocean floor metamorphism on the major elements. In particular, the previous chapter seems to intimate that any Na enrichment happened at the albite-chlorite late-stage metamorphic event.

Thus generally the effects of ocean floor metamorphism must have been, as expected, very weak, because no distinct enrichment or depletion of characteristic elements can be demonstrated. This does not preclude the possibility that single specimens may have been influenced more strongly. The particularly Na-rich Fe-Ti gabbro specimen Hi 27-2-96 (rightmost data point in the Ca - Mg - Na plot fig 6.15) however probably gained its excess Na some other way:

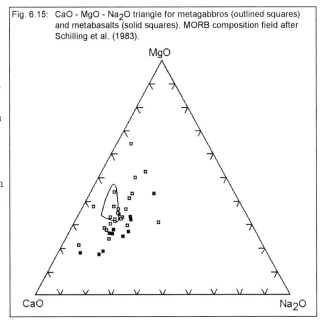

Fig. 6.15: CaO - MgO - Na$_2$O triangle for metagabbros (outlined squares) and metabasalts (solid squares). MORB composition field after Schilling et al. (1983).

it was collected close to a major thrust plane at the Spitzi Flue pass.

6.10 Preliminary summary

A large part of the magmatic differentiation history of the gabbroic rocks from the ZSF has been already summarized above in chapter 6.4 on the results of the log-ratio method. These remarks are repeated here with some additions. Also some results on the effects of metamorphism on the rocks are presented.

Gabbroic magmas were derived by different but small degrees of partial melting from a fertile mantle source containing the following minerals:

– Olivine; this is not surprising and was already seen from the REE diagrams;
– Spinel, because normal mantle olivine has too low Ni contents to explain the correlation between MgO and Ni alone;
– cpx, considering the falling REE curves;
– probably Ti-clinohumite, as is attested by the negative P and Nb anomalies.

These findings make it quite probable that the metagabbros and the serpentinites today found together in the ZSF are indeed related as mantle residuum and extracted melt.

This magma differentiated by forming cumulates. In the Allalin gabbro, cumulate layers as in layered intrusions can still be observed despite the polymetamorphic history (Meyer 1983a). Plagioclase cumulates are most easily proved from the positive Eu- and Sr-anomalies. But there were also layers enriched in olivine and in cpx, as discussed in chapter 6.4. The distinct grouping of the gabbros according to Sc/Cr ratios is most easily explained by cpx enrichment: gabbros with spidergram curves rising from Sc to Cr are controlled by olivine plus spinel, whereas gabbros with curves flat from Sc to Cr are enriched in cpx, thus containing more Sc.

Some of the Fe-Ti gabbros could have formed from magma depleted in Sr and Eu, squeezed off from the plagioclase cumulates. These are having small negative Eu- and Sr-anomalies and are similar to potential basaltic educts. In fact, a metabasaltic rock from a dike enclosed in metagabbros (the purported sheeted dike complex of Barnicoat and Fry 1986) has very similar trace element patterns as this type of Fe-Ti gabbros (see subsequent chapter). Yet, as the vast majority of metabasalts from the ZSF does not seem to be related magmatically to the metagabbros, as will be shown in the next chapter, there is a major volume problem associated with the gabbros: although a representative sampling of the different gabbro types was not intended, it is clear that Fe-Ti gabbros are generally very scarce within the ZSF (cf. Ganguin 1988) and thus the liquids squeezed off from the cumulates must have disappeared somewhere.

Those Fe-Ti gabbros not showing any Eu- or Sr anomaly could stem from melts derived by very small degrees of partial melting, being therefore strongly enriched in incompatible elements and crystallizing early Fe-Ti oxides.

Whether oceanic metamorphism resulted in any chemical changes can not be shown with the data at hand. Devolatilization during prograde HP metamorphism as well cannot be proven to have caused metasomatism, but any effects would have been very minor due to the small amounts of water involved. Retrograde metamorphism, caused by influx of extraneous fluids, resulted in some enrichment in Na (albitization), as well as in some redistribution of K and Ba, elements characteristic for sheet silicates, which were formed extensively at this stage. The scale of this redistribution may have been quite small (mm to cm) to explain the observed enhanced correlations.

7   Geochemistry of the metabasalts

7.1   General description of the metabasalt geochemistry

With the metabasalts, some specimens were collected from rocks displaying a pillow structure with the hope that this would help in disentangling the possible effects of ocean floor metamorphism from other processes. This topic has been discussed in the literature (Bearth & Stern 1971; Widmer 1996), but results seem to be not very well founded. However, the number of appropriate specimens collected was quite small; also, it can not be stated with certainty which parts were coming from pillow cores and which from pillow rims, as this had to be decided on petrographic grounds and not on observations of the relictic pillow structure. Nevertheless, these specimens were divided for the chemical analysis and the parts are tentatively designated with an R and C, respectively. One analysis seems to represent interpillow matrix and is designated with an I.

The metabasalts are chemically much more homogeneous than the metagabbros. On the whole, they compare well with averages for tholeiitic basalts given in the literature (Wedepohl 1969, p. 238). Minimal and maximal values and geometric means (calculated excluding specimen Hi 24-96 for reasons to be explained later) are given in table 7.1.

SiO_2 varies between 44 and 51 %, Al_2O_3 between 14 and 18 %. According to its geometric mean of 11 %, CaO is the next most important oxide, followed by total Fe (calculated as Fe_2O_3) with 9,5 %. CaO varies considerably; the maximum is about 15 %, the minimum 6 %. This is the reason that about as many specimens are lower in CaO than in Fe_2O_3 than the other way round. The variation of Fe_2O_3 is smaller: between 7 % and 11 %. Somewhat over two thirds of the Fe exists still in the form of FeO. The variation in Fe contents is largely offset by opposite variations in MgO contents. This oxide has a geometric mean of 5.31 %, the minimum and maximum being about 3 % and 8.5 %. These relatively low MgO contents are characteristic for tholeiites.

Conspicuous are the high Na_2O contents (1.77 % minimum – 3.84 % geometric mean – 4.89 maximum) and the low K_2O contents (0.02 % – 0.2 % – 1.53 %). According to the Na_2O values, the rocks might be classed as alkaline basalts, but the low K_2O testifies against this. Widmer (1996) has argued that the high Na contents are caused by oceanic metamorphism. This question will be taken up below. Part of the Na enrichment might also be caused by the late stage albitization.

Tab. 7.1: Minimal, maximal and geometric mean of element/oxide contents of metabasalts			
	Min	Max	geom. Mean
SiO_2	44.2	51.1	48.4
TiO_2	1.020	2.088	1.777
Al_2O_3	13.8	18.1	15.4
Fe_2O_3 (t)	7.32	10.91	9.58
Fe_2O_3	2.22	3.09	2.59
FeO	4.5	7.7	6.3
MnO	0.120	0.236	0.175
MgO	3.27	8.5	5.35
CaO	6.21	14.71	10.72
Na_2O	1.77	4.89	3.62
K_2O	0.02	1.53	0.17
P_2O_5	0.179	0.337	0.310
H_2O (t)	1.83	5.42	2.85
S	0.00	0.13	0.04
CO_2	0.02	5.94	0.27
Mg#	49	68	56
Li	3.5	42.7	13.0
Sc	21	35	30.1
V	170	249	225.1
Cr	115	309	141.4
Co	27	45	34.5
Ni	66	336	88.1
Cu	35	59	45.4
Zn	40	94	64.6
Ga	14	23	16.5
Rb	0	22	1.1
Sr	140	738	294.0
Y	26.51	36.93	34.5
Zr	108	205	163.2
Nb	2.65	7.86	5.68
Sn	2.48	6.43	3.62
Ba	2.89	86.14	12.82
La	4.43	11.29	7.20
Ce	13.25	34.04	24.23
Pr	2.11	4.92	3.65
Nd	10.68	23.73	17.76
Sm	3.09	5.71	4.89
Eu	0.95	1.77	1.53
Gd	3.54	6.63	5.41
Tb	0.62	0.96	0.86
Dy	4.37	6.72	5.28
Ho	0.94	1.37	1.21
Er	2.57	3.72	3.17
Tm	0.39	0.59	0.50
Yb	2.37	3.59	3.04
Lu	0.37	0.54	0.45
Ta	0.58	2.51	1.40
Th	0.10	1.47	0.26
U	0.05	8.45	0.38

The remaining minor oxides TiO_2 and P_2O_5 and the trace element contents are showing only very minor variation. The ranges are comparable to normal tholeiites.

Volatiles are varying strongly. All specimens are more or less hydrated, with H_2O contents between 1.8 % and 5.4 %. CO_2 varies between 0 % and almost 6 %.

The minimal and maximal values and geometric means of all oxides and elements are to be found in table 7.1.

Because the oxidation state of Fe is certainly not primary, the Mg# was calculated assuming that 15 % of the total Fe were primarily in the form of Fe_2O_3. The Mg# is in the range of 49 to 68 with a geometric mean of 55. This is much lower than for the gabbros, even the Fe-Ti gabbros, and points to some fractionation prior to eruption, if Fe- and Mg contents have not been altered during metamorphism.

Specimen Hi 24-96 is lying outside of the range of the other metabasalts for several oxides and elements: it is much higher in incompatible elements like Ti, P, V, Y and the REE and also in total Fe_2O_3, and it is lower in Cr, Ni, Sn and SiO_2.

The AFM triangle (fig. 7.1) shows much scatter. Although the specimens are supposed to have been primarily tholeiitic basalts, the AFM diagram by no means reflects this; most specimens are lying in the calc-alkaline field according to Irvine & Baragar (1971). Again this can be ascribed to the metamorphic addition of Na_2O.

Table 7.2 contains the CIPW norm minerals for the metabasalts. According to these, 5 specimens are alkali-olivine-basalts, 5 specimens are olivine tholeiites and 2 specimens are quartz tholeiites according to the classification in the basalt tetrahedron after Yoder.

The normative nepheline is again due to the high Na_2O contents, and thus the norm minerals do not adequately reflect the modal mineralogy of the protolith.

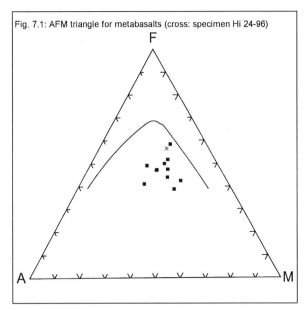

Fig. 7.1: AFM triangle for metabasalts (cross: specimen Hi 24-96)

7.2   REE diagrams and scattergrams

As with the other rock types, some of the trace element data is summarized graphically in the form of normalized scattergrams (fig 7.2 to 7.4). These diagrams have been calculated water-free. The normalization coefficients are to be found in Tab. 5.1. For the REE diagrams, all values are well above the l.o.d. In the other diagrams, the l.o.d. is marked with a heavy line.

The **REE diagram** shows curves rising from La to Pr, but generally falling towards the HREE. Specimen Hi 25-96 R has a curve rising slightly from Eu to Lu. The spread between the curves is slightly larger in the LREE than in the HREE. Most specimens are occupying a narrow range between 10 and 40 times chondritic values. For other basaltic specimens from the ZSF, Himmelheber (1996) reports contents up to 70 times chondritic. Specimen Hi 24-96 has distinctly higher REE contents. This specimen comes from a basaltic dike enclosed in metagabbros (the purported sheeted dike complex of Barnicoat & Fry 1986). This specimen also shows a distinct negative Eu anomaly, whereas with the other metabasalts either no anomaly is present or else it is very weak and seems to be only apparent, due to overestimated Gd contents.

These curves are quite similar to other published REE curves from the ZSF (Pfeifer et al. 1989). The general type of the patterns resembles T-MORB (cf. Sun et al. 1979 and particularly Humphris et al. 1984).

With the **mantle-normalized spidergram** after Thompson (1982), again specimen Hi 24-96 is clearly distinguished from the other specimens. It generally has much higher trace element contents than the other metabasalts. An other distinguishing trait are its large negative anomalies at Th, Nb, Sr and Zr.

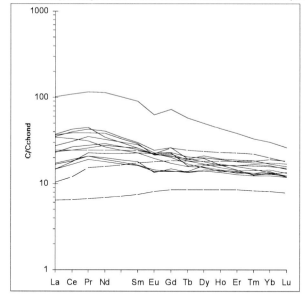

Fig. 7.2: REE diagram for metabasalts. Dashed lines from top to bottom: T-MORB after Humpris et al (1984), average N-MORB after GERM, T-MORB after Sun et al. (1979).

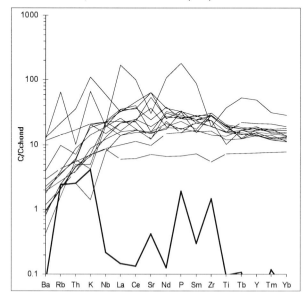

Fig. 7.3: Chondrite-normalized spidergram for metabasalts after Thompson (1982). Bold line: l.o.d. Dashed lines from top to bottom: T-MORB after Humphris et al (1984), average N-MORB after GERM, T-MORB after Sun et al. (1979). Values below the l.o.d are set to ½ l.o.d.

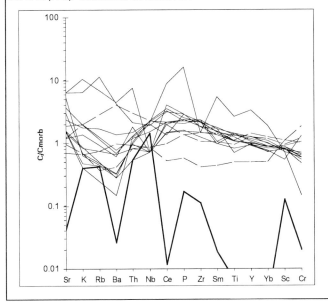

Fig. 7.4: MORB-normalized spidergram after Peirce (1983) for metabasalts. Bold line: l.o.d. Dashed lines from top to bottom: T-MORB after Humphris et al (1984), average N-MORB after GERM, T-MORB after Sun et al. (1979). Values below the l.o.d are set to ½ l.o.d.

The other specimens are showing subparallel curves, rising strongly through the LIL elements and falling slightly with decreasing degree of incompatibility in the REE and HFSE.

There are no major anomalies, except at Sr, where positive as well as negative anomalies are found. With some specimens (Hi 25-96 R, Hi 29-4-96 C), there seem to be minor positive anomalies at P and Zr. The P anomaly is close to the bandwidth of the analytical error, but the Zr anomaly seems to be real. The interpillow matrix analysis Hi 29-4-96 I has positive anomalies at K and Rb. Also high in K, Ba and Rb is specimen Hi 29-4-96 R.

The **MORB-normalized spidergram** after Pearce (1983) contains much the same information as the mantle-normalized spidergram. The special character of specimen Hi 24-96 extends also to its Cr content. Whereas most specimens have curves falling slightly from Sc to Cr, and 2 rising curves (Hi 2-96 and Hi 22-96), Hi 24-96 shows a steep drop.

How can these findings be interpreted? It is clear that specimen Hi 24-96 is different from the other metabasalts and probably has a different origin. The majority of specimens can be compared to the T-MORB after Humphris et al. (1984). The T-

MORB after Sun et al (1979) has lower trace element contents. The generally falling REE curves can be explained by derivation from a mantle source containing clinopyroxene, that is, from a fairly fertile source. The rise from La to Nd is probably caused by olivine in the mantle residue. The lack of a negative Eu-anomaly proves that no plagioclase fractionation has taken place in the development of the basaltic magma. This means that the metabasalts from the ZSF can not be related to the metagabbros. Naturally the question arises where the fractionated magma belonging to the cumulate metagabbros has remained. In the geodynamic framework of Lemoine et al. (1987), the gabbros are supposed to originate by decompression melting of the lithospheric mantle which is rising due to extension by simple shear. The melt is supposed to have stayed more or less in place. As an afterthought to the chapter on the metagabbros it is considered here how the Eu anomalies of the normal metagabbros can be balanced. This is a volumetric problem. The highly evolved Fe-Ti gabbros, having a negative Eu-anomaly, are fairly restricted in outcrop area. Ganguin (1988, p. 49) reports two small outcrops in the Täsch area (at Sparrenflue and at the foot of the Mellich glacier). On the other hand, the analyses presented here show that there are also Fe-Ti gabbros in the Allalin body. The metagabbros of the present study have on the average a positive Eu-anomaly with a size around 3 (measured additively as Eu_N - $Sqrt(Sm_N{*}Gd_N)$ to allow volumetric considerations), whereas Fe-Ti gabbros typically reach values of -12. This would call for a ratio of 4 to 1 to balance Eu contents. Perhaps also some evolved magma escaped into the continental crust of what is now the Siviez-Mischabel nappe, at that time lying above the ZSF according to the ideas of Stampfli & Marthaler (1990).

The similarity of the specimens to T-MORB is also seen in the mantle-normalized spidergram, as far as the HFSE and REE are concerned. The LILE are much lower than is common for T-MORB and are relatively close to the pattern of N-MORB ac-

Tab. 7.2: CIPW norms of metabasalts

	CM 1-98	Hi 2-96	Hi 22-96	Hi 24-96	Hi 25-96 R C	Hi 25-96 R	Hi 29-2-96 C	Hi 29-2-96 R	Hi 29-4-96 C	Hi 29-4-96 R	Hi 29-4-96 I	Hi 31-96
ap	0.7	0.5	0.5	4.7	0.7	0.8	0.6	0.7	0.6	1.0	0.9	0.6
ilm	3.7	3.2	2.6	7.5	4.3	3.8	3.8	3.1	4.1	3.6	2.1	3.9
or	0.4	0.1	0.8	0.6	1.3	0.4	1.8	0.6	1.7	10.8	5.9	1.6
ab	38.3	31.4	47.2	27.1	37.1	28.4	30.2	15.8	33.5	8.6	22.3	30.1
an	28.3	30.7	21.9	25.3	23.6	27.3	26.5	31.0	21.1	19.9	34.9	21.6
pl	66.6	62.1	69.1	52.4	60.7	55.8	56.7	46.8	54.7	28.5	57.3	51.7
mag	2.2	2.1	2.4	3.0	2.5	2.3	2.3	2.0	2.3	1.9	1.8	2.2
di	15.0	20.9	9.7	16.5	18.4	24.5	24.7	36.6	23.4	38.1	24.2	27.3
hy	6.8	6.2	5.7		7.6	12.1	9.2	0.6				
qtz						0.3		9.8				
ol	4.6	4.9	9.4	13.5	4.6		1.0		9.4	0.7	4.5	7.3
ne				1.7					3.8	16.3	3.3	5.3

118

cording to GERM. This is surprising, as the degree of incompatibility was supposed by the authors of this diagram to increase towards the left. Thus a melt produced from normal mantle would be supposed to display curves falling from left to right. This problem occurred also with the metagabbros and was discussed already in chapter 6.2. As stated there, the rise through the LILE is probably mainly caused by the normalization coefficients, which were overestimated by Thompson et al. (1984). Also these elements are not differing greatly in their degree of incompatibility. According to modern estimates of distribution coefficients, a mantle-derived melt would have rising curves with Ba_N being about 15 times lower than K_N. This view is encouraged by the pattern of T-MORB after Humphris et al. (1984) (the T-MORB after Sun et al (1979) shows a different behavior, particularly in the right part of the diagram) and by the pattern of the average composition of MORB with 8 % MgO according to GERM.

However, there are also distinct differences to normal and to T-MORB. The rise from Ba through La is much steeper than for normal MORB and still somewhat steeper than for the T-MORB after Humphris et al. (1984). In the right half of the diagram, the meta-MORBs from the ZSF display a slight drop not seen in T- and N-MORB. This means the source was somewhat more fertile than a T-MORB source as far as HSFE and REE are considered, but significantly depleted in LILE. This peculiarity of the mantle source is here tentatively ascribed to the probably subcontinental character of the mantle source. Subcontinental mantle is supposed to be very heterogeneous, but it seems plausible that it is particularly depleted in LILE, considering the concentration of this group of elements in the continental crust.

This interpretation depends on the assumption that LILE contents have not been strongly overprinted by ocean floor metamorphism and by later metamorphic processes. This will be discussed in greater detail below; here attention is drawn to the relatively good parallelism of the curves, lending some credibility to this assumption.

Hitherto the metabasalts of the ZSF have been interpreted in the literature as N-MORB (Beccaluva et al. 1984, on the basis of petrographic triangles) or T-MORB (Pfeifer et al. 1989). In the light of the scattergrams presented here, this tenet has to be questioned. Geochemically, the ZSF metabasalts are very similar to modern plume-related oceanic tholeiites. There is however also a strong similarity to flood basalts from the North Atlantic tertiary volcanic province, which are broadly contemporaneous to the ZSF, being related to the opening of the North Atlantic. The REE diagrams and spidergrams of some basalts from East Greenland, Scotland (Skye, Mull) and Ireland are presented in fig. 7.5 and 7.6 (data are from the Georoc database). The similarity to the patterns of the ZSF (fig. 7.2, 7.3) is striking. Eruption of these lavas accompanied the early phases of continental breakup. In the

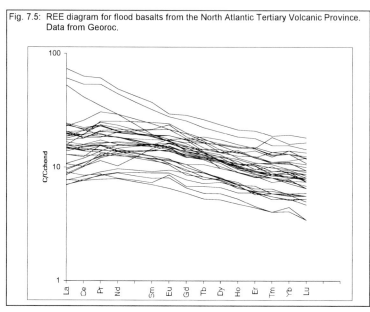

Fig. 7.5: REE diagram for flood basalts from the North Atlantic Tertiary Volcanic Province. Data from Georoc.

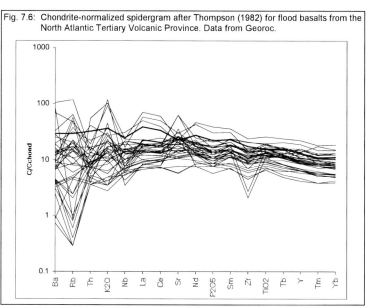

Fig. 7.6: Chondrite-normalized spidergram after Thompson (1982) for flood basalts from the North Atlantic Tertiary Volcanic Province. Data from Georoc.

literature there are diverging opinions as to their mantle source. Most authors prefer to relate these basalts to the early Icelandic plume (e.g. Scarrow & Cox 1995, Kerr 1995), whereas others invoke an additional contribution from fusible components of the subcontinental lithosphere (e.g. Holm et al. 1992, Gariépy et al. 1983). According to Thompson & Morrison (1988), the lithospheric part of the mantle source is residual uppermost asthenosphere from the late paleozoic (Pangaea break-up) magma extraction event, which after a small degree of melt extraction became attached to the Scottish lithosphere. Due to its young age, this source is not showing a continental isotopic signature.

It is interesting to note that other flood basalt provinces, e.g. Deccan, Karroo and Columbia River, are having different trace element patterns.

Considering the modern geodynamical interpretation of the Ligurian-Piemonte Ocean as being the result of passive rifting according to Wernicke's (1981, 1985) simple shear model, resulting in a "lithospheric ocean" (cf. chapter 1), there is no longer a need to stick to a oceanic character of the basaltic rocks. True oceanization may very well never have occurred in the ZSF.

The features seen in the MORB-normalized spidergram after Pearce (1983) tend to corroborate the interpretations made above. The patterns here resemble N-MORB closely for the LILE, though the curves are lying somewhat higher, and through the REE and HFSE they are slightly enriched over T-MORB with a steeper drop. Why the estimates of N-MORB contents of K, Rb and Ba are so much lower according to GERM as compared to Pearce's (1983) assumptions (his normalization coefficients) is not known.

There is no similarity to other oceanic basalt types, as within plate basalts and volcanic arc basalts (cf. the figures in Pearce 1982, p. 527).

Specimen Hi 24-96, coming from a basaltic dike enclosed in metagabbroic rocks, would preferably be explained as the fractionated lava belonging to the cumulate metagabbros. The presence of a negative Eu-anomaly pleads in favor of this. The metagabbros are, as we have seen above in chapter 6.2, characterized by negative anomalies at Nb and P. Specimen Hi 24-94 now is also having a negative Nb anomaly, whereas its P anomaly is positive. It also has a distinct negative Zr anomaly. The common negative Nb anomaly could be explained by Nb being retained in the mantle residue, thus leaving both metagabbros and their related lavas depleted in Nb.

On the other hand, as these dikes did obviously crystallize much more finely grained than the enclosing metagabbros, the latter must already have been cooled down at the time of emplacement, implying a younger age for the dikes. Crosscutting relationships can not be observed in the field due to the strong deformation and boudinage.

However the relationships between metagabbros and enclosed metabasaltic dikes may be, it is clear that the dike basalts are not related to the other metabasalts. In contrast to Specimen Hi 24-96, specimen CM 1-98, coming from the basalt-gabbro layered complex of Spitzi Flue, is not displaying any peculiarities as compared to the main body of metabasalt data.

In the geodynamical model after Lemoine et al. (1987) and Stampfli & Marthaler (1990) the basalts are thought to be of later origin than the peridotites and gabbros. Can the higher incompatible trace element contents of the dike basalt interpreted in that context? Perhaps it is pertinent here that metabasalts from the Tsaté nappe, believed to be of earlier origin than the metabasalts of the ZSF by Stampfli & Marthaler (1990), contrary to expectations, seem on the average to be slightly less enriched than the latter, according to analyses carried out by Gernuks (1999) (Fig. 7.7)

7.3 Ocean-floor metamorphism

In his study on the effects of oceanic metamorphism on the basalts from the ZSF, Widmer (1996) has argued that there have been two major processes of metasomatic alteration: spilitization with the addition of Na, and chloritization with the addition of Mg. He discusses also two other processes, rodingitization and epidotization, which are found less frequently. Rodingitization will be discussed here in chapter 8. Of these processes, addition of MgO can be related to experimental results (e.g. Seyfried & Bischoff 1979, 1981, Mottl & Holland 1981) and has been found also in samples from the Middle Atlantic Ridge (Humphris & Thompson 1978), whereas it was not possible experimentally to create rocks enriched in Na_2O except with fluids having high Na/Ca ratios, but not with seawater.

Already Bearth (1967) noticed that the two different HP metamorphic parageneses, the eclogites and the glaucophanites, are often found as the cores and the rims of basaltic pillows (see also Oberhänsli 1980, 1982, Barnicoat 1988). As already mentioned in chapter 2.1, this difference has been ascribed in the literature to various causes, as both parageneses obviously underwent the same conditions of metamorphism. Among the different interpretations are the following: differences in the activity of H_2O, caused by different degrees of hydration of cores and rims prior to HP metamorphism (Barnicoat 1988); differences in Na_2O content (Bearth & Stern 1971); differences in MgO content (Widmer 1996). Another possibility would be that the glaucophanites represent a lower degree of metamorphism, because the pillow rims were in some way more susceptible to retrograde overprint. This seems however not likely, as it would hardly result in such coherent relationships.

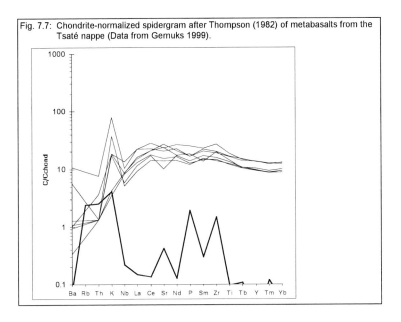

Fig. 7.7: Chondrite-normalized spidergram after Thompson (1982) of metabasalts from the Tsaté nappe (Data from Gernuks 1999).

To elucidate possible effects of oceanic metamorphism, Widmer (1996) plotted all available analyses from the literature, in addition to his own, (totaling 144) in diverse triangular diagrams. In the $Al_2O_3 - 3*Na_2O - 10*TiO_2$ diagram it turns out that both eclogites and glaucophanites are variably enriched in Na_2O. Thus the thesis of Bearth & Stern (1971) is refuted; it was founded on too small a data base. This is also seen in the $MgO - CaO - Na_2O$ diagram, where however another effect is discerned: the glaucophanites are displaced towards the MgO apex. By adding density contour lines to this diagram (fig. 7.8), Widmer could demonstrate this alteration trend of MgO enrichment, coupled to the glaucophanite-eclogite distinction. A second trend of Na_2O enrichment is also seen from the density contour lines. This trend may have been enhanced by further addition of Na_2O during the late metamorphic albitization, although this effect has been demonstrated only in the vicinity of the late-stage albite veins (Mueller 1989).

Igneous differentiation of basalts would also result in variable MgO contents and in trends similar to the purported oceanic alteration trend. However, the range spanned would be much smaller, as is evidenced by the field of MORB compositions from the middle Atlantic ridge after Schilling et al. (1985), also shown, following again Widmer (1996), in fig. 7.8.

Most of the specimens analyzed for the present study fall within the 5 % of specimens per 1 % area density contour line of Widmer (fig. 7.8). The three specimens outside this contour line (Hi 29-2-96 R, Hi 29-4-96 I, Hi 29-4-96 R) are close to the CaO apex. One of these specimens represents probably interpillow matrix; all three are rich in modal carbonate and thus are probably enriched in CaO.

Most specimen points are seen to follow the chloritization (MgO addition) trend. However, petrographically all analyzed specimens are strongly retrogressed, so a distinction in former glaucophanites and eclogites can not be drawn. It even turns out that criteria which are given in the literature for distinguishing these two types in retrogressed specimens do not lead to correct results. For example, specimen Hi 22-96 is closest to the MgO apex and thus has been presumably most strongly chloritized. It is thoroughly retrogressed to a prasinite. This shows that Widmer (1996, p. 23) was in error when he stated that eclogites retrogress to prasinites, glaucophanites to ovardites. However, also the presence of chlorite, a criterion given by Ganguin (1988, p. 56) for the exclusion of an eclogitic provenance, seems not to be diagnostic, as all analyzed specimens contain chlorite, including those plotting on the more Ca-rich end of the diagram.

Spilitization (exchange for Na against Ca) seems to have been not so important with the specimens analyzed here; only Hi 31-96 and Hi 29-4-96 C are lying on the spilitization branch outlined by the density contour lines in fig. 7.8.

Those data points belonging to pillow rims and cores have been joined by a line in fig. 7.8. It turns out, contrary to expectations, that the rims are displaced towards the CaO apex, whereas the cores are closer to the MgO apex. As already remarked above, two of these specimens are rich in modal carbonate. Specimen Hi 25-96 R is very rich in epidote. Enrichment of pillow

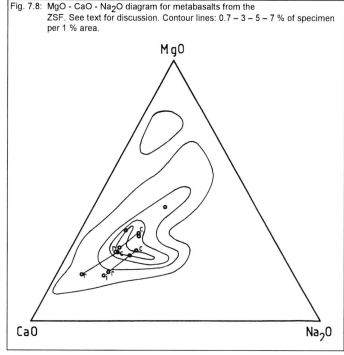

Fig. 7.8: MgO - CaO - Na$_2$O diagram for metabasalts from the ZSF. See text for discussion. Contour lines: 0.7 – 3 – 5 – 7 % of specimen per 1 % area.

MgO

CaO

Na$_2$O

rims in carbonate has been often observed and is usually ascribed to sedimentary carbonate in the interpillow matrix (Barnicoat & Fry 1986).

Another way to elucidate the effects of ocean floor metamorphism is to use metamorphic phase diagrams. Following (and modifying) Schliestedt (1986) and Widmer (1996), here a projection of the rock and mineral chemistry to the plane $(Al,Fe^{III})_2O_3 - Na_2O - (Ca,Mg)O_2$ is used. As garnet and epidote are part of almost all HP-metamorphic metabasalts, these phases are used for projection. According to Schliestedt (1986), the exact composition of the garnet does not significantly affect the projection. For a detailed discussion of this projection see also Ganguin (1988, p. 249). Following Widmer (1996), in fig. 7.9 a garnet and an epidote analysis from Ganguin (1988) was used for the projection (specimen numbers: grt: 216/01; ep: 216/01). Furthermore, constant activity of SiO_2 and H_2O is assumed. The details of the projection are contained in tab. 7.3.

The phase diagram represents the metamorphic conditions of the late HP stage; approximate phase compositions after Ganguin's (1988) analyses are also shown. Again it is seen that the two specimens Hi 29-4-96 R and Hi 29-2-96 R are enriched in Ca (probably secondarily; both are containing modal carbonate) and classify as outliers. The other specimens can be neatly classified as former glaucophanites and eclogites. Again contrary to expectations, analyses representing pillow rims are not classifying as glaucophanites, and analyses representing pillow cores not as eclogites. Secondary carbonate explains this in two cases, but not for specimen Hi 25-96. In general, the distribution of the specimens is the same as in the MgO-Na2O-CaO diagram: the 4 specimens plotting most closely to the MgO apex in fig. 7.6 (CM 1, Hi 2-96, Hi 22-96, Hi 25-96 C) are now classified as former glaucophanites, and the two very CaO rich specimens Hi 29-4-96 R and Hi 29-2-96 R are again outliers in the opposite direction.

To sum up, with the specimens studied here, oceanic metamorphism has resulted in two effects: enrichment of MgO in exchange for CaO and addition of CaO in the form of probably sedimentary carbonate.

Is there any influence of the oceanic metamorphism on the trace element contents of the specimens? There is no peculiarity to be seen in the spidergrams and REE diagram curves of the group of former glaucophanites (Mg enriched) as compared to the former eclogites. Neither is there a conclusive difference in the behavior of the Ca-enriched specimens. Two specimens show a distinct enrichment in LILE in the spidergram (fig. 7.4): Hi 29-4-96 R and Hi 29-4-96 I. One of these is also deviant in the MgO-CaO-Na2O diagram and in the metamorphic phase diagram discussed above (figs 7.8, 7.9). The other CaO enriched specimen however show perfectly normal behavior. Neither does there emerge a clear picture when pillow cores are compared to pillow rims. Rims can both have higher and lower trace element contents compared

Table 7.3: construction of metamorphic mole-fraction phase diagram

Matrix A: Composition of projection phases and of system components (cation numbers)

	grt	ep	qtz	H_2O	Na_2O	$(Ca,Mg)O_2$	$(Al,Fe^{III})_2O_3$
Si	3	3	1	0	0	0	0
$Al+Fe^{III}$	2	3	0	0	0	0	2
Fe^{II}	1.54	0	0	0	0	0	0
Mg	0.84	0	0	0	0	1	0
Ca	0.59	2	0	0	0	1	0
Na	0	0	0	0	2	0	0
H	0	1	0	2	0	0	0

The coordinates of projected phase X then can be calculated from the composition of specimen/mineral Y with the matrix equation

$A^{-1} * Y = X$.

Matrix A-1:

0	0	0.65	0	0	0	0
0	0	0.08	-0.5	0.5	0	0
1	0	-2.19	1.5	-1.5	0	0
0	0	-0.04	0.25	-0.25	0	0.5
0	0	0	0	0	0.5	0
0	0	-0.545	1	0	0	0
0	0,5	0.771	0.75	-0.75	0	0

Therefore the components are calculated thus:

$Na_2O = 0.5 * Na$
$(Ca,Mg)O_2 = -0.545 * Fe^{II} + Mg$
$(Al,Fe^{III})_2O_3 = 0.5 * (Al + Fe^{III}) - 0.771 Fe^{II} + 0.75 * Mg - 0.75 * Ca$

to pillow cores. Neither does the sign of the Sr anomaly seem to be affected by oceanic metamorphism. To sum up, although oceanic metamorphism may well result in
some alteration of trace element contents, and in particular sometimes a strong enrichment in LILE is observed, the general depletion in LILE and the curves rising
from Ba to K in the spidergram are not affected.

Another way to study element changes is possible by using the analyses of pillow
cores and rims. It would be supposed that pillow rims have been altered more
strongly than cores. The method of Gresens (1967) can not be used, as no rock den-
sities were measured. The alternative method of Grant (1986) leads to clumsy diagrams and is replaced here with the following, mathematically equivalent, bar diagrams of $\log(C_{core}/C_{rim})$ for each element (fig. 7.10). The crossing of the abscissa
with the horizontal axis has been positioned at the value of the TiO_2 and Zr ratios,
which are equal or almost equal in the 3 cases studied and are therefore assumed to
have been constant. In the case of specimen Hi 25-96, where the water contents of
core and rim are about equal, this point is close to zero. In the case of the other two

specimens, the rim has a much higher water content, therefore the abscissa crosses at a higher value of the logarithmized core-to-rim ratio.

The ratios of V, Al_2O_3, $Fe_2O_3(t)$, Cr and Nb (where they could be calculated) are close to the Ti and Zr values (and curiously also MnO), lending support to the assumption that the latter were immobile.

There are similarities as well as differences between the three cases. CaO and Sr are strongly enriched in the rims. To a lesser degree, this holds also for $SiO2$, Al_2O_3, P_2O_5 and Ga, probably due to metasomatic enrichment from a sedimentary component in the interpillow matrix. K_2O and Ba behave congruently; they can be either strongly depleted or

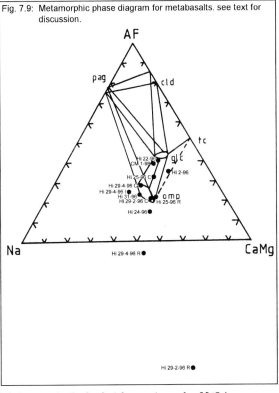

Fig. 7.9: Metamorphic phase diagram for metabasalts. see text for discussion.

strongly enriched in the rims. Na_2O behaves similarly, but less extremely. MgO is either slightly depleted in the rims or immobile. The REE are not at all immobile. They can be either enriched or depleted in the rims; LREE are more mobile than HREE. Co, Ni and Zn are either immobile or slightly depleted in the rims. On the whole, most changes are minor. K and Ba, the elements most strongly affected by metasomatic alteration, are altered for less than a factor of around 6 in the most strongly altered specimen pair Hi 29-4-96 C and R.

All this lends support to the above contention that there was no major mobilization and thus the spidergrams and REE diagrams can be reliably used for petrogenetic interpretations.

7.4 Results of the log-ratio method: correlation structure

Table 7.4 contains the log-ratio variances and log-ratio correlation coefficients for the metabasalts, calculated without specimen Hi 24-96. Due to the rather homoge-

Fig. 7.10: Element concentration changes between pillow cores and rims

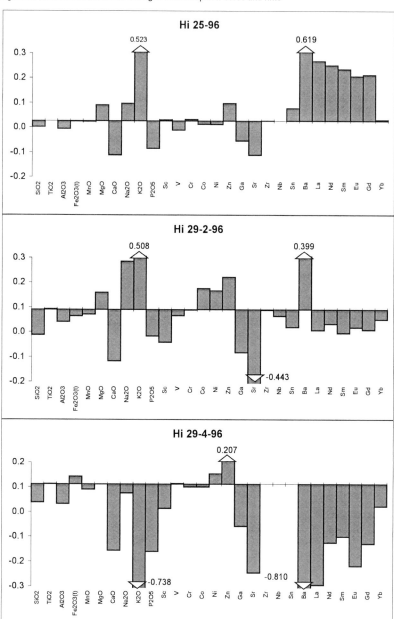

neous chemical composition of the specimens, the log-ratio variances are generally quite small. So it is easier to see which elements do not correlate with each other than which ones do. The group K, Ba and Sr seems not to correlate with any of the other elements. They have probably been mobile during metamorphism, as already discussed above. The smallest log-ratio variance of these elements with any other element are those between K and Sr (0.254), K and Ba (0.288) and Sr to Ca (0.249), all three rather high values. The group Mg, Fe, Cr, Co, Ni is standing also for the most part alone, except for the low log-ratio variances of Fe and Mg to Al and Si. Ca and Na seem not to correlate well with either the ferromagnesians and the incompatible elements.

Because of the generally indistinct picture seen from the log-ratio variances, a detailed comparison to the log-ratio correlation coefficient and to element-element plots is made.

Many of the small log-ratio variances are disclaimed by insignificant and sometimes even negative log-ratio correlation coefficients. Si and Al loose many of their correlations to other elements. In particular, Al is found no longer correlated to the ferromagnesians, the Ti-V group, Ca and Na or any of the incompatible HFSE and REE. Si looses its correlations to Ca and Na, the HFSE and LREE, but remains correlated to the HREE. According to the log-ratio correlation coefficients, the following picture emerges (fig. 7.11): there is the group of ferromagnesian elements plus Si (Si, Mg, Fe, Cr, Co, Ni). This group is also correlated to Ti, V and the HREE. Si is also correlated to Al. Another group of coherently correlated elements is formed by Ti, V, Zr and the HREE, to which more loosely also Sc is attached. Sc furthermore is correlated to Ca. Ca is weakly correlated to Sr. The HREE are not correlated to the LREE. K is correlated to Ba; both have negative correlations to the HREE, the ferromagnesians, Ti, V, Si and Al.

Fig. 7.12 shows some of these element-element relationships as plots. It turns out that some seemingly good correlations do not look very impressive after all. The Si-Al relationship is good, but Si-Mg and Si-Ti aren't valid correlations. The surprisingly good correlations of Si with the HREE however seem real (shown: Si-Yb). The correlations among the ferromagnesians are borne out by the plots (shown here are Cr-Ni and Fe-Co) and also of the ferromagnesians to the HREE and to Ti. The case of Mg-Ti is special: the plot resembles a typical tholeiitic fractionation trend, caused by oli vine fractionation during the first part of the liquid development, followed by Fe-Ti oxides.

The Fe-Mg relationship, although less well defined, might perhaps also be interreted as a curved tholeiitic trend. But it turns out that those data points marking the low-Mg low Fe/Ti end of the purported trends are those specimens representing pillow rims and interpillow matrix which were already found above (chapter 7.3

on ocean floor metamorphism) to be relatively depleted in Mg, the depletion being probably caused by addition of Ca.

The correlations of the HFSE Ti, V and Zr among each other and to the HREE are also valid. K-Ba, despite its relatively low correlation values, displays a clear relationship. The Ca-Sr correlation, although real, is by far weaker.

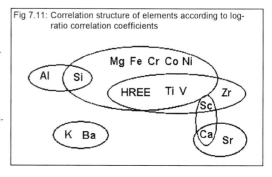

Fig 7.11: Correlation structure of elements according to log-ratio correlation coefficients

As there are many ZSF metabasalt analyses to be found in the literature, it is possible to compare these findings to a larger sample. For this purpose 38 analyses from Widmer (1996) were selected, as the analyses in the older literature are lacking in trace element data. Widmer's data are XRF analyses; most of his specimens were eclogites. Only two specimens are glaucophanites, one specimen is a prasinite. Most of the specimens come from two moraine blocks below Pfulwe pass. They were selected to have only minor retrograde overprint.

Tab. 7.4: log-ratio variance (lower left) and log-ratio correlation coefficient (upper right) matrices for metabasalts. Log-ratio correlation coefficients significant at the 5 % level and log-ratios < 0.1 are in bold print.

	SiO₂	Al₂O₃	Ga	Fe₂O₃(t)	MnO	MgO	Cr	Co	Ni	TiO₂	V	Zn	CaO	Na₂O	P₂O₅
SiO2		**0.767**	**0.745**	**0.832**	**0.746**	**0.733**	**0.751**	**0.793**	0.594	0.441	**0.665**	0.317	0.075	0.142	-0.084
Al2O3	**0.008**		0.415	**0.569**	0.401	**0.597**	**0.589**	**0.557**	0.453	0.074	**0.505**	0.094	-0.001	0.266	-0.018
Ga	**0.016**	**0.032**		0.390	**0.776**	0.209	0.422	0.476	0.176	0.174	0.331	-0.030	**0.518**	-0.483	0.166
Fe2O3(t)	**0.009**	**0.021**	**0.041**		**0.687**	**0.784**	**0.664**	**0.834**	**0.626**	**0.703**	**0.804**	**0.640**	-0.206	0.386	-0.182
MnO	**0.023**	**0.044**	**0.020**	**0.027**		0.434	0.489	**0.752**	0.288	0.479	**0.565**	0.228	0.126	-0.196	-0.077
MgO	**0.076**	**0.091**	0.136	**0.061**	0.109		**0.864**	**0.855**	**0.819**	0.450	0.542	0.497	-0.465	**0.711**	-0.468
Cr	**0.084**	0.103	0.118	**0.086**	0.110	**0.037**		**0.876**	**0.856**	0.175	0.409	0.455	-0.313	0.456	**-0.621**
Co	**0.019**	**0.033**	**0.044**	**0.014**	**0.024**	**0.043**	**0.046**		**0.761**	0.469	**0.610**	0.541	-0.288	0.425	-0.455
Ni	0.178	0.199	0.242	0.162	0.226	**0.080**	**0.065**	0.124		**0.081**	0.185	**0.640**	**-0.621**	**0.663**	-0.494
TiO2	**0.052**	**0.074**	**0.083**	**0.032**	**0.060**	0.111	0.173	0.119	0.282		**0.837**	0.363	0.103	0.231	0.052
V	**0.015**	**0.021**	**0.042**	**0.011**	**0.036**	**0.092**	0.119	**0.030**	0.238	**0.022**		0.312	0.215	0.197	-0.053
Zn	**0.081**	**0.097**	0.128	**0.011**	0.109	0.111	0.129	**0.066**	0.141	**0.098**	**0.085**		-0.430	0.443	-0.281
CaO	**0.068**	**0.069**	**0.044**	0.101	**0.091**	0.261	0.252	0.130	**0.043**	**0.064**	**0.064**	0.202		**-0.761**	-0.442
Na2O	**0.085**	**0.074**	0.164	**0.070**	0.152	**0.064**	0.124	**0.073**	0.136	**0.094**	**0.064**	**0.092**	0.229		0.157
P2O5	**0.058**	**0.051**	**0.059**	**0.077**	**0.092**	0.227	0.265	0.119	0.363	**0.094**	**0.064**	0.155	**0.041**	0.157	
Zr	**0.043**	**0.047**	**0.068**	**0.032**	**0.056**	0.160	0.216	**0.073**	0.321	**0.023**	**0.020**	**0.096**	**0.069**	0.114	**0.039**
Sn	0.109	**0.078**	0.143	0.120	0.157	0.290	0.340	0.177	0.465	0.147	**0.094**	0.232	0.116	0.160	**0.063**
Sc	**0.023**	**0.023**	**0.035**	**0.039**	**0.051**	0.140	0.153	**0.061**	0.310	**0.051**	**0.014**	0.137	**0.030**	0.124	**0.053**
La	0.152	0.128	0.166	0.186	0.194	0.325	0.375	0.221	0.486	0.208	0.167	0.312	0.144	0.203	**0.088**
Nd	**0.085**	**0.069**	0.100	0.105	0.106	0.229	0.252	0.131	0.389	0.137	**0.089**	0.226	0.101	0.164	**0.076**
Sm	**0.053**	**0.046**	**0.062**	**0.073**	**0.070**	0.185	0.197	**0.094**	0.350	0.102	**0.056**	0.184	**0.066**	0.153	**0.066**
Eu	**0.054**	**0.051**	**0.055**	**0.078**	**0.068**	0.175	0.187	**0.089**	0.341	0.100	**0.061**	0.194	**0.063**	0.153	**0.066**
Gd	**0.050**	**0.046**	**0.061**	**0.068**	**0.068**	0.176	0.198	**0.089**	0.337	**0.089**	**0.053**	0.179	**0.068**	0.137	**0.057**
Tb	**0.027**	**0.028**	**0.034**	**0.041**	**0.041**	0.144	0.151	**0.056**	0.281	**0.071**	**0.033**	0.120	**0.052**	0.124	**0.048**
Ho	**0.015**	**0.017**	**0.039**	**0.016**	**0.036**	0.107	0.134	**0.039**	0.234	**0.044**	**0.015**	**0.095**	**0.069**	**0.088**	**0.043**
Tm	**0.020**	**0.032**	**0.028**	**0.013**	**0.029**	**0.094**	0.140	**0.037**	0.233	**0.029**	**0.013**	**0.083**	**0.066**	**0.048**	**0.060**
Yb	**0.016**	**0.033**	**0.032**	**0.011**	**0.030**	**0.093**	0.111	**0.029**	0.213	**0.028**	**0.013**	**0.057**	**0.068**	0.104	**0.061**
Lu	**0.016**	**0.023**	**0.035**	**0.014**	**0.032**	0.106	0.124	**0.034**	0.222	**0.043**	**0.016**	**0.071**	**0.068**	**0.097**	**0.048**
K2O	1.624	1.550	1.624	1.684	1.748	2.061	2.067	1.749	2.029	1.714	1.651	1.517	1.427	1.516	1.291
Sr	0.437	0.412	0.378	0.529	0.519	0.809	0.730	0.579	0.915	0.573	0.470	0.610	0.249	0.630	0.320
Ba	1.264	1.165	1.283	1.351	1.427	1.587	1.523	1.367	1.579	1.451	1.305	1.277	1.120	1.143	1.071
H2O(t)	0.114	0.135	0.083	0.163	0.133	0.251	0.181	0.150	0.300	0.241	0.173	0.209	0.144	0.280	0.199
S	0.972	0.992	0.929	0.980	1.041	1.163	0.910	0.989	1.000	1.143	0.999	0.876	0.927	1.202	1.047
CO2	4.704	4.654	4.503	4.856	4.706	5.662	5.208	4.893	5.522	4.965	4.720	4.808	4.173	5.078	4.465

The results of the log-ratio analysis of this sample is presented in table 7.5. Despite the differing specimen selection criteria and the smaller list of analyzed ele- ments, there are common points to be noticed. According to the log-ratio correlation coefficients, there is a distinct group of incompatible elements, encompassing Ti, V, P, Zr, Hf, Y and surprisingly Fe. There is a group of loosely correlated magnesian elements: Mg, Ni, and Cr, but Fe is having significant negative correlations to this group. Mg is also correlated to Si and Al. Si is correlated to Na. Ca and K are for the most part uncorrelated to any other element. Ca seems to have some relationship to V and Co. K has significant negative correlations to Fe and Sc.

A difference to the result of the log-ratio analysis of my sample is that in Widmer's sample there is a distinct group of incompatible element, whereas in my sample, the Mg group elements are correlated to the Fe-Ti group elements and also more loosely to the HREE. The only truly incompatible elements left over are P and Zr on the one hand, the LREE and Sn on the other.

This major difference serves to illustrate the problems of doing statistics with small samples. The pseudo-tholeiitic trends discussed above are causing a fake correlation between Mg group elements and some incompatible elements. However, this refutation of some of the correlations makes the interpretation quite straightfor-

Table 7.4 continued

	Zr	Sn	Sc	La	Nd	Sm	Eu	Gd	Tb	Ho	Tm	Yb	Lu	K2O	Sr	Ba	H2O	S
SiO2	0.179	-0.260	0.435	**-0.617**	-0.391	-0.153	-0.116	-0.203	0.081	0.522	**0.613**	**0.669**	0.528	**-0.865**	-0.385	**-0.857**		
Al2O3	0.025	0.143	0.401	-0.388	-0.179	-0.064	-0.152	-0.194	-0.073	0.420	0.304	0.261	0.266	**-0.698**	-0.275	-0.556		
Ga	0.022	-0.339	0.423	-0.406	-0.242	0.038	0.171	-0.018	0.334	0.237	0.579	0.505	0.353	-0.575	0.062	**-0.616**		
Fe2O3(t)	0.488	-0.197	0.271	**-0.699**	-0.413	-0.258	-0.311	-0.272	0.013	**0.690**	**0.782**	**0.811**	**0.742**	**-0.795**	**-0.689**	**-0.886**		
MnO	0.340	-0.290	0.327	-0.447	-0.113	0.112	0.151	0.087	0.427	0.529	**0.664**	**0.650**	**0.602**	**-0.699**	-0.420	**-0.807**		
MgO	0.018	-0.464	0.106	-0.535	-0.369	-0.240	-0.137	-0.222	-0.063	0.428	0.524	0.531	0.424	**-0.721**	**-0.756**	**-0.619**		
Cr	-0.290	**-0.618**	0.110	**-0.670**	-0.411	-0.211	-0.116	-0.280	0.027	0.255	0.252	0.470	0.366	**-0.677**	-0.514	-0.488		
Co	0.097	-0.500	0.133	**-0.707**	-0.429	-0.261	-0.168	-0.279	0.034	0.419	0.538	**0.638**	0.526	**-0.731**	**-0.678**	**-0.707**		
Ni	-0.281	-0.593	-0.330	-0.572	-0.519	-0.515	-0.438	-0.506	-0.314	0.152	0.207	0.327	0.264	-0.404	-0.551	**-0.741**		
TiO2	**0.803**	-0.083	0.462	-0.406	-0.275	-0.129	-0.074	-0.022	0.067	0.581	**0.744**	**0.759**	0.579	**-0.780**	-0.455	**-0.827**		
V	**0.673**	0.042	**0.711**	-0.594	-0.271	-0.045	-0.105	-0.080	0.106	**0.655**	**0.775**	**0.755**	**0.630**	**-0.780**	-0.455	**-0.827**		
Zn	0.237	-0.449	-0.281	**-0.797**	**-0.744**	**-0.683**	**-0.735**	**-0.740**	-0.329	0.107	0.345	**0.602**	0.443	-0.136	-0.466	-0.296		
CaO	0.217	0.076	**0.673**	-0.041	-0.022	0.204	0.265	0.133	0.278	-0.031	0.225	0.184	0.032	-0.072	0.582	-0.116		
Na2O	-0.037	-0.079	-0.295	-0.255	-0.367	-0.525	-0.486	-0.437	**-0.634**	0.031	**0.617**	0.013	-0.085	-0.163	-0.595	-0.098		
P2O5	0.425	0.439	0.086	0.295	0.042	-0.058	-0.023	0.016	-0.061	0.122	0.085	0.026	0.060	0.182	0.320	-0.067		
Zr		0.384	0.386	-0.167	-0.079	-0.050	-0.178	0.007	0.026	0.539	**0.609**	0.534	0.490	-0.261	-0.270	-0.571		
Sn	0.068		0.198	**0.733**	0.838	0.616	0.432	**0.665**	0.282	0.426	0.011	-0.397	0.021	-0.028	0.101	-0.032		
Sc	0.035	0.079		-0.219	0.051	0.371	0.313	0.265	0.351	0.349	0.407	0.385	0.235	-0.588	0.111	-0.499		
La	0.138	0.042	0.127		0.803	0.555	0.543	**0.686**	0.159	-0.140	-0.436	**-0.739**	-0.481	0.447	0.249	0.477		
Nd	0.084	0.022	0.064	0.031		0.892	0.758	0.920	0.585	0.268	-0.062	-0.473	-0.068	0.044	-0.028	0.174		
Sm	0.064	0.045	0.032	0.059	0.010		0.888	0.946	0.836	0.345	0.128	-0.186	0.108	-0.207	-0.036	-0.030		
Eu	0.074	0.062	0.036	0.060	0.019	0.006		0.916	0.818	0.233	0.141	-0.125	0.060	-0.215	-0.079	-0.041		
Gd	0.056	0.041	0.034	0.047	0.009	0.003	0.005		0.752	0.362	0.194	-0.217	0.073	-0.166	-0.109	-0.049		
Tb	0.043	0.067	0.023	0.086	0.030	0.010	0.011	0.010		0.470	0.458	0.275	0.486	-0.376	-0.243	-0.280		
Ho	0.024	0.059	0.024	0.108	0.045	0.028	0.034	0.024	0.013		**0.939**	0.600	0.860	-0.760	**-0.697**	**-0.795**		
Tm	0.025	0.101	0.032	0.159	0.079	0.050	0.051	0.043	0.024	0.006		0.825	0.943	**-0.968**	-0.504	**-0.944**		
Yb	0.029	0.136	0.031	0.185	0.106	0.066	0.065	0.062	0.029	0.018	0.010		0.851	**-0.660**	-0.522	**-0.801**		
Lu	0.026	0.086	0.030	0.136	0.064	0.039	0.043	0.036	0.014	0.004	0.005	0.008		**-0.670**	**-0.683**	**-0.760**		
K2O	1.478	1.424	1.563	1.119	1.358	1.442	1.450	1.417	1.436	1.551	1.753	1.618	1.544		0.569	**0.900**		
Sr	0.448	0.396	0.359	0.354	0.408	0.391	0.402	0.399	0.397	0.465	0.490	0.489	0.474	0.906		0.564		
Ba	1.253	1.100	1.186	0.817	0.982	1.049	1.055	1.050	1.081	1.209	1.370	1.309	1.218	0.254	0.691			
H2O(t)	0.218	0.320	0.140	0.301	0.247	0.186	0.183	0.195	0.151	0.188	0.169	0.156	0.177	1.535	0.288	1.181		
S	1.111	1.311	1.019	1.428	1.212	1.119	1.135	1.151	1.007	1.003	1.054	0.913	0.922	2.511	1.186	2.033	1.005	
CO2	4.617	4.374	4.485	4.267	4.357	4.407	4.506	4.449	4.467	4.705	4.696	4.784	4.692	2.673	2.789	2.820	4.022	4.413

ward. The basalts have seen some olivine fractionation, resulting in positive correlations among Mg, Ni and Cr. Fe and Ti are enriched in the more evolved members of the series, forming Fe-Ti minerals, which are also incorporating some HREE. This explains also the correlation of Si to the HREE, because Si also is enriched in the more evolved rocks. Ca is strongly affected by ocean floor metamorphism, making it uncorrelated to most other elements. In particular, there probably was no plagioclase fractionation, as then Ca would have some relationship to some of the potential fractionation indices. K is also strongly affected by ocean floor metamorphism and probably also the later stages. Its relationship to Ba is caused by their both entering preferentially sheet silicates.

It seems that the log-ratio analysis in the case of the metabasalts did not offer much information which could not be gained from the study of the most traditional element-element plots. Perhaps some strange correlations like the one between Si and the HREE might have gone unnoticed. However, as previous studies have focus sed their sampling on specimens with minor retrograde overprint, it seemed an open and interesting question whether the major traits of the magmatic petrogenesis remain unchanged in such a long history of metamorphic alteration. The answer in the case of the ZSF is that not much was changed after the impact of oceanic metamorphism, with the possible exception of the K and Ba contents.

The effects of ocean floor metamorphism can be also studied using the log-ratio adaptation of Beach & Tarney (1978) method already introduced in chapter 6.8. We have to separate two samples of metabasalt specimens, differing in the degree to which they have been subjected to ocean floor metamorphic alteration, and calculate

Tab. 7.5: log-ratio correlation coefficients (upper right) and log-ratios (lower left) of metabasalt specimens of Widmer (1996). Log-ratio correlation coefficients significant at the 5 % level and log ratios < 0.05 are in bold print.

	SiO2	TiO2	Al2O3	Fe2O3	MnO	MgO	CaO	Na2O	K2O	P2O5	Y	Zr	V	Cr	Ni	Co	Zn	Hf	Sc
SiO2		**-0.402**	**0.755**	-0.020	**-0.539**	**0.744**	-0.172	**0.419**	-0.312	-0.252	**-0.699**	**-0.518**	0.203	0.275	**0.442**	-0.293	0.241	**-0.425**	**0.346**
TiO2	**0.055**		**-0.401**	**0.562**	0.198	**-0.521**	-0.155	0.117	-0.096	**0.697**	**0.445**	**0.894**	0.057	**-0.744**	**-0.676**	-0.071	**0.345**	**0.441**	0.253
Al2O3	**0.010**	0.069		-0.196	**-0.328**	**0.624**	-0.225	0.034	0.084	-0.123	**-0.672**	**-0.521**	0.174	0.226	0.282	**-0.458**	0.100	-0.247	0.243
Fe2O3	**0.018**	0.018	**0.032**		0.263	-0.160	-0.069	0.078	**-0.350**	**0.356**	**0.328**	**0.513**	0.192	**-0.520**	**-0.428**	-0.170	**0.359**	**0.350**	0.321
MnO	0.067	0.046	0.072	**0.029**		**-0.484**	0.159	**-0.451**	0.143	0.178	**0.659**	0.222	0.068	-0.293	**-0.398**	-0.009	**-0.415**	0.101	-0.010
MgO	0.076	0.208	0.079	0.135	0.214		-0.322	0.152	-0.243	**-0.341**	**-0.701**	**-0.684**	0.042	**0.391**	**0.586**	0.091	0.157	**-0.468**	0.126
CaO	0.118	0.133	0.136	0.099	0.106	0.284		**-0.501**	-0.095	**-0.390**	0.077	-0.158	**0.421**	0.059	-0.061	**0.327**	**-0.571**	0.081	0.010
Na2O	0.061	0.090	0.095	0.076	0.149	0.168	0.249		-0.285	0.156	-0.185	0.195	-0.209	-0.084	0.099	-0.099	**0.662**	-0.209	0.147
K2O	0.596	0.578	0.539	0.568	0.524	0.777	0.663	0.717		0.140	0.019	-0.099	-0.221	-0.224	-0.098	-0.242	-0.318	0.017	**-0.344**
P2O5	0.148	0.065	0.151	0.103	0.125	0.319	0.287	0.160	0.575		0.322	**0.653**	-0.006	**-0.676**	**-0.629**	**-0.381**	0.324	0.183	0.190
Y	0.331	0.194	0.362	0.221	0.156	0.601	0.307	0.362	0.754	0.246		**0.594**	0.041	**-0.344**	**-0.476**	-0.156	**0.433**	**0.441**	0.233
Zr	0.185	**0.041**	0.093	0.093	0.116	0.394	0.235	0.150	0.688	0.078	0.156		0.041	**-0.646**	**-0.785**	-0.156	**0.433**	**0.441**	0.233
V	0.238	0.285	0.245	0.239	0.267	0.356	0.213	0.379	0.937	0.385	0.550	0.345		-0.266	**-0.459**	**-0.345**	-0.158	-0.302	**0.804**
Cr	0.414	0.629	0.421	0.494	0.543	0.384	0.511	0.541	1.189	0.863	0.904	0.841	0.867		**0.691**	0.182	-0.230	-0.191	-0.313
Ni	0.178	0.339	0.196	0.243	0.308	0.146	0.321	0.262	0.807	0.523	0.664	0.564	0.670	0.232		**0.383**	-0.131	-0.177	**-0.521**
Co	0.160	0.157	0.195	0.135	0.155	0.224	0.145	0.216	0.775	0.328	0.376	0.270	0.491	0.481	0.211		-0.080	0.133	**-0.439**
Zn	0.053	0.053	0.069	**0.046**	0.117	0.150	0.223	**0.044**	0.687	0.117	0.337	0.097	0.335	0.565	0.292	0.188		0.151	0.187
Hf	0.056	**0.029**	0.062	**0.023**	0.052	0.203	0.110	0.119	0.552	0.121	0.223	0.090	0.323	0.511	0.265	0.134	0.067		-0.36
Sc	**0.030**	**0.042**	**0.041**	**0.027**	0.062	0.138	0.121	0.091	0.646	0.123	0.294	0.114	0.139	0.546	0.326	0.206	0.068	0.077	

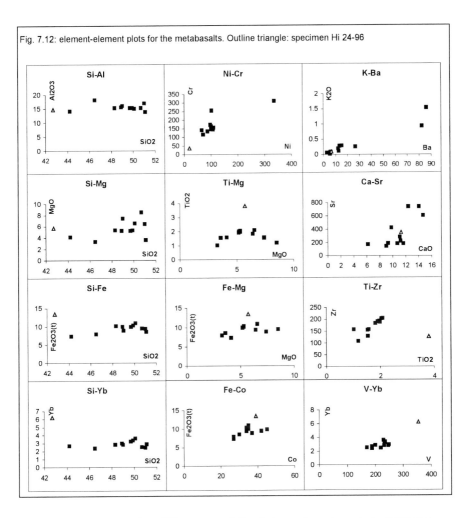

Fig. 7.12: element-element plots for the metabasalts. Outline triangle: specimen Hi 24-96

their correlation matrices. When a correlation between two elements is distinctly lower in the sample with the larger oceanic overprint, one of these elements has probably been mobile. The opposite – retained correlations – however does not prove immobility.

Here this method is applied to the pillow rims and the pillow cores, because these have presumably been altered in different degrees by ocean floor metamorphism. The results are somewhat tentative, as there were only 3 analyses each of pillow cores and rims, due to sampling insufficiencies during the field campaign.

Tab. 7.6: log-ratio variances (lower left) and log-ratio correlation coefficients (upper right) for pillow cores and rims. Significant values in bold face.

Pillow Cores

	SiO2	Al2O3	Ga	Fe2O3(t)	MnO	MgO	Cr	Co	Ni	TiO2	V	Zn	CaO	Na2O	P2O5	Zr
SiO2		**0.934**	-0.342	**0.994**	-0.365	**0.995**	0.335	-0.482	**0.993**	**0.937**	0.670	**0.951**	**-0.966**	0.673	**0.878**	**0.925**
Al2O3	0.001		-0.655	**0.967**	-0.673	**0.896**	-0.022	-0.762	**0.969**	**1.000**	**0.891**	**0.999**	**-0.995**	**0.892**	0.650	**1.000**
Ga	0.023	0.025		-0.440	**1.000**	-0.251	0.770	**0.988**	-0.449	-0.650	**-0.927**	-0.616	0.573	**-0.925**	0.149	-0.674
Fe2O3(t)	0.002	0.003	0.037		-0.462	**0.980**	0.234	-0.572	**1.000**	**0.968**	0.745	**0.978**	**-0.988**	0.747	0.823	**0.960**
MnO	0.052	0.056	**0.007**	0.070		-0.275	0.755	**0.992**	-0.471	-0.668	**-0.936**	-0.635	0.592	**-0.934**	0.125	-0.692
MgO	0.014	0.018	0.058	**0.006**	0.092		0.424	-0.396	**0.978**	**0.899**	0.596	**0.917**	**-0.937**	0.599	**0.920**	**0.884**
Cr	**0.006**	**0.007**	**0.006**	0.013	0.024	0.029		0.664	0.224	-0.016	-0.474	0.027	-0.082	-0.471	0.745	-0.049
Co	0.022	0.023	**0.000**	0.036	**0.009**	0.060	**0.006**		-0.580	-0.758	**-0.973**	-0.729	0.691	**-0.972**	-0.004	-0.779
Ni	**0.004**	**0.005**	0.042	**0.000**	0.077	**0.005**	0.016	0.041		**0.971**	0.752	**0.980**	**-0.990**	0.754	0.817	**0.963**
TiO2	0.003	0.003	0.042	**0.001**	0.079	**0.010**	0.017	0.040	**0.001**		**0.888**	**0.999**	**-0.995**	**0.889**	0.655	**0.999**
V	0.003	0.001	0.033	0.006	0.068	0.024	0.013	0.029	0.008	0.004		0.867	-0.839	**1.000**	0.233	**0.902**
Zn	0.013	0.015	0.070	**0.005**	0.114	**0.006**	0.035	0.069	0.004	0.004	0.015		**-0.999**	0.869	0.687	**0.997**
CaO	0.013	0.011	**0.009**	0.025	0.029	0.053	**0.006**	**0.006**	0.030	0.025	0.013	0.051		-0.841	-0.725	**-0.991**
Na2O	0.026	0.022	0.094	0.019	0.049	0.030	0.055	0.089	0.018	0.012	0.016	**0.010**	0.058		0.237	**0.904**
P2O5	0.003	0.007	0.021	**0.004**	0.044	**0.009**	**0.006**	0.022	**0.005**	0.008	0.013	0.017	0.021	0.041		0.630
Zr	**0.002**	**0.002**	0.040	**0.001**	0.077	0.011	0.015	0.038	**0.001**	0.003	0.006	0.023	0.013	**0.008**		
Sc	0.003	0.001	0.035	0.005	0.072	0.022	0.014	0.032	0.006	0.003	0.000	0.013	0.015	0.015	0.013	0.002
La	0.046	0.058	0.055	0.043	0.064	0.027	0.042	0.062	0.042	0.055	0.073	0.057	0.081	0.111	0.025	0.057
Nd	0.060	0.072	0.054	0.059	0.055	0.045	0.048	0.062	0.060	0.074	0.091	0.080	0.088	0.141	0.035	0.075
Sm	0.048	0.059	0.048	0.047	0.053	0.034	0.039	0.055	0.047	0.060	0.076	0.065	0.077	0.121	0.026	0.061
Eu	0.080	0.093	0.054	0.084	0.046	0.072	0.058	0.064	0.085	0.101	0.114	0.113	0.098	0.180	0.052	0.101
Gd	0.045	0.056	0.043	0.045	0.048	0.033	0.035	0.050	0.045	0.057	0.072	0.064	0.071	0.118	0.024	0.058
Tb	0.046	0.057	0.039	0.047	0.043	0.037	0.034	0.047	0.048	0.060	0.073	0.069	0.069	0.123	0.025	0.061
Ho	0.018	0.025	0.037	0.015	0.056	**0.008**	0.020	0.041	0.015	0.022	0.035	0.025	0.048	0.063	**0.006**	0.023
Tm	0.011	0.017	0.030	**0.009**	0.077	**0.007**	0.013	0.033	**0.009**	0.012	0.025	0.020	0.037	0.053	**0.002**	0.016
Lu	0.030	0.039	0.040	0.028	0.053	0.018	0.027	0.046	0.027	0.038	0.052	0.041	0.060	0.087	0.013	0.039
K2O	0.047	0.039	0.043	0.065	0.070	0.111	0.042	0.035	0.072	0.059	0.034	0.094	0.016	0.077	0.069	0.055
Sr	0.325	0.297	0.387	0.342	0.468	0.427	0.358	0.362	0.349	0.312	0.265	0.347	0.281	0.249	0.395	0.306
Ba	0.015	0.012	0.010	0.027	0.031	0.057	**0.008**	**0.007**	0.032	0.027	0.013	0.053	0.000	0.058	0.024	0.024
H2O(t)	0.055	0.044	0.079	0.067	0.122	0.113	0.064	0.068	0.072	0.057	0.034	0.083	0.035	0.053	0.085	0.053
S	0.024	0.025	**0.001**	0.040	**0.010**	0.066	**0.008**	**0.000**	0.045	0.043	0.031	0.073	**0.005**	0.091	0.026	0.041
CO2	4.717	4.621	4.559	4.858	4.624	5.210	4.663	4.483	4.906	4.776	4.529	4.998	4.323	4.679	4.932	4.742

Pillow Rims

	SiO2	Al2O3	Ga	Fe2O3(t)	MnO	MgO	Cr	Co	Ni	TiO2	V	Zn	CaO	Na2O	P2O5	Zr
SiO2		**0.918**	-0.804	**0.979**	**0.901**	0.795	**1.000**	**0.977**	0.844	**0.890**	**0.953**	0.763	0.831	0.098	0.622	**0.904**
Al2O3	0.007		0.502	**0.979**	0.654	**0.971**	**0.923**	**0.981**	0.888	**0.998**	**0.998**	**0.995**	**0.957**	0.542	0.485	**0.882**
Ga	0.021	0.051		0.666	**0.982**	0.278	0.795	0.659	0.360	0.443	0.586	0.229	**0.999**	-0.513	0.034	0.472
Fe2O3(t)	0.016	0.019	**0.061**		0.794	**0.902**	**0.982**	**1.000**	**0.935**	**0.964**	**0.995**	**0.878**	0.701	0.298	0.768	**0.972**
MnO	0.033	0.071	0.015	0.048		0.453	**0.895**	0.787	0.528	0.603	0.727	0.406	**0.990**	-0.344	0.220	0.628
MgO	0.045	0.022	0.128	0.022	0.130		0.804	**0.906**	**0.996**	**0.984**	**0.942**	**0.999**	0.324	0.681	**0.969**	**0.978**
Cr	0.003	0.012	0.027	0.007	0.027	0.041		**0.980**	0.852	0.896	**0.957**	0.772	0.823	0.112	0.633	**0.910**
Co	0.006	0.007	0.046	0.003	0.047	0.022	0.003		**0.939**	**0.967**	**0.996**	**0.883**	0.694	0.308	0.774	**0.974**
Ni	0.065	0.044	0.155	0.024	0.139	0.006	0.052	0.032		**0.996**	**0.967**	**0.991**	0.404	0.616	**0.945**	**0.992**
TiO2	0.019	0.007	0.080	0.009	0.085	0.006	0.016	0.006	0.015		**0.986**	**0.974**	0.486	0.542	**0.911**	**0.999**
V	0.009	0.005	0.056	0.004	0.059	0.015	0.006	0.001	0.026	0.002		**0.923**	0.624	0.395	0.830	**0.991**
Zn	0.073	0.044	0.172	0.036	0.167	0.004	0.065	0.040	0.003	0.017	0.032		0.275	0.718	**0.981**	**0.966**
CaO	0.025	0.031	0.028	0.079	0.076	0.103	0.044	0.051	0.148	0.068	0.054	0.146		-0.472	0.082	0.514
Na2O	0.222	0.148	0.364	0.216	0.424	0.105	0.239	0.194	0.139	0.139	0.173	0.100	0.231		**0.840**	0.514
P2O5	0.035	0.010	0.102	0.046	0.136	0.020	0.045	0.030	0.048	0.017	0.023	0.036	0.049	0.081		**0.897**
Zr	0.015	0.005	0.072	0.008	0.008	0.008	0.013	0.004	0.019	0.000	0.001	0.021	0.061	0.145	0.016	
Sc	0.013	0.018	0.023	0.056	0.061	0.079	0.027	0.033	0.117	0.047	0.035	0.117	**0.002**	0.218	0.038	0.041
La	0.231	0.209	0.236	0.348	0.364	0.322	0.283	0.284	0.414	0.286	0.278	0.380	0.110	0.277	0.182	0.276
Nd	0.063	0.071	0.050	0.142	0.118	0.169	0.091	0.104	0.227	0.125	0.106	0.221	**0.009**	0.280	0.088	0.116
Sm	0.060	0.075	0.037	0.138	0.098	0.178	0.085	0.102	0.233	0.129	0.107	0.232	0.010	0.317	0.101	0.119
Eu	0.087	0.092	0.074	0.175	0.153	0.196	0.120	0.132	0.261	0.151	0.134	0.250	0.019	0.283	0.102	0.141
Gd	0.059	0.066	0.048	0.135	0.114	0.161	0.086	0.097	0.212	0.117	0.100	0.212	**0.007**	0.271	0.082	0.109
Tb	0.024	0.030	0.027	0.078	0.074	0.103	0.043	0.050	0.146	0.067	0.053	0.145	0.000	0.233	0.049	0.060
Ho	**0.004**	**0.003**	0.043	**0.008**	0.051	0.022	**0.004**	0.001	0.006	0.006	0.001	0.043	0.040	0.181	0.022	**0.004**
Yb	**0.008**	**0.006**	0.054	**0.004**	0.056	0.016	0.006	0.000	0.027	0.003	0.001	0.033	0.055	0.178	0.025	0.002
Lu	0.005	0.005	0.050	0.004	0.053	0.019	0.004	0.000	0.004	0.000	0.004	0.000	0.037	0.182	0.025	0.002
K2O	3.384	3.190	3.519	3.686	3.957	3.355	3.561	3.492	3.620	3.378	3.437	3.432	2.934	2.459	2.921	3.369
Sr	0.641	0.653	0.547	0.858	0.734	0.886	0.723	0.757	0.127	0.798	0.761	0.992	0.414	0.865	0.651	0.777
Ba	3.453	3.280	3.541	3.790	3.993	3.493	3.638	3.586	3.774	3.493	3.536	3.590	2.970	2.638	3.030	3.479
H2O(t)	0.285	0.340	0.172	0.428	0.258	0.536	0.325	0.373	0.618	0.440	0.390	0.627	0.173	0.756	0.403	0.421
S	0.530	0.555	0.621	0.374	0.463	0.441	0.458	0.447	0.348	0.453	0.454	0.408	0.785	0.874	0.636	0.463
CO2	6.005	6.206	5.439	6.619	5.826	6.962	6.194	6.389	7.287	6.632	6.450	7.280	5.369	7.222	6.362	6.560

Tab. 7.6 continued

	Sc	La	Nd	Sm	Eu	Gd	Tb	Ho	Tm	Lu	K2O	Sr	Ba	H2O(l)	S	CO2
SiO2	0.714	0.707	0.573	0.622	0.396	0.601	0.548	0.830	0.848	0.729	-0.872	-0.435	-0.995	-0.494	-0.611	-0.865
Al2O3	0.917	0.409	0.243	0.302	0.042	0.276	0.215	0.577	0.603	0.437	-0.641	-0.086	-0.966	-0.152	-0.853	-0.629
Ga	-0.902	0.422	0.574	0.523	0.728	0.546	0.598	0.239	0.208	0.394	-0.161	-0.697	0.435	-0.648	0.953	-0.176
Fe2=3(t)	0.784	0.628	0.483	0.536	0.296	0.513	0.457	0.767	0.787	0.853	-0.816	-0.338	-1.000	-0.399	-0.691	-0.807
MnO	-0.912	0.400	0.554	0.502	0.711	0.525	0.578	0.215	0.184	0.371	-0.137	-0.679	0.457	-0.629	0.960	-0.151
MgO	0.644	0.771	0.649	0.694	0.481	0.674	0.626	0.880	0.895	0.791	-0.915	-0.519	-0.981	-0.575	-0.533	-0.909
Cr	-0.420	0.903	0.964	0.946	0.998	0.955	0.972	0.803	0.784	0.889	-0.753	-0.994	-0.239	-0.985	0.541	-0.763
Co	-0.957	0.279	0.442	0.386	0.614	0.411	0.468	0.088	0.056	0.249	-0.008	-0.579	0.568	-0.524	0.988	-0.023
Ni	0.791	0.620	0.474	0.527	0.286	0.504	0.448	0.760	0.781	0.645	-0.810	-0.328	-1.000	-0.390	-0.699	-0.801
TiO2	0.914	0.415	0.250	0.309	0.049	0.283	0.221	0.583	0.609	0.443	-0.646	-0.092	-0.967	-0.158	-0.850	-0.634
V	0.998	-0.051	-0.224	-0.164	-0.417	-0.191	-0.253	0.143	0.175	-0.020	-0.222	0.376	-0.742	0.314	-0.997	-0.207
Zn	0.896	0.453	0.291	0.349	0.092	0.323	0.263	0.617	0.642	0.481	-0.678	-0.135	-0.977	-0.200	-0.826	-0.667
CaO	-0.870	-0.501	-0.343	-0.400	-0.146	-0.375	-0.315	-0.659	-0.683	-0.528	0.717	0.189	0.987	0.254	0.794	0.707
Na2O	0.998	-0.047	-0.221	-0.160	-0.413	-0.187	-0.249	0.147	0.179	-0.016	-0.225	0.373	-0.744	0.311	-0.997	-0.211
P2O5	0.292	0.959	0.895	0.921	0.787	0.910	0.882	0.996	0.998	0.968	-1.000	-0.813	-0.825	-0.850	-0.158	-1.000
Zr	0.927	0.384	0.217	0.277	0.016	0.251	0.189	0.555	0.582	0.413	-0.620	-0.059	-0.958	-0.125	-0.867	-0.608
Sc		0.010	-0.165	-0.104	-0.361	-0.131	-0.194	0.203	0.234	0.041	-0.281	0.319	-0.781	0.256	-0.990	-0.266
La	0.072		0.985	0.994	0.990	0.929	0.990	0.979	0.981	0.974	1.000	-0.963	-0.944	-0.632	-0.964	-0.967
Nd	0.090	0.003		0.985	0.998	0.979	0.999	1.000	0.932	0.920	0.979	-0.900	-0.987	-0.487	-0.996	0.299
Sm	0.075	0.001	0.001		0.965	1.000	0.996	0.953	0.943	0.989	-0.925	-0.976	-0.540	-0.988	0.240	-0.931
Eu	0.115	0.013	0.004	0.008		0.972	0.985	0.840	0.822	0.917	-0.794	-0.999	-0.301	-0.994	0.485	-0.803
Gd	0.072	0.001	0.001	0.000	0.007		0.998	0.944	0.933	0.985	-0.915	-0.981	-0.517	-0.992	0.266	-0.921
Tb	0.073	0.003	0.001	0.001	0.006	0.000		0.921	0.908	0.972	-0.887	-0.992	-0.461	-0.998	0.327	-0.894
Ho	0.034	0.007	0.016	0.009	0.032	0.009	0.011		0.999	0.987	-0.997	-0.863	-0.770	-0.895	-0.066	-0.998
Tm	0.025	0.013	0.022	0.014	0.038	0.013	0.015	0.001		0.981	-0.999	-0.846	-0.790	-0.880	-0.098	-0.999
Lu	0.051	0.002	0.006	0.003	0.018	0.002	0.004	0.002	0.005		-0.971	-0.934	-0.656	-0.955	0.097	-0.974
K2O	0.037	0.168	0.178	0.162	0.190	0.154	0.151	0.115	0.097	0.135		0.820	0.819	0.856	0.146	1.000
Sr	0.268	0.617	0.660	0.620	0.704	0.608	0.608	0.493	0.454	0.551	0.176		0.342	0.998	-0.447	0.828
Ba	0.015	0.087	0.094	0.082	0.104	0.077	0.074	0.052	0.040	0.064	0.013	0.272		0.404		0.810
H2O(l)	0.036	0.200	0.221	0.200	0.245	0.192	0.191	0.135	0.115	0.164	0.010	0.119	0.032		0.006	0.064
S	0.033	0.071	0.070	0.063	0.072	0.058	0.054	0.047	0.038	0.053	0.030	0.349		0.064		0.131
CO2	4.556	5.582	5.601	5.529	5.579	5.478	5.444	5.273	5.144	5.396	3.831	2.871	4.284	3.794	4.415	

	Sc	La	Nd	Sm	Eu	Gd	Tb	Ho	Yb	Lu	K2O	Sr	Ba	H2O(l)	S	CO2
SiO2	0.974	-0.999	-0.226	0.039	-0.641	-0.164	0.838	0.970	0.959	0.965	-0.924	-0.872	-0.958	-0.271	0.989	-0.629
Al2O3	0.804	-0.901	-0.594	-0.361	-0.893	-0.542	0.553	0.987	0.993	0.990	-0.696	-0.995	-0.765	-0.461	0.967	-0.886
Ga	0.918	-0.828	0.398	0.626	-0.058	0.455	0.998	0.635	0.602	0.621	-0.970	-0.411	-0.941	0.355	0.705	-0.043
Fe2O3(t)	0.907	-0.970	-0.419	-0.165	-0.783	-0.361	0.710	0.999	0.996	0.998	-0.827	-0.954	-0.880	-0.461	0.999	-0.774
MnO	0.976	-0.918	0.220	0.469	-0.244	0.281	0.992	0.768	0.740	0.756	-0.998	-0.574	-0.988	0.174	0.825	-0.229
MgO	0.636	-0.769	-0.770	-0.575	-0.975	-0.729	0.336	0.919	0.935	0.926	-0.503	-0.990	-0.587	-0.799	0.878	-0.972
Cr	0.971	-0.998	-0.240	-0.652	-0.178		0.830	0.973	0.963	0.969	-0.918	-0.879	-0.954	-0.285	0.991	-0.640
Co	0.903	-0.967	-0.429	-0.175	-0.790	-0.371	0.702	1.000	0.999	0.999	-0.821	-0.957	-0.875	-0.419	0.998	-0.780
Ni	0.700	-0.821	-0.713	-0.503	-0.952	-0.667	0.415	0.949	0.962	0.955	-0.575	-0.998	-0.655	-0.745	0.915	-0.948
TiO2	0.763	-0.870	-0.646	-0.421	-0.921	-0.596	0.496	0.974	0.983	0.978	-0.647	-0.999	-0.721	-0.681	0.948	-0.915
V	0.859	-0.940	-0.511	-0.363	-0.843	-0.455	0.633	0.998	1.000	0.999	-0.765	-0.980	-0.826	-0.550	0.988	-0.835
Zn	0.596	-0.736	-0.802	-0.616	-0.985	-0.763	0.286	0.897	0.915	0.905	-0.457	-0.981	-0.545	-0.829	0.852	-0.983
CaO	0.936	-0.854	0.353	0.588	-0.106	0.412	1.000	0.671	0.639	0.658	-0.981	-0.454	-0.956	0.309	0.738	-0.091
Na2O	-0.131	-0.057	-0.992	-0.991	-0.867	-0.998	-0.461	0.337	0.377	0.354	0.290	-0.572	0.192	-0.984	0.247	-0.835
P2O5	0.427	-0.589	-0.904	-0.758	-1.000	-0.875	0.094	0.793	0.819	0.804	-0.275	-0.925	-0.370	-0.923	0.733	-1.000
Zr	0.783	-0.885	-0.821	-0.392	-0.908	-0.570	0.524	0.981	0.988	0.984	-0.671	-0.998	-0.743	-0.657	0.958	-0.901
Sc		-0.982	0.001	0.265	-0.449	0.065	0.940	0.889	0.869	0.881	-0.987	-0.739	-0.998	-0.045	0.928	-0.436
La	0.138		0.069	0.186	-0.080	0.609	0.123	-0.860	-0.959	-0.946	-0.954	0.939	0.852	0.969	0.231	-0.981
Nd	0.020	0.069		0.186	0.965	0.893	0.998	0.342	-0.456	-0.494	-0.472	-0.164	0.673	-0.063	0.999	-0.370
Sm	0.020	0.091	0.002		0.007	0.742	0.979	0.578	-0.205	-0.247	-0.222	-0.419	0.454	0.393	0.913	-0.749
Eu	0.034	0.047	0.002	0.007		0.862	-0.118	-0.808	-0.833	-0.819	-0.226	0.625	-0.126	0.913	-0.311	0.870
Gd	0.017	0.071	0.000	0.002	0.003		0.008	0.401	-0.399	-0.438	-0.415				0.746	-0.103
Tb	0.002	0.113	0.010	0.010	0.020	0.008		0.680	0.648	0.667	-0.983	-0.485	-0.959	0.298	0.746	-0.103
Ho	0.024	0.253	0.087	0.087	0.113	0.082	0.039		0.999	1.000	-0.803	-0.965	-0.859	-0.497	0.996	-0.799
Tm	0.035	0.283	0.108	0.108	0.135	0.101	0.053	0.001	1.000		-0.777	-0.975	-0.837	-0.534	0.991	-0.824
Lu	0.032	0.275	0.101	0.101	0.128	0.095	0.049	0.001	0.000		-0.792	-0.970	-0.850	-0.513	0.994	-0.810
K2O	3.047	1.940	2.740	2.869	2.591	2.750	2.947	3.384	3.458	3.445		0.619	0.995	-0.118	-0.856	0.284
Sr	0.475	0.164	0.301	0.313	0.257	0.312	0.417	0.710	0.766	0.749	1.822		0.695	0.707	-0.936	0.929
Ba	0.384	1.948	2.592	2.872	2.597	2.764	2.983	3.474	3.540	3.540	0.028	1.701		-0.017	-0.904	0.379
H2O(l)	0.203	0.204	0.118	0.101	0.115	0.124	0.173	0.351	0.390	0.376	3.038	0.183	2.935		-0.413	0.919
S	0.707	1.443	0.955	0.926	1.046	0.939	0.779	0.488	0.448	0.458	6.205	2.317	6.413	1.409		-0.739
CO2	5.562	4.672	4.966	4.916	4.851	5.013	5.375	6.289	6.452	6.396	5.866	3.102	5.163	3.682	9.427	

Table 7.6 contains the log-ratio variances and the log-ratio correlation coefficients of both the pillow cores and the pillow rims, and also the quotients between log-ratio variances (core - rim) (calculated as already described in chapter 6.8 with the larger variance as the denominator and a plus or minus sign to indicate which of the two samples has the larger variance) and the difference of the Fisher-transforms of the log-ratio correlation coefficients. This latter measure is calculated as $(LN((1+\rho_1)/(1-\rho_1)/2 - LN((1+\rho_2)/(1-\rho_2)/2)$ and has been chosen following again a proposal of Beach & Tarney (1978), because it is always normally distributed and can therefore be tested with standard procedures. However, due to the low sample size, the sample error is infinite. The F-test for the quotients of the log-ratio variances is however possible; it turns out that a coefficient is significant at the 5 % level if it is larger than 9.28.

Contrary to expectations, Ca and Mg do not show significantly different correlations between pillow cores and rims. Correlations deteriorating towards the rims are those with the elements K, Na, Ba and Sr (this latter is consistent, although not significant), and (at a lower level of significance) those of Mg, Ni and Zn to the LREE. The correlation of K to Ba does not deteriorate towards the rim. The case of La is dubious; it is not clear why only La of all REE should have been preferentially mobile. Perhaps there were some analytical problems.

Considering the extremely small sample size, these results are not considered any further here.

Tab. 7.6 continued: quotients between log-ratio variances (core – rim; lower left; calculated with the larger variance as the denominator and a plus or minus sign to indicate which of the two samples has the larger variance); difference of the Fisher-transforms of the log-ratio correlation coefficients (upper right)

	SiO2	Al2O3	Ga	Fe2O3(t)	MnO	MgO	Cr	Co	Ni	TiO2	V	Zn	CaO	Na2O	P2O5	Zr
SiO2		0.117	-1.466	0.661	-1.859	1.953	-4.603	-2.751	1.608	0.291	-1.052	0.838	-3.226	0.718	0.640	0.127
Al2O3	-11.7		-1.335	-0.240	-1.599	-0.650	-1.633	-3.335	-0.457	2.346	-1.565	1.790	-3.563	0.904	-0.608	0.244
Ga	1.068	-2.050		-1.276	2.039	-0.542	-0.065	1.775	-0.860	-1.251	-2.306	-0.952	-3.083	-1.057	0.116	-1.332
Fe2O3(t)	-7.935	-5.579	-1.664		-1.581	0.810	-2.110	-5.948	3.507	0.069	-2.001	0.888	-3.429	0.660	0.150	-0.184
MnO	1.557	-1.267	-2.243	1.458		-0.770	-0.461	1.676	-1.098	-1.505	-2.624	-1.182	-1.977	-1.332	-0.098	-1.591
MgO	-3.237	-1.197	-2.208	-3.537	-1.417		-0.656	-1.924	-0.908	-0.954	-1.064	-2.088	-2.052	-0.140	-0.493	-0.860
Cr	2.093	-1.707	-4.596	1.868	-1.144	-1.419		-1.495	-1.035	-1.467	-2.427	-0.998	-1.249	-0.624	0.216	1.575
Co	3.661	3.339	-113.0	11.0	-5.318	2.772	1.976		-2.393	-3.029	-5.220	-2.317	-0.005	-2.452	-1.035	-3.214
Ni	-18.5	-8.620	-3.702	-112.5	-1.809	-1.322	-3.214	1.291		-0.975	-1.065	-0.369	3.059	0.264	-0.633	-0.804
TiO2	-6.548	-2.534	-1.907	-10.8	-1.064	1.731	1.005	7.340	-15.0		-1.073	1.679	-3.551	0.813	-0.748	-0.028
V	-2.675	-4.821	-1.692	1.311	1.169	1.584	2.067	46.0	-3.474	1.427		-0.287	1.548	5.685	-0.950	-1.214
Zn	-5.422	-2.995	-2.455	-6.798	-1.467	1.536	-1.838	1.716	1.121	-3.895	-2.124		-3.883	0.425	-1.474	1.237
CaO	-1.906	-2.741	-3.136	-3.132	-2.628	-1.945	-6.796	-8.629	-4.903	-2.656	-4.156	-2.845		-0.711	-1.001	-3.295
Na2O	-8.575	-6.676	-3.861	-11.5	-2.840	-3.542	-4.372	-2.177	-7.925	-11.2	-10.59	-10.31	-4.018		-0.981	0.926
P2O5	-10.2	-1.499	-4.839	-10.7	-3.081	-2.175	-8.011	-1.342	-8.918	-1.971	-2.123	-2.271	-1.985			-0.716
Zr	-6.364	-2.507	-1.817	-7.881	-1.016	1.411	1.137	9.7	-12.9	-3.143	1.870	-3.808	-2.644	-11.28	-1.925	
Sc	-4.025	-15.3	1.552	-11.2	1.178	-3.628	-1.940	-1.028	-18.0	-17.7	-433.8	-8.965	6.962	-15.00	-2.905	-21.97
La	-4.983	-3.615	-4.316	-8.138	-5.672	-11.9	-6.770	-4.558	-9.8	-5.180	-3.789	-6.699	-1.358	-2.498	-7.376	-4.885
Nd	-1.061	1.015	1.069	-2.386	-2.145	-3.794	-1.905	-1.662	-3.813	-1.685	-1.174	-2.764	9.77	-1.992	-2.534	-1.536
Sm	-1.253	1.268	1.282	-2.943	-1.853	-5.281	-2.184	-1.843	-4.977	-2.151	-1.409	-3.568	7.627	-2.632	-3.924	-1.949
Eu	-1.098	1.018	-1.361	-2.090	-3.356	-2.728	-2.065	-2.061	-3.057	-1.500	-1.170	-2.208	5.188	-1.566	-1.984	-1.396
Gd	-1.314	-1.180	-1.124	-3.013	-2.381	-4.822	-2.426	-1.941	-4.830	-2.047	-1.389	-3.308	9.826	-2.297	-3.478	-1.860
Tb	1.907	1.902	1.459	-1.650	-1.747	-2.748	-1.239	-1.075	-3.051	-1.109	1.398	-2.105	3458.0	-1.887	-1.974	1.019
Ho	4.518	8.607	-1.177	1.954	1.095	-2.718	4.825	44.6	-2.575	3.966	34.56	-1.694	1.186	-2.886	-3.777	6.349
Yb	1.302	2.863	-1.825	2.526	-1.104	-2.314	2.351	80.1	-2.799	5.332	636.00	-1.611	-1.492	-3.382	-10.92	9.011
Lu	4.587	7.782	-1.238	6.252	-1.568	-2.078	6.070	208.9	-1.106	9.7	306.59	1.127	1.193	-2.081	-1.921	15.78
K2O	-71.669	-81.749	-82.556	-56.669	-56.540	-30.176	-84.620	-100.664	-50.402	-57.100	-101.255	-36.654	-186.291	-31.748	-42.084	-61.165
Sr	-1.971	-2.196	-1.415	-2.503	-1.568	-1.588	-2.020	-2.087	-2.941	-2.555	-2.873	-2.862	-1.471	-3.479	-2.081	-2.537
Ba	-234.910	-269.486	-337.838	-138.682	-128.356	-61.742	-448.566	-507.873	-117.011	-129.551	-263.712	-67.312	-29696	-45.484	-125.929	-142.748
H2O(t)	-5.163	-7.694	-2.812	-6.353	-2.108	-4.735	-5.042	-5.447	-8.525	-7.700	-11.537	-7.534	-14.264	-1.974	-4.756	-7.892
S	-21.924	-22.644	-513.074	-9.454	-47.419	-6.700	-59.653	-1718.885	-7.700	-10.483	-14.889	-5.549	-150.067	-9.556	-24.364	-11.403
CO2	-1.273	-1.343	-1.193	-1.363	-1.260	-1.336	-1.328	-1.425	-1.485	-1.389	-1.424	-1.457	-1.242	-1.544	-1.290	-1.383

7.5 Preliminary summary

It has been shown that the metabasalts from the ZSF can be explained as the product of low degrees of melting of subcontinental mantle, slightly differentiated by olivine and probably spinel fractionation. Contrary to the metagabbros, there are no signs of the presence in the mantle residue of any exotic phase retaining incompatible elements like Nb or Ti.

Oceanic metamorphism resulted in exchange of Mg for Ca; also Na, K, Ba and Sr were influenced. K and Ba may have been redistributed to some extent during this or the later metamorphic stages, when sheet silicates were formed.

The long standing interpretation of the ZSF metabasalts as MORB, normal or enriched, has to be questioned in the light of the evidence presented here. Of course the relictic pillow structures are proof of a subaquatic extrusion, but this can as well result from the melting of mantle which is denuded of the continental crust overlaying it by simple shear rifting according to the model of Wernicke (1981, 1985) as adapted to the Alpine situation by Stampfli & Marthaler (1990). From this it follows that the basalts of the ZSF are representing very early stages of oceanization. The metabasalts of the Tsaté nappe, which are generally somewhat less evolved, are then representing an yet earlier stage. This is in keeping with Stampfli & Marthalers (1990) assumption that these rocks were scraped off from parts of the Ligurian ocean which were closer to the southern continental margin

Table 7.6 continued

	Sc	La	Nd	Sm	Eu	Gd	Tb	Ho	Yb	Lu	K_2O	Sr	Ba	$H_2O(l)$	S	CO_2
SiO_2	-1.266	4.769	0.882	0.689 n	1.178	0.859	-0.599	-0.902	-0.681	-1.094	0.273	0.877	-1.065	-0.263	-3.288	-0.573
Al_2O_3	0.459	1.910	0.932	0.690	1.478	0.891	-0.404	-1.848	-2.115	-2.156	0.100	2.888	-1.013	0.590	-3.314	0.663
Ga	-3.060	1.831	0.233	-0.154	0.982	0.121	-2.816	-0.506	-0.485	-0.311	1.935	-0.425	2.212	-1.142	0.986	-0.134
$Fe_2O_3(t)$	-0.457	2.831	0.974	0.764	1.359	0.944	-0.393	-2.875	-2.108	-2.743	0.035	1.518	-4.729	0.076	-4.482	-0.089
MnO	-3.745	1.999	0.401	0.042	1.137	0.295	-2.090	-0.797	-0.765	-0.598	3.424	-0.174	3.030	-0.916	0.775	0.081
MgO	0.013	2.043	1.794	1.510	2.711	1.744	0.386	-0.206	-0.250	-0.553	-1.005	2.074	-1.642	0.442	-1.959	0.599
Cr	-2.549	5.077	2.249	1.769	4.209	2.062	0.932	-1.044	-0.926	-0.657	0.599	-1.541	1.628	-2.143	-2.071	-0.245
Co	-3.403	2.336	0.933	0.584	1.787	0.826	-0.364	-4.090	-3.249	-3.463	1.152	1.244	1.996	-0.071	-0.893	1.024
Ni	0.205	1.887	1.408	1.139	2.152	1.360	0.040	-0.827	-0.922	-1.116	-0.471	3.247	-4.077	0.550	-2.425	0.709
TiO_2	0.549	1.775	1.023	0.768	0.978		-0.320	-1.497	-1.666	-1.771	0.002	3.923	-1.138	0.671	-3.070	0.807
V	2.202	1.685	0.335	0.106	0.789	0.298	-1.006	-3.325	-4.454	-3.830	0.782	2.682	0.221	0.944	-5.801	0.995
Zn	0.784	1.430	1.405	1.083	2.542	1.339	-0.026	-0.737	-0.797	-0.974	-0.332	2.200	-1.621	0.983	-2.438	1.561
CaO	-3.038	0.718	-0.726	-1.098	-0.040	-0.832	-5.456	-1.604	-1.591	-1.377	3.218	0.681	4.425	-0.061	0.137	0.972
Na_2O	3.690	0.010	2.509	2.515	0.793	3.212	0.244	-0.203	-0.216	-0.386	-0.528	1.042	-1.154	2.746	-3.458	0.991
P_2O_5	-0.156	2.613	2.939	2.587	5.465	2.880	1.290	1.998	2.356	0.943	-4.848	0.487	-0.784	0.350	-1.094	0.987
Zr	0.584	1.806	0.948	0.699	1.530	0.904	-0.391	-1.690	-1.898	-1.974	0.088	3.317	-0.968	0.661	-3.241	0.772
Sc		2.371	-0.168	-0.376	0.107	-0.196	-1.935	-1.213	-1.090	-1.339	2.215	1.278	2.423	0.307	-4.318	0.194
La	-1.913		2.243	2.943	0.945	2.525	3.567	4.263	3.968	6.030	-3.709	-3.039	-2.818	-2.236	2.465	-2.724
Nd	4.537	-26.89		1.475	0.848	0.611	3.866	2.168	2.131	2.778	-1.309	-3.343	-0.469	-6.820	0.697	-2.978
Sm	3.744	-104.0	-2.513		1.061	2.018	2.432	2.069	2.012	2.846	-1.180	-2.688	-0.267	-4.400	0.357	-2.643
Eu	3.411	-3.610	1.754	1.073		0.823	2.559	2.343	2.360	2.723	-1.390	-5.511	-0.726	-4.443	1.501	-6.011
Gd	4.147	-50.45	9.82	-17.54	2.437		3.024	2.198	2.151	2.889	-1.327	-3.064	-0.445	-5.652	0.594	-2.927
Tb	38.92	-39.80	-8.287	-16.61	-3.528	-30.96		0.768	0.746	1.326	0.971	-2.227	1.440	-3.748	-0.625	-1.338
Ho	1.425	-34.96	-5.612	-9.29	-3.549	-9.240	-3.679		0.285	-2.212	-2.111	0.710	0.270	-0.899	-3.123	-2.328
Tm	-1.434	-22.57	-4.976	-7.500	-3.546	-7.872	-3.607	-1.282		-2.063	-2.697	0.950	0.138	-0.779	-2.779	-2.939
Lu	1.622	-148.5	-15.98	-37.61	-7.016	-38.37	-13.04	3.923	40.58		-1.025	0.401	0.470	-1.323	-2.784	-1.039
K2O	-82.38	-11.57	-15.36	-17.71	-13.65	-17.84	-19.54	-29.43	-35.84	-25.43		0.433	-1.830	1.397	1.424	4.612
Sr	-1.774	3.772	2.191	1.980	2.738	1.952	1.457	-1.439	-1.685	-1.358	-10.35		-0.502	2.527	1.226	-0.466
Ba	-200.5	-22.47	-29.25	-34.92	-25.02	-36.05	-40.10	-67.02	-88.71	-55.03	-2.078	-6.254		0.445	2.335	0.727
H2O(l)	-5.685	-1.020	1.877	1.984	2.138	1.545	1.106	2.601	3.400	-2.299	-300.7	-1.544	-91.93		0.032	-0.274
S	-21.38	-20.46	-13.56	-14.72	-14.60	-16.30	-14.47	-10.31	-11.73	-8.650	-205.9	-6.633	-1064	-22.10		1.081
CO2	-1.221	1.195	1.128	1.125	1.150	1.093	1.013	-1.193	-1.254	-1.185	-1.531	-1.081	-1.205	1.030	-2.135	

than the materials in the ZSF. Later basalts with a truly oceanic character are either not preserved or never existed, because the oceanization never reached that stage. This then would speak in favor of a very small ocean, closer to 100 than to 1000 km in width.

8 Geochemistry of rodingites and miscellaneous specimens

8.1 Metarodingites. Geochemical properties.

Rodingites are rocks of diverse mineralogy, supposed to have originated as gab-broic dikes enclosed in a peridotite matrix, and which have been variably but strongly metasomatized due to the large chemical contrast to the matrix during serpentinization of the latter. In particular, they are enriched in Ca and depleted in alkaline elements. The rodingites of the ZSF have undergone the same metamorphic conditions as the other units and thus have to be described as metarodingites. They are well known among mineral collectors for beautiful specimens of several mine-rals.

According to Ganguin (1988, p. 39), the minerals are in order of abundance: Ca-rich garnet, chlorite, diopside, vesuvianite, clinozoisite/epidote, calcite, titanite and apatite. Frequently the metarodingites are practically monomineralic. They are fine-grained. At the contact to the enclosing serpentinites there are usually found reaction zones, often developed as chloritoschists.

The metarodingite specimens of this study all come from the Lichenbretter serpen-tinite complex, from a restricted area in the valley of the Gorner glacier, in front of its present termination. These are the specimens Hi 10-96, Hi 12-1-96 and 12-2-96 (impure; contains also part of the reaction zone), Hi 13-3-96 and probably also Hi 8-96. This latter specimen comes from a defunct talc mine. They consist mostly of Chl and Ves; in Hi 12-2-96 also uvarovite with a core of spinel (according to Ganguin 1988:34 a magmatic relic), pyroxene, amphibole and magnetite. Hi 8-96 is made up of chlorite, diopside and opaques.

The major element chemistry of the metarodingites resembles in many respects that of the metagabbros. Table 8.1 contains the data. It has to be taken into account that the metarodingites are by far more hydrated than the metagabbros. Less mobile elements like Al_2O_3, TiO_2 and P_2O_5 are well within the normal range of the meta-gabbros. This holds also of $Fe_2O_3(tot)$. Specimen Hi-8-96 resembles more closely a Fe-Ti-gabbro (or perhaps a metabasalt), Hi 10-96 is an intermediary type, whereas the remaining specimens Hi 12-1-96 and Hi 13-3-96 resemble normal gabbros.

SiO_2 and particularly Na_2O and K_2O are lower in the metarodingites than in the metagabbros and seem to have been lost during metasomatic alteration. MgO is part-ly in the normal range of the metagabbros, partly it seems to have been enriched. For 3 of the 4 specimens, CaO is well outside the normal range of the metagabbros (maximal and minimal values can be checked in table 6.1) and seems to be partly enriched, partly depleted. In specimen Hi 12-1-96 it is in the normal range.

Tab. 8.1: Major element chemistry of metarodingites, compared to metagabbro geometric mean values

	Hi 8-96	Hi 10-96	Hi 12-1-96	Hi 13-3-96	normal gabbros (geometric means)	Fe-Ti-gabbros	metabasalts (w/out Hi 24-96)
SiO_2	32.3	34.8	36.6	34.5	48.0	46.4	49.0
TiO_2	1.146	0.316	0.078	0.044	0.344	3.005	1.658
Al_2O_3	15.6	15.6	13.8	18.6	17.6	15.4	15.4
Fe_2O_3 (t)	10.44	8.04	4.66	3.69	5.55	11.51	9.28
Fe_2O_3	0.64	7.01	2.66	2.38			
FeO	8.8	0.9	1.8	1.2			
MnO	0.398	0.209	0.063	0.141	0.100	0.190	0.190
MgO	26.13	13.05	26.1	14.67	9.32	6.38	6.38
CaO	4.23	23.8	10	23.01	11.35	10.28	10.28
Na_2O	0.02	0.05	-0.01	0.04	2.88	3.60	3.60
K_2O	-0.04	0.01	0	0.01	0.08		0.096
P_2O_5	0.115	0.020	0.007	0.012	0.026	0.156	0.156
H_2O (t)	11.53	5.99	9.19	7.25	2.94	1.81	0.26

Depletion in SiO_2 and alkaline elements is expected due to the large chemical contrast between mafic and ultramafic rocks. The same holds for MgO enrichment. Perhaps also the 2 specimens with MgO contents in the normal range had initially a lower content and thus also have been enriched. CaO enrichment is almost the defining property of rodingites, but specimen Hi 8-96 must have been depleted. This remains somewhat unclear.

The trace element chemistry is discussed with the help of REE- and spidergrams (fig. 8.1 to 8.3). Generally they are bearing out the similarity to the metagabbros; the differences to the metabasalts become clearer. The REE patterns are dropping from the LREE to the HREE. The REE contents are generally rather low, much lower than those of the metabasalts, except for Hi 8-96, which was seen already above to have either Fe-Ti gabbro characteristics or affinities to the metabasalts. This specimen has a negative Eu anomaly, in keeping with its affiliation to the Fe-Ti metagabbros. The other specimens have positive Eu anomalies and thus resemble the normal metagabbros. Perhaps there is a negative Ce anomaly, but this remains unclear due to the low analytical quality of the data for the 2 specimens Hi 12-1-96 and Hi 13-3-96 with the lowest Ce contents.

The same characteristics are seen in the chondrite normalized spidergram after Thompson (1982, fig. 8.2). There is a steep rise in the mobile LILE and a moderate drop in the HFSE/REE right part of the diagram, where again the curves are subparallel. The negative Nb anomaly resembles the metagabbros. In contrast however to this group of rocks, the metarodingites display huge negative Sr anomalies, except for specimen Hi 13-3-96, having a positive anomaly. These are to be explained by the Ca metasomatism. Specimens Hi 10-96 and Hi-13-3-96 have been enriched in Ca. Of these specimens, the latter has a positive Sr anomaly, whereas the formers negative anomaly is not as large as with those specimens which have been depleted in

140

Ca. At any rate, the primary Ca carrier plagioclase has been completely removed during metasomatism and thus Sr may have escaped. The particularly low Ba contents are probably also in part due to mobilization.

In the MORB-normalized spidergram after Pearce (1983), the steep rise from Yb over Sc to Cr is seen, which has been also observed in most metagabbros.

The metarodingites analyzed are thus most probably derived from gabbroic proto-liths, which have been metasomatized during serpentinization of enclosing ultrama-fites, probably in the ocean floor metamorphic stage. However, the number of speci-mens studied is very low. It would be worthwhile to study the metarodingites of the ZSF in more detail.

8.2 Chloritoschist. Geochemical properties.

Specimen Hi 17-96 comes from a 2 m wide shear zone within a body of metagabbro-ic rocks, comprising "Zone 6" of the little map in Bearth (1973). This specimen is composed mainly of chlorite with some talc and opaques. Its major element chemis-try with very low SiO_2 (29.7 %) and high MgO (28.7 %) resembles the serpentinites, except for the high Al_2O_3 content of 17.9 %. There is also a great similarity to mi-croprobe analyses of chlorites from the ZSF (cf. Ganguin 1988, Meyer 1983a). The trace elements however are more resembling a leukogabbro (fig. 8.1 to 8.3). Per-haps this specimen is the result of a tectonic mixture between metagabbro and ser-pentinite. It might also be part of a former reaction rim between metabasic rocks and metasediments. In the vicinity of the specimen location there are calcschists to be found. Chloritoschists are frequently observed as reaction rocks between metaba-sites and serpentinites (Bearth 1967, Ganguin 1988).

8.3 HP Veins. Geochemical properties.

2 Veins from the HP metamorphic stadium have been analyzed, one omphacite vein (Hi 32-4-96) and one Mg-chloritoid vein containing also talc and minor rt, omp, white mica and chl (Hi 32-10-96). The omp vein's major element chemistry closely resembles microprobe omphacite analyses from the ZSF (cf. Ganguin 1988, Meyer 1983a). The trace element contents of both veins are very low, partly below the l.o.d. (fig 8.1 to 8.3). Relatively high are the contents of TiO_2, for which thus some mobilization must be assumed.

Fig. 8.1: REE diagram for metarodingites and other specimens (solid lines: metarodingites; dashed line: chloritoschist; dotted lines: veins; bold line: l.o.d). Values below the l.o.d have been set to ½ l.o.d.

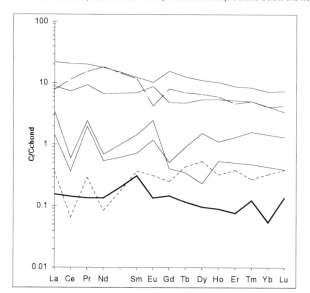

Fig. 8.2: Chondrite-normalized spidergram for metarodingites and other specimens (solid lines: metarodingites; dashed line: chloritoschist; dotted lies: veins; bold line: l.o.d.) . Values below the l.o.d have been set to ½ l.o.d.

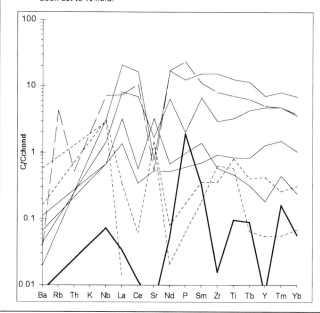

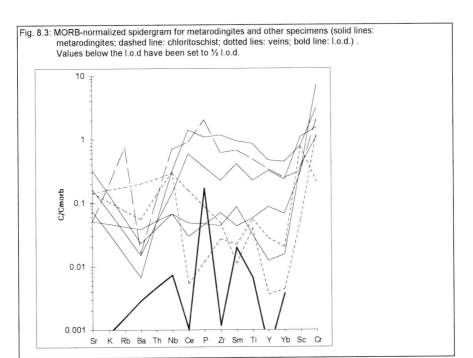

Fig. 8.3: MORB-normalized spidergram for metarodingites and other specimens (solid lines: metarodingites; dashed line: chloritoschist; dotted lies: veins; bold line: l.o.d.). Values below the l.o.d have been set to ½ l.o.d.

9 Magmatic modeling

Magmatic modeling of cumulate rocks is only rarely undertaken. Notable exceptions are the works of Bédard (2001) and Casey (1997). A reason for this neglect can be sought in the fact that there have to be assessed even more variables than with simple melt-residue relationships. In the case at hand there is the further problem that the rocks have been metamorphosed, so that the primary mineral chemistry can not be ascertained. Besides, the specimens have also probably been metasomatized to an unknown degree. Therefore a variety of methods is used to arrive at tentative results.

9.1 The equilibrium distribution method (EDM) after J. H. Bédard

This method is used to determine the composition of interstitial melt in a series of cumulate rocks. It is developed in the papers of Bédard (1994, 2001). For a valid application of this method the following assumptions have to be thought to hold true:

- the cumulate minerals and the melt from which they were precipitating were in equilibrium (no fractional crystallization);
- the system was sealed fast, so that no differences between the intercumulus melt and the melt from which the cumulus phases crystallized could develop before in-situ crystallization of the former;
- there are no post-cumulus metasomatic effects.

For an application of the EDM the modes of the rocks are needed. In the case at hand these are determined from the CIPW norm in the sequence of steps described in Bédard (2001, electronic supplement).

The norm is recalculated to arrive at a more realistic picture of the rock modes. Small amounts of qtz, crn and ne are simply converted to feldspar or pyroxene (whichever is predominating). Also kalifeldspar is added to plagioclase, and opx, which is minor or absent in the rocks studied, to cpx, in which it has been dissolved at the temperature of cumulate crystallization. For the normal metagabbros, apatite, magnetite and ilmenite are left only in very small amounts after allowance is made for solution of these components in feldspar and cpx (ap left: < 0.11 %; mag + ilm left: < 1,33 % except for Hi 32-11-96: 2,55 %). These small amounts are neglected. The result is a simple 3-mineral composition for each specimen (Table 9.1). The "normal" group of Fe-Ti gabbros is treated according to the same scheme, whereas for the "extreme" Fe-Ti gabbros some magnetite and ilmenite remains in the recalculated norm mineral composition.

Table 9.1: recalculated modal compositions of gabbroic specimens

	CM 2-98	Hi 4-96	Hi 16-96	Hi 18-96	Hi 19-96	Hi 24-96	Hi 30-1-96
feldspar	66.83	62.36	71.67	73.01	61.76	63.36	65.33
cpx	24.35	30.00		9.10	28.13	23.59	31.75
olivine	8.82	7.64	28.33	17.88	10.12	13.05	2.92

	Hi 30-3-96	Hi 30-6-96	Hi 32-1-96	Hi 32-8-96	Hi 32-10-96	Hi 32-11-96	Hi 27-1-96
feldspar	60.16	57.87	59.73	59.73	65.28	52.66	14.37
cpx	23.43	28.23	28.04	28.04	15.04	13.30	73.57
olivine	16.40	13.90	12.23	12.23	19.68	21.33	12.06

	Hi 27-2-96	Hi 30-4-96	Hi 30-5-96	Hi 32-12-96	Hi 32-2-96	Hi 32-4-96	Hi 32-5-96	Hi 32-6-96
feldspar	73.12	69.62	67.09	57.10	48.51	50.00	47.16	54.33
cpx	13.24	7.23	25.60	38.74	28.32	26.70	26.87	24.23
olivine	13.64	23.15	7.31	4.16	7.94	13.64	8.36	5.62
ilm + mgt					15.23	9.67	17.61	15.81

From these modal compositions the amount of trapped melt has to be estimated. It is assumed that there are only two cumulus phases. The process of cumulate formation is conceptually reversed. The phases are diminished in the proportions of some melting model. As soon as one (or two for the extreme Fe-Ti gabbros) phase vanishes, we arrive at a minimal amount of trapped melt. The melting model used here is taken from Bédard (2001, electronic supplement). Feldspar, cpx and ol are supposed to enter the melt in the proportions 0.4286 : 0.4286 : 0.1428. It turns out that for most specimens cpx is the first phase to disappear. For Hi 4-96 and Hi 30-1-96 olivine disappears first. This is in agreement with the observation from relictic textures that for most rocks pl and ol are the cumulus phases, whereas cpx is forming xenomorphic crystals (Meyer 1983a).

With the assumed trapped melt fraction the concentration of trace elements in the trapped melt and in the cumulus minerals can be calculated from the following equations:

$$C_{WR} = \phi_{fs}\, C_{fs} + \phi_{cpx}\, C_{cpx} + \phi_{ol}\, C_{ol} + \phi_{TM}\, C_{TM}$$
$$C_{TM} = C_{fs} / D_{fs} = C_{cpx} / D_{cpx} = C_{ol} / D_{ol}$$

(C: concentration of an element; WR: whole rock, ϕ: fraction; TM: trapped melt; D: distribution coefficient mineral/melt)

According to Bédard (op. cit.), for incompatible trace elements the concentrations obtained are not very sensitive to errors in the modal composition.

9.2 Double-logarithmic diagrams

This method was developed by Treuil and is used here as in Pearce & Norry (1979) and Pearce (1980). Two trace elements are plotted against each other on logarithmic scales. When magmatic modeling results are plotted on such a diagram, the shape of the curves is independent of the starting compositions due to the logarithmic scaling. This makes it easy to fit fractionation paths etc. to observed data.

Table 9.2: Distribution coefficients after Bédard (2001) and other sources used for magmatic modeling

	cpx	opx	pl	ol	spl
K	0.007		0.036	0.006	
Ba	0.00068		0.9	0.01	
Nb	0.008		0.03	0.002	
La	0.0536		0.124	0.003	
Ce	0.0858		0.117	0.004	
Sr	0.1283		1.3	0.008	
P	0.13		0.075	0.008	
Nd	0.1873		0.068	0.01	
Sm	0.291		0.058	0.02	
Zr	0.26		0.01	0.02	
Ti	0.34		0.043	0.02	
Eu	0.3288		0.71	0.02	
Tb	0.404		0.031	0.035	
Y	0.412	0.2	0.026	0.007	0.002
Yb	0.43		0.0097	0.05	
Lu	0.433		0.008	0.05	
Sc	3.9		2	0.16	
Cr	3.8	1.9	0.02	1.25	77

In this study only the plot of Cr against Y, representing a typical compatible respectively incompatible trace element, is used. In contrast to Pearce & Norry (1979), Pearce (1980), not only melt compositions, but also restite and fractionating phases are plotted. The distribution coefficients are to be found in table 9.2; they are taken from Bédard (2001), where the pertaining literature is to be found, and from other sources, as specified below.

Two magmatic processes have been modeled: mantle melting and fractionation of individual minerals. The model parameters for mantle melting are the following:

Table 9.3: Model parameters for mantle melting

Source composition:

ol	opx	cpx	spl
0.55	0.245	0.18	0.025

proportion of phases entering the melt:

	ol	opx	cpx	spl
a)	0.10	0.20	0.68	0.02
b)	-0.068	0.390	0.562	0.114

The problem of selecting proportions of the phases entering the melt is rarely discussed in depth. Model a) in table 9.3 is taken from Johnson et al. (1990), where it is claimed to represent the proportions of a peritectic melt of a 4-phase lherzolite (p. 2668/2669). Whereas this is in itself contradictory, as a peritectic would involve some phase being gained by the peritectic reaction and thus having a negative proportion going into the melt, the model is producing reasonable results, if fractionated melting is assumed. Model b) was calculated in the following way: The composition of the first formed melt from the Kilborne Hole lherzolite at 3 GPa after Takahashi (1986; quoted after Phillpotts 1990) was approximated by a mixture of the phases ol, opx, cpx and spl, each with a composition typical for spl-lherzolites. For this the average compositions after BVSP (1981, p. 286) were taken. A system of 4 linear equations was set up and solved to match the concentrations of SiO_2, Al_2O_3, MgO and CaO between melt and four-phase mixture. The proportions given above are the exact solution of this system of equations, rounded to 3 decimals. Of course, this calculation will give deviant values for the other oxides. Table 9.4 compares the composition of the calculated mixture and the starting composition.

Table 9.4: Comparison lherzolite melt – 4-phase mixture

	Melt	Model
SiO_2	46.9	46.9
TiO_2	0.9	0.5
Al_2O_3	11	11.0
FeO_t	7.8	6.9
MnO	0.2	0.1
MgO	19.2	19.2
CaO	12.2	12.2
Na_2O	1.2	0.5

It turns out that the incompatible elements Ti and Na are somewhat too low according to the model, but this is not surprising. As incompatible elements, Ti and Na are not only coming from molten minerals, but are also extracted from the solids left. Model FeO is also somewhat too low; this is explained by Fe partitioning preferentially into the melt compared to MgO with all minerals containing Fe and Mg in solid solution.

The negative fraction of olivine entering the melt is reflecting the peritectic reaction of opx to melt plus olivine. This reaction however exists only at depths less than 0.5 GPa (Philpotts 1990, p. 157, 469)

The melting proportions resulting are very similar to the model of Kinzler (1997).

When the same calculations are made for the Kilborn Hole lherzolite melt at 1 GPa, an unrealistically high negative ol fraction is found. Probably Al_2O_3 is behaving slightly incompatible, entering the melt more strongly than would be supposed from its contents in the phases entering the melt alone. This means that model b) probably reflects melting conditions at lower pressure than 3 GPa.

The possibility that the mantle source did perhaps contain titanian clinohumite, as discussed above in chapter 5 as an explanation of the characteristics of HFSE element contents, is not relevant for the modeling of Cr and Y contents, as the effect of Ti-clinohumite on the bulk partition coefficients would be very minor due to the low amount of Ti-clinohumite possibly involved, in conjunction with the very low D values for Cr and Y (cf. data in Weiss 1997, pp. 151, E2, J6, K3).

The formulas used for modeling are the familiar ones:

Batch melting and equilibrium crystallization:

$$C_L / C_O = 1 / (D + F - FD)$$

$$C_S / C_O = D / (D + F - FD)$$

Rayleigh melting:

$$C_S / C_O = (1-F)^{(1/D-1)} \text{ (restite, in equilibrium with the melt increments)}$$

$$C_L / C_O = (1 - (1-F)^{1/D}) / F \text{ (pooled melt)}$$

Rayleigh crystallization:

$$C_L / C_O = F^{(D-1)} \text{ (pooled melt)}$$

$$C_S / C_O = (1 - F^D) / (1 - F) \text{ (averaged solid)}$$

(C_S : concentration in solid; C_L: in melt; C_O: original source concentration; F: fraction of melt produced / of melt remaining; D: (bulk) distribution coefficient). The case of fractional (Rayleigh) melting is dual to fractional crystallization)

The compositions of restites and melts, calculated again from the modal model compositions and the average phase oxide contents after BVSP (1981), compare favorably with the compositions of the studied serpentinites and gabbros (cf. table 9.5). The higher Al and lower Mg and Fe content of the gabbros as compared to the melt reflects the presence of cumulate plagioclase and perhaps also some fractionation.

Figure 9.1 shows the result of modeling the melting of a four-phase spinel lherzolite. Two of the used D values are different from those given in Bédard (2001). For Y in ol this author uses 0.033, whereas here 0.007 is taken. The former value would result in degrees of melting higher than 70 % for the most Y depleted serpentinites and batch melting. GERM (as in spring 2001) is listing the following values for basalts: 0.00494-0.0125: Beattie (1994); 0,0036: Nielsen et al (1992); 0.0070-0.018: Kennedy et al (1993); all three from experimental work. For Cr in spl, Bédard uses the 500 given by Peirce & Parkinson (1993). Here the experimental value of Ringwood (1970) of 77 is considered as more appropriate.

Table 9.5: Comparison between actual rocks and model

Serpentinite (calculated water-free)			Model a); batch melting		Model a); fractional melting	
	from	to	at 21.89 % melt	at 30.81 % melt	at 5 % melt	at 25 % melt
SiO$_2$	42.78	45.78	42.87	41.45	43.8	42.11
Al$_2$O$_3$	1.11	2.22	1.26	0.7	3.29	2.56
FeO$_t$	6.67	11.11	12.62	13.49	11.7	13.01
MnO	0.11	0.11	0.21	0.22	0.19	0.21
MgO	42.22	44.44	41.29	43.95	37.5	41.58
CaO	0.01	1.83	1.75	0.19	3.52	0.53
Mg-gabbro (geometric mean)						
SiO$_2$	48.00		48.70	50.27	50.21	
Al$_2$O$_3$	17.60		11.42	9.73	6.06	
FeO$_t$	5.55		7.15	6.75	6.75	
MnO	0.10		0.12	0.12	0.11	
MgO	9.32		19.93	20.05	21.99	
CaO	11.35		12.67	13.08	14.88	

Both models lead to subhorizontal trends for the restites and to curved trends
for the melts. Whereas the scaling of the restites trend is very different for
fractional
and batch melting, the melts trends are practically indistinguishable.

With the given model parameters for model a, cpx is melting out at 26.47 %
melt produced. Opx is melting out at 57.1 %. With model b, spl is the first phase to
melt out at about 21.89 % melt production, followed by cpx at 30.81 % and opx at
40.73 %. The calculation has been continued up to complete melting. The
proportions of the phases entering the melt were kept the same (except for
renormalizing to 1) after any phase being molten out.

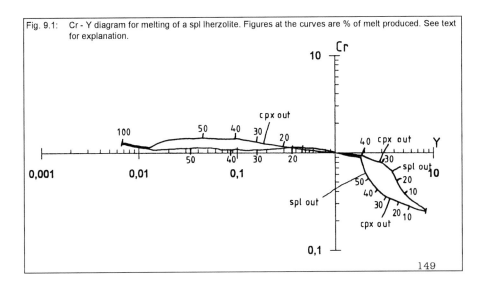

Fig. 9.1:   Cr - Y diagram for melting of a spl lherzolite. Figures at the curves are % of melt produced. See text for explanation.

As spl is the phase most strongly retaining Cr in the residue, Cr contents of the melt increase steeply as soon as spl is getting scarce. The development of the restite follows a subhorizontal path; slight undulations (more clearly visible for model b) are caused by changing bulk Cr distribution coefficients.

Figure 9.2 shows the effects of equilibrium crystallization of the phases ol, pl, opx, cpx, ilm. The endpoints at 0 % melt left are of course having only a mathematical meaning, as the amount of any one phase able to crystallize from a melt is very limited. As is evident from the distribution coefficients, the crystallizing solids are depleted in the incompatible Y. As Cr is compatible in all phases except pl, Cr is enriched in the solids except for pl. The curves for the melts have the same shape as those for the solids. This is different from fig. 9.1, where the bulk distribution coefficient changes during progressive melting.

In figure 9.3 the same modeling is done with the formulas for Rayleigh fractionation. The curves for the averaged solids are in shape similar to those of batch melting, but the degrees of fractionation, where points of similar composition are

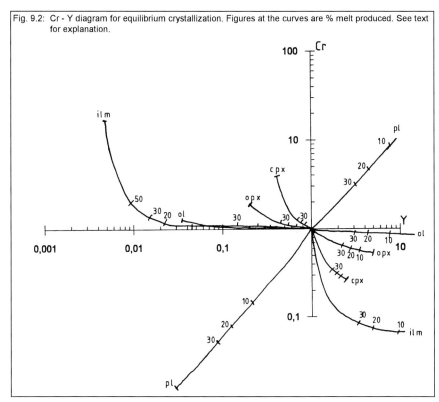

Fig. 9.2: Cr - Y diagram for equilibrium crystallization. Figures at the curves are % melt produced. See text for explanation.

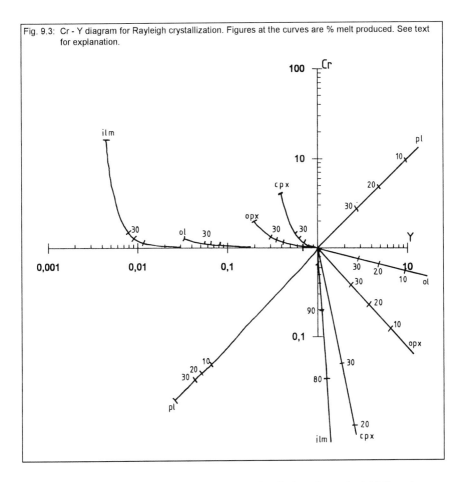

Fig. 9.3: Cr - Y diagram for Rayleigh crystallization. Figures at the curves are % melt produced. See text for explanation.

reached, are quite different. The pooled melt is developing along straight lines in the double-logarithmic scaling.

9.3   The studied specimens in the Cr-Y diagram

Figure 9.4 displays the analyzed specimens in the Cr-Y diagram. The lines of restite and melt development by batch melting have been superimposed on the datapoints to obtain a good fit between model and data.

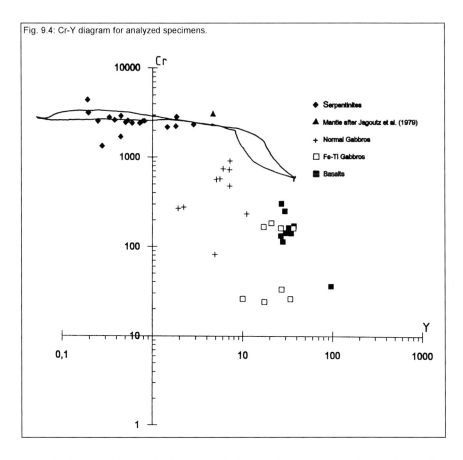

Fig. 9.4: Cr-Y diagram for analyzed specimens.

To obtain this fit the model has to start at a mantle composition slightly depleted in Cr compared to the mantle composition after Jagoutz et al (1979). The serpentinites are forming a subhorizontal array, which is in keeping with the model. With fractional melting, regardless of whether model parameters a or b are used, the serpentinites represent restites after a melt extraction between ca. 5 % and 22 %. For batch melting and model b), these figures are ca. 8 % and 37 %. If batch melting is used with a model with a positive value for the fraction of ol entering the melt, the most depleted serpentinites would require unrealistically high degrees of melt extraction of around 50 %.

On the other hand, below in the chapter on spidergram modeling fractional melting is found not to be realistic either. Most probably the real process was somewhere in between both extremes, although closer to batch melting.

Two serpentinite specimens are lying somewhat below the linear array. It is tempting to speculate that perhaps these specimens were formed under lower pressure conditions in the plagioclase lherzolite stability field. From fig. 9.2 it is clear that pl in the source would displace the restite curve towards lower Cr contents. There is no independent test of this assumption, as we do not have any information on the Eu anomaly of these 2 specimens. Of course it is also possible that the serpentinites are coming from an inhomogeneous mantle source. A third possibility – these specimens might be cumulates produced by fractionation of ol + spl for basalt or Fe-Ti gabbro production – is discussed below.

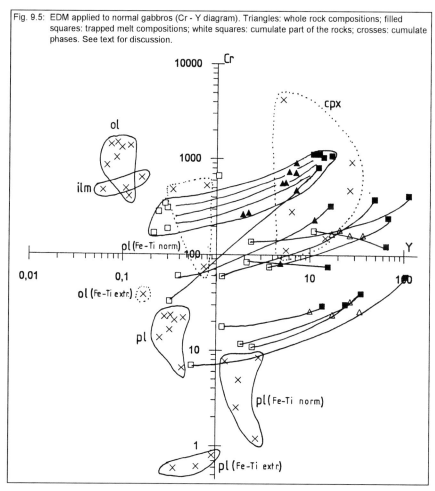

Fig. 9.5: EDM applied to normal gabbros (Cr - Y diagram). Triangles: whole rock compositions; filled squares: trapped melt compositions; white squares: cumulate part of the rocks; crosses: cumulate phases. See text for discussion.

Generally, the serpentinites in the more depleted left part of the diagram are having also smaller Al_2O_3 contents, thus corroborating the idea that the linear array indeed is reflecting progressive depletion by progressive melt extraction.

The Cr-Y diagram does not make it possible to decide whether batch melting or fractional melting was operating. However, above in chapter 5 it was argued from the REE patterns that the restite must be containing some cpx. This would exclude batch melting. REE modeling, which will shed some light on this question, is tried below. Also it has been argued that mantle rock is not capable of retaining any larger amounts of melt (e.g. Johnson et al. 1990). On the other hand, the situation of the mantle below the Ligurian-Piemonte ocean as a "lithospheric ocean" (Lemoine et al. 1987) was perhaps quite different from middle oceanic ridges with their comparably fast rates of mantle ascent and concomitant intense tectonism, a factor facilitating melt escape.

The normal gabbro specimens are lying to the left of and below the melt composition curve, as is to be expected for cumulate rocks, enriched in plagioclase and olivine (cf. the composition of fractionating solids in fig. 9.2 and 9.3). Below some more detailed modeling will be tried.

The Fe-Ti gabbros are forming two clusters at rather Cr-poor compositions. These two clusters do not exactly correspond to the two groups of Fe-Ti gabbros distinguished above in chapter 6.5, "normal" and "extreme". The more Cr rich specimens are all "normal", but one of the Cr-poor specimens (Hi 27-2-96) is also normal and not extreme. Both groups call for some fractionation; 10 respectively 20 % fractionation of olivine plus spinel in the proportions 4:1 would be sufficient (see below). Incidentally, the restites belonging to such a fractionated melt are lying on a curve passing through the most Y-depleted end of the serpentinites (cf. fig. 9.3), thus offering an alternative explanation for the vertical spread of the serpentinite data points at the Y-poor end. As will be seen below, for most normal gabbros no fractionation has to be invoked, although this was discussed above in chapter 6. Notable exceptions are specimens Hi 30-1-96 and Hi 30-6-96.

The metabasalts are forming a rather tight subvertical cluster, with the exception of specimen Hi 24-96, which probably has a different history, as already discussed above in chapter 7. The metabasalt specimens of this study are all coming from a very restricted area; when other analyses from the literature are plotted, they are showing more spread in the Y direction, perhaps indicating derivation from different primary melts or different fractionation paths. The metabasalts analyzed here can very well be explained by fractionation of ca. 10 % of ol+spl, similar to the more Cr-rich group of Fe-Ti gabbros. Fractionation of these phases has been already assumed from the REE diagrams (chapter 7), as well as the lack of plagioclase fractionation, which is endorsed by the Cr-Y diagram.

To gain a clearer picture of the formation of cumulate gabbros, Bédard's technique of EDM has been applied; the results are plotted in fig. 9.5. Compositions of trapped melt and cumulate rock in equilibrium are joined by mixing lines, on which the whole rock data points are lying. Also included in the diagram are the compositions calculated for the cumulate phases making up the cumulate part of the rocks. The original melts, from which both the cumulate phases and the trapped melt evolved, have to be somewhere in between the data points for whole rocks and for trapped melt, if a closed system is presumed.

It is seen that the curve of primary melts for model a) passes between the whole rock data points and the corresponding trapped melt points for most normal gabbros. So it is very well possible that they have developed from primary melts, derived directly from the mantle. The degree of melting as taken from the diagram is rather high with 30 to 40 %; this may be caused by distribution coefficients which are not quite adequate. There are no rocks analyses corresponding to the trapped melt compositions, although for reasons of mass balance a considerable amount of melt must have evolved to comparable compositions. This corresponds to the problem, already noted above in chapter 6, that there are no rocks with a negative Eu-anomaly balancing all those plagioclase cumulates with their positive anomalies. The only rocks with a negative Eu anomaly are some Fe-Ti gabbros and the metabasalt Hi 24-96. But according to the EDM results discussed here, they are not the appropriate candidates for this mass-balance problem.

For the normal Fe-Ti gabbros the EDM calculation has been performed neglecting the modal ilmenite and magnetite, because the amounts involved are minor (less than 6.5 %) and calculation of ilmenite as a cumulus phase results in unrealistic ilmenite compositions containing up to 3400 ppm Cr. For the extreme Fe-Ti gabbros the situation is different; here some ilmenite has been calculated as cumulus phase.

According to the calculations, there are some specimens where cpx and pl have been the cumulus phases; modal ol is vanishing as the first phase upon "backstripping".

9.4   Modeling spidergrams

Using the parameters gained from the study of Cr-Y diagrams as starting point, it has been tried to model the trace element distributions of chondrite- and MORB normalized spidergrams. The distribution coefficients used are as in table 9.6.

Figure 9.6 shows the result of the calculation of restite and melt compositions after 15, 25 and 40 % batch melting of a source corresponding to bulk silicate earth after GERM (see table 9.6). Using fractional melting results in extremely low contents of the restite for the elements from Ba to La (around 10^{-10} to 10^{-16}), resulting in extremely steep curves, and thus seems not realistic. The general appearance of the model restite is reasonably similar to the serpentinites (cf. fig. 5.3 to 5.4), although the curves are still a bit too steep and the range is a bit too small. These little inconsistencies are again probably due to distribution coefficients which are not quite appropriate. The positive Nb and Ti anomalies in the chondrite normalized spidergram are more pronounced in the serpentinites than in the model restites, as is the negative Zr anomaly. In chapter 5 above these anomalies have been explained by the presence of some exotic residual phase in the mantle, possible Ti-clinohumite. This phase has not been used for modeling as not enough is known about its distribution coefficients. Also the serpentinites are having a pronounced negative Sr anomaly, which might be explained by melting in the plagioclase lherzolite field or by Sr loss during serpentinization.

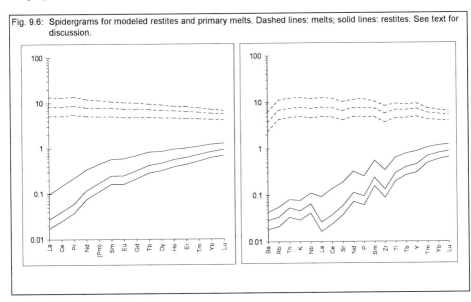

Fig. 9.6: Spidergrams for modeled restites and primary melts. Dashed lines: melts; solid lines: restites. See text for discussion.

Tab. 9.6: bulk silicate earth composition after GERM

Ba	Rb	Th	K	Nb	La	Ce	Sr	Nd	P	Sm	Zr	Ti	Eu	Tb	Y	Tm	Yb	Lu	Sc	Cr
6.6	0.6	0.08	2400	0.66	0.65	1.68	20	1.25	90	0.41	10.5	1200	0.15	0.1	4.3	0.068	0.44	0.068	16	2625

The situation is less favorable for the gabbros. Figure 9.7 shows calculated spidergrams for primary melts, melts after some olivine fractionation and for the po-

Fig. 9.7: Spidergrams for cumulus minerals and fractionated melts. Bold line: bulk silicate earth; solid lines: upper two with ragged shape: plagioclase; lower two ragged: olivine; topmost two: primary melt after 15 / 20 % melting; dotted lines at the top: fractionated melts after 10 / 20 % ol + spl fractionation; dashed line: ilmenite; lower dashed-dotted line: magnetite; upper dashed-dotted line: fractionated melt after plagioclase fractionation.

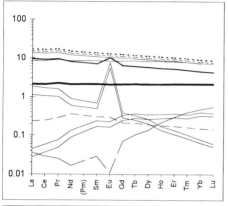

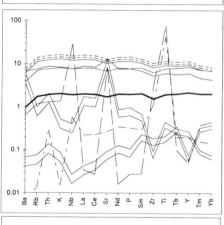

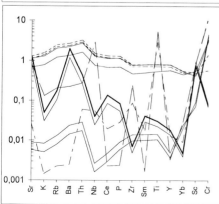

Fig. 9.8: Spidergrams for modeled gabbros

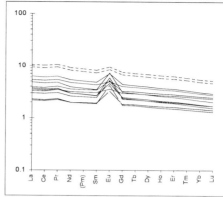

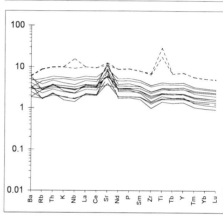

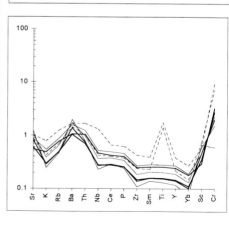

tential cumulus phases olivine, plagioclase, ilmenite and magnetite. The starting composition was again the bulk silicate earth composition after GERM. The melt curves are again primary melts after 15 and 25 % bulk melting and a curve derived from the 15 % primary melt after fractionation of 10 resp. 20 % of a mixture of olivine (90 %) and spinel (10 %). All three curves are subhorizontal and subparallel except for the compatible elements Sc and Cr, which are influenced by the loss of cpx from the model for higher degrees of melting. There are two curves each for olivine and

plagioclase resulting from 20 % crystallization of these minerals from the two primary melts. The plagioclase curves are generally falling from left to right in the chondrite-normalized spidergram and the REE diagram, whereas the olivine curves are rising. Aside from these general trends, both curves are showing some anomalies. For ilmenite and magnetite only one curve is shown, representing solids derived from the primary melt by 10 % crystallization.

From the generally falling curves of the gabbro's spidergrams (fig. 6.6) it is clear that plagioclase is the dominant cumulus phase. This is also in accordance with the falling curves of the REE diagrams (fig. 6.5). Figure 9.8 shows the spidergrams for mixtures of cumulus minerals and melt in equilibrium to them in a variety of proportions as specified in table 9.7.

The correspondence between the model and the actual metagabbros is less good than for the serpentinites. The general parallelism of the curves and the falling trend in the right part of the chondrite normalized spidergram are faithfully reproduced by the model, as is the general range of the normal gabbros.

Table 9.7: proportions of components in model gabbros

the curves in fig. 9.8 have been computed with these parameters. Starting point primary melts were derived by 15 % (solid lines) resp. 25 % (broken lines) of mantle melting. Dotted lines: with ilmenite fractionation.

cumulate ol formed by ... % fractionation	20	20	10	20	20			
cumulate pl formed by ... % fractionation	20	5	10	20	5	25	10	10
parts of ol in mixture	4	2	5	4	8			
parts of pl in mixture	6	7	3	6	2	6	4	4
parts of ilm/mgt (produced by 5 % frct)							4	4
parts of trapped melt in mixture	2	3	2	5	5	1	10	3

The most conspicuous difference between the model and the actual rocks is to be found in the LILE part of the chondrite-normalized spidergram with its drop from K towards the left to Ba. The fact that Ba contents are low both in the metagabbros and in the serpentinites makes it very likely that the Ba depletion is inherited from the source. This can be taken as indicative of subcontinental mantle as source. Another point of difference is the negative P anomaly which is not reproduced in the model. In the MORB-normalized spidergram the differentiation into 3 different groups according to the ferromagnesian element contents is also not very well reproduced. In the model as in the real gabbros, most curves are rising from Sc to Cr. The model curve falling slightly from Sc to Cr belongs to the calculation without any cumulate olivine. There are metagabbro analyses with similar curves. However, there is no correspondence in the model to those gabbros having a steep drop from Sc to Cr. A further difference is in the positive Sr anomaly, which is much stronger in the actual rocks than in the model. The Eu anomaly on the contrary is modeled well.

The Fe-Ti gabbros are even less well modeled. The general enrichment is much less for the most enriched model Fe-Ti gabbros than for the real rocks. Also there are no model Fe-Ti gabbros with a negative Eu anomaly. To model the more Cr-depleted and incompatible element-enriched Fe-Ti gabbros, the parameters have to be chosen similar to those for metabasalt modeling below. That is, we have to assume either a primary melt derived at a smaller melt percentage or a high degree of ol + spl fractionation.

Figure 9.9. displays the metabasalt modeling effort. A primary melt of 10 %, fractionating 10 to 30 % olivine + spinel in the proportions 9:1, results in a fairly good approximation to the observed curves (fig. 7.2 to 7.4). The generally falling trend for the REE and HFSE is well reproduced. The model values for K, Ba, Rb and Th are much too high, as is particularly clear from the MORB-normalized spidergram. A similar deviation was found above also for the cumulate gabbros and the restite and was ascribed to a depletion of the source.

Tab. 9.8: Model parameters for basalts

% melt extraction	10	10	10	15	20	20	20	
% fractionation	40	30	20	20	30	20	10	of ol and spl in proportions 9:1

The shape of the curve from Sc to Cr is not very sensitive to the percentage of primary melt formation, but depends primarily on the degree of ol + spl formation and on the proportions of these phases fractionating. This latter parameter is not very likely to vary much in nature. As most analyzed specimens show a slight drop of their curves from Sc to Cr, this would imply that most specimens underwent about 25 % fractionation. Two specimens display rising curves, corresponding to 15 to 20 % fractionation.

On the whole, the modeling of the metabasalts, metagabbros and serpentinites of the Zermat-Saas Fee ophiolite zone can be considered successful. However, it remains a mystery where the liquids complementary to the cumulate gabbros have remained.

Fig. 9.9: Spidergrams for modeled basalts. Curves from top to bottom modeled according to parameters in table 9.8.

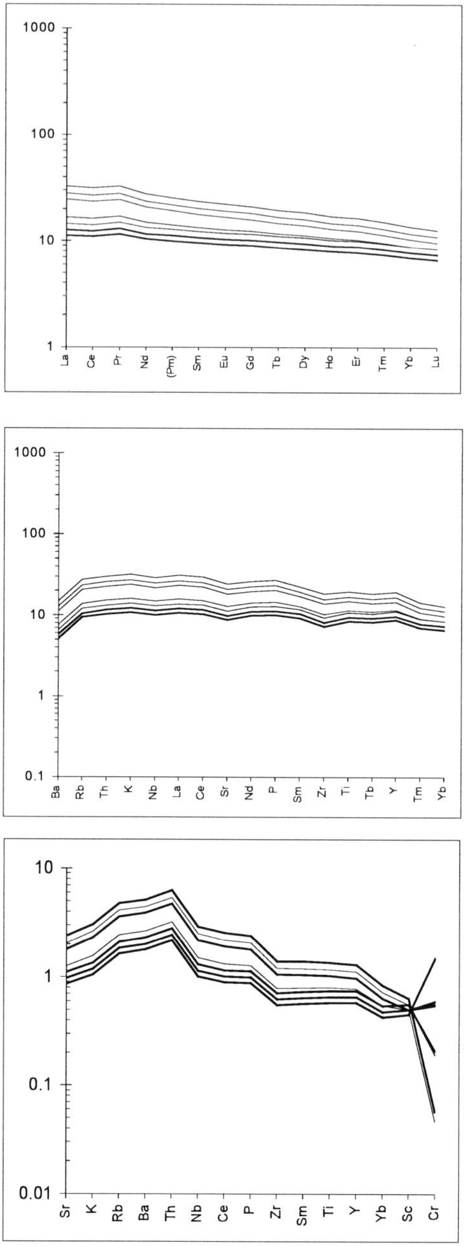

10 Assessment of the potential of Aitchisons log-ratio method for the study of magmatic and metamorphic rocks

Before asking what the log-ratio method can do for geochemical studies, we have to ask what statistics in general might do. Perhaps we are tempted to say: not very much. Most questions of petrogenesis can very well be addressed with the "classical" methods: petrographical observation, diagrams (element-element plots, Harker diagrams, triangle diagrams and trace element spidergrams) and model calculations. However, this may be too rash a view.

Firstly, element-element plots and triangular diagrams are essentially statistical. They both have to do with correlations in the broadest sense and also with specimens forming clusters and groups.

Furthermore, I see three different fields on which statistical methods can serve as tools in geochemical studies:
- detection and study of relationships between petrogenetic processes and chemical composition;
- classification: similarity and difference between specimens;
- comparison between different measures for properties.

Statistics can be helpful in disentangling petrogenetic processes by several means. Correlations between elements will to some extent reflect geochemical processes, so that the presence of a correlation indicates that a certain process has probably been operating. This kind of study can be done with correlation matrices. Another method is the study of changes in the element correlations, when subsamples are dealt with that can be distinguished on some non-geochemical parameter. This can be achieved by studying the difference between correlation matrices, as e.g. proposed by Beach & Tarney (1978).

Or we can study the correlations between geochemical composition and other parameters, like e.g. petrography, position in the stratigraphical column, distance from some interface (like sediment/water or host rock/intrusion) or distance from a hotspot along strike of a MOR. These correlations or trends will ultimately be also related to some petrogenetic processes.

Statistics can also be useful for purposes of classification by indicating differences within a population or by the detection of outliers.

It may be interesting to have some appreciation of the diversity within a suite of rocks. This can be achieved with principal component analysis. Diversity may be used for classificatory purposes and for outlier detection, or it may be related to geochemical processes again.

Finally, statistics can be used to compare different measures. If the results of different measures, applied to the same sample, correlate well, it can be assumed that they are measuring the same property.

Now everything that can be done with statistics can be done better with the proper statistical methods. This is where Aitchison's log-ratio method comes into the play. We have discussed the merits and some criticisms of this method above (chapter 4) from a more theoretical point of view. Here is the place to summarize the results arrived at in the present study and so to assess the potential of the method from a practical viewpoint.

For all the uses of statistics listed above there are examples of a successful application in this study. Correlations have been used to confirm processes of magmatogenesis that were detected by diagrammatic methods. Different groups (up to 4) of mutually correlating elements could be identified in the several rock groups and related to petrogenetic processes. For example, in the normal metagabbros elements hosted by plagioclase were correlated with each other; with the Fe-Ti gabbros, a group of compatible elements could be distinguished, which contained also Ti and V, indicating that the crystallization of some phase hosting them was an important petrogenetic process. This study of correlation can be done in a very differentiated way, as the degree of correlation between element pairs can be directly compared on a homogeneous scale. The scale independence of the log-ratios is particularly valuable in this context.

Also the lack of correlation was found to be significant, as with the serpentinites, where Ca did align neither with the compatible nor incompatible elements. This was interpreted as indicating that Ca was influenced less by magmatic processes, but rather by serpentinization. Also with the serpentinites, Zr and Ti were not among the incompatible elements, because they are probably buffered in the restite by some exotic phase.

Some strange correlations, as between H_2O and the compatible elements for the serpentinites, were detected. This correlation might have gone unnoticed, if we had not had the complete matrix of log-ratio variances. This particular correlation could be interpreted as being due to the mineralogical expression of the chemical composition and to the serpentinization reaction.

It should be noted that not all element pairs that are being well correlated according to the log-ratio statistics are appearing as such in element-element plots, e.g. Si-Mg for the serpentinites (cf. fig. 5.4). This is due to the unit-sum constraint and demonstrates another point where geochemical studies can profit from this method, as compared to simple inspection of element-element plots.

Himmelheber (1996) tried to prove that the tetrad effect (see e.g. Masuda & Ikeuchi 1979, Irber & Bau 1995, McLennan 1994) of the REE was effective in metabasalts from the ZSF, using log-ratio variance matrices. The results obtained there could not be reproduced with the specimens studied here; however, this might be caused by the small sample size and by analytical dishomogeneity of the REE data.

Changes of correlation were studied with a modified matrix difference technique after Beach & Tarney (1978) for the metagabbros and the metabasalts. With the metagabbros, an effort was made to study the effects of the retrograde metamorphism. Processes of albitization and sheet silicate formation not obvious from the raw chemical data could be proved. With the metabasalts it was tried to find differences between pillow cores and rims, as the latter should have been influenced more strongly by sea water. However, this was not successful due to the very small sample size. In general, this method is particularly useful for studies of element mobility.

Relationships between chemical composition and other variables have not played a major role in this study. A somewhat similar case is the discussion of the relationship between the Fe oxidization state and the chemical composition as a whole. For the serpentinites it was possible to find such a relationship by plotting the pca specimen coordinates against $\log(Fe^{II}/Fe^{III})$. It was found that the degree of oxidization is mainly determined by the completeness of the serpentinization process. This made it possible to estimate the degree of primitiveness of the specimens, both from a magmatic and from a metamorphic point of view. Combining this estimate with the REE diagrams, it could be demonstrated that the REE had not suffered major mobilization.

Pca has been used here also for classification and outlier detection purposes. For the serpentinites, 2 specimens could thus be excluded as outliers. For the Fe-Ti gabbros a separation into two groups could be confirmed. Aitchison (1986) describes other, more specialized statistical tools for outlier detection and for relating composition to other parameters, but for heuristical purposes, as in this study, the proposed method, using pca, seems sufficient and also intuitive.

The main use of pca in this study however was for relating the variability of composition among a group of rock specimens to potential processes. This was quite successful for the serpentinites and the metagabbros. For the former, the combination of correlation structure and pca interpretation showed that the process of melt extraction and of serpentinization are leading to very similar changes in the chemical composition.

For the metagabbros, the processes extracted from the trace element diagrams – partial melting and cumulate formation – were found to be mirrored in the pca. For the metabasalts, pca did not lead to stable results, due to the rather small and chemically too homogeneous sample.

As the first factors of the pcas of the normal and of the Fe-Ti metagabbros were interpreted as indicative of magmatic differentiation, it was tested whether the specimen coordinates could be used as differentiation index. This was done by plotting the coordinates against other differentiation indices, like Mg#, content of incompatible elements, as Ti, Zr and Y, and also against some potential indices of the degree of plagioclase cumulate formation, like the content of normative plagioclase and the size of the Sr and Eu anomalies. It was found that the pca coordinates can serve very well as a measure of differentiation. No effort was made to quantify the degree of agreement between the different measures, as no appropriate statistical tool was known.

On the whole, the log-ratio method was found to be practicable, leading to sensible results, which are in accordance to what would be expected from a petrological point of view. There were even some unexpected phenomena disclosed. As with all statistical methods, it is important to make a careful selection of the specimens that are to be included in the sample. For example, the picture gained much clearness after the normal gabbros were separated from the Fe-Ti gabbros. The appropriate specimen selection depends of course on the questions posed. It could be interesting to have sedimentary and magmatic rocks in one sample, if the question centers on similarity and difference. If the interest is in petrological processes, such a sample will not be very helpful.

The use of statistical methods in this study was somewhat hampered by relatively small sample sizes. An example of what can be done with the log-ratio method with a larger sample can be found in Himmelheber & Sheraton (in prep.), reproduced here in appendix 4. But even with small samples as used here, in combination with other methods it was possible to arrive at robust results.

11 Results and open questions

This chapter deals with questions of the magmatic history of the rocks of the ZSF. Those results obtained in this study concerning the metamorphic history are not dealt with here.

The most important finding of this study is that the ZSF does not resemble normal oceanic lithosphere. The points of departure from the model of a normal ophiolite will be recapitulated briefly here.

According to the magmatic modeling, the serpentinites can be explained as residual mantle after extraction of ca. 5 to 22 % melt in the spinel lherzolite field. The mode of melting seems to have been closer to batch melting than to fractional melting, judging e.g. from the linear, only slightly LREE depleted REE patterns. This is one peculiarity, as for MORB production usually depletion by fractional melting is envisaged as the most plausible mechanism (e.g. Johnson et al. 1990). Another peculiarity are the positive Ti and Nb anomalies and a negative Zr anomaly. These can be most plausibly be explained by assuming that the serpentinites represent subcontinental mantle. The Ti and Nb anomalies are explained by assuming that this mantle contains some exotic phase, perhaps Ti-clinohumite (see above chapter 5.3, p. 50/51 and the discussion in Weiss 1997: 3f). The Zr anomaly remains somewhat enigmatic; it may have been a characteristic of the mantle source prior to melt extraction.

The metagabbros have been ol + pl cumulates, sometimes also cpx + pl cumulates. Some highly differentiated members of this group are containing also Fe-Ti minerals as cumulate phases. The chemical data are compatible with the assumption that these rocks were derived from those melts which belong to the restites now represented by the serpentinites. In particular the gabbros are displaying Ti and Nb anomalies complementary to the serpentinites. That they are also having a negative Zr anomaly then would speak in favor of a primary lack of Zr in the mantle source.

For the metabasalts, on the other hand, a magmatic relationship to either the metagabbros or to the serpentinites can be excluded, arguing from the lack of the aforementioned characteristic trace element anomalies. Furthermore, the lack of a negative Eu anomaly precludes a relationship to the gabbros as fractionated melt to cumulate.

From what kind of mantle source could the basalts have been derived? It must have been rather fertile, even more fertile than a T-MORB source, as the REE and HFSE are rather enriched. On the other hand, the LILE are strongly depleted. This peculiar combination of enrichment and depletion might again point to subcontinental mantle, but this time lacking the exotic phase which has been present when the gabbroic melts were extracted.

These findings are not compatible with a MOR genetic model, but can perhaps be fitted into the framework of a simple shear rifting model as proposed by Wernicke (1981, 1985). Lemoine et al. (1987) applied this model to the Alpine situation, motivated by field evidence from several ophiolites in the Alps, Sardinia and Corsica, which is showing that the gabbros and the basalts are derived from two distinct magmatic events. In their view, in a first step, the gabbros are produced from a lherzolite by decompression melting. This is effected by unroofing of the lower plate through extension of the lithosphere along a detachment fault. It is hard to conceive how the lithospheric mantle could have melted in this situation to produce the gabbros, as it would not have the necessary temperature. Thermal modeling (Ruppel et al. 1988) showed that a linear decrease of both T and P is to be expected under those circumstances. If the gabbros were produced from the underlying asthenospheric mantle instead, there would be either no magmatic link between gabbros and serpentinites, or we would face the problem of why the deeper asthenospheric mantle has been preserved as serpentinites, but not the shallower lithospheric mantle. However that may be, the later basaltic magmatism can be explained as the product of the asthenosphere, upwelling in response to the extension. On the whole, the lithospheric ocean has been very narrow in this view. True oceanization, which would set in as soon as the first strip of asthenosphere is unroofed by the retreating upper plate, is not supposed to have occurred. This puts an upper limit to the width of the ocean at around three times the thickness of the lithosphere prior to extension, judging from the geometry of a listric detachment fault, that is, at ca. 450 to 600 km.

Stampfli & Marthaler (1990), also taking Wernicke's model as starting point, believe that the ZSF rocks are of a truly oceanic nature. In their model, the unroofed subcontinental lithosphere was only a narrow strip, upon continuing extension followed by an oceanic development. The restite part of the present-day ophiolite then definitely would be former asthenospheric mantle. In this model it is difficult to see how the serpentinites, metagabbros and metabasalts could escape having a magmatic relationship. Also the mantle responsible for melt production would be assumed to loose its subcontinental character as extension is going on and new fresh mantle is welling up from the depth. Thus it seems that the model of Stampfli & Marthaler is not compatible with the geochemical findings presented here.

Besides these problems related to finding a genetic model, there remain also the following open questions:

1. cumulate problem: material balance calls for some rocks having a negative Eu anomaly corresponding to the positive anomaly of most metagabbros, or more

166

generally, there are no rocks found in the field representing the evolved liquids left over from cumulus production. The Fe-Ti gabbros were found not to be suitable candidates;

2. why do we not find restites and cumulates belonging to the metabasalts?
At present no solution to these question can be presented.

For comparison we turn to the northern Apennines, where another part of the Ligurian Tethys has been obducted. There the situation might seem more favorable for an elucidation of the magmatic relationships, because the rocks there have suffered less metamorphic overprint. There are even unserpentinized peridotites to be found. Indeed a lot of work on geochemical and geochronological aspects has been done there.

Piccardo et al. (1992), summarizing the older literature and presenting new data, distinguish three different types of peridotite according to their REE patterns:

- slightly depleted (around 1 times chondritic concentration) with REE curves rising slightly and almost linearly from LREE to HREE; curves of differently depleted rocks are subparallel;
- more strongly depleted, with REE curves rising approximately linearly from 0.1 .. 0.3 in the LREE to 0.5 .. 1 times chondritic concentration in the HREE; curves are subparallel;
- strongly fractionated with strongly convex REE patterns, rising from 0.1 .. 0.5 to 1 .. 2 times chondritic concentrations; curves are converging towards the HREE.

These three types are ascribed by Piccardo et al. (1992) to different units, namely the External Ligurides spinel lherzolites, the Internal Ligurides peridotites, and the Erro-Tobbio spinel lherzolites, respectively.

These findings are interpreted by the authors on the background of a simple shear extension model as follows:

The lherzolites from the External Ligurides represent relatively fertile subcontinental lithospheric mantle, whereas the peridotites from the Internal Ligurides would correspond to the former asthenosphere. In both units equilibrium melting is assumed as melt extraction mechanism. For the Erro-Tobbio lherzolites a progressive fractional melting model is assumed.

This interpretation is also strengthened by geochronological results of Rampone et al. (1995). These authors found that the External Ligurides (EL henceforth) peridotites underwent a melt extraction event in proterozoic times (2.4 to 0.78 Ga). An event of sub-solidus decompression occurred around 165 Ma, which is the assumed time of oceanic development of the Ligurian-Piemonte ocean and thus is in accordance with the simple shear mechanism.

The gabbros and basalts both of the IL and EL are assumed by Piccardo et al. (1992), drawing on older literature, to be derived from the upwelling asthenosphere. The development goes from transitional basalts, sometimes found in the EL, to N-MORB tholeiites found both in the EL and IL. The cumulate gabbros can be derived from the same type of melts as the basalts. Those basalts from the EL termed N-MORB by these authors are having REE patterns broadly similar to those described for the ZSF in this study (that is, they are rather more strongly enriched than N-MORBs), except for a negative Eu anomaly which seems to be present in some specimens.

A slightly different picture is presented by Rampone et al. (1998). They address the question of a genetic link between peridotites on the one hand, gabbros and basalts on the other hand. They find it likely that the gabbros and basalts derive from the same parental magma. Both are dated at around 164 Ma and are having a similar ε_{Nd} (164) of 8.6. The peridotite is dated at 270 Ma and has an extremely depleted ε_{Nd} (164) of 11.9 to 14.8. Thus these authors differ from Piccardo et al. (1992) on the time of depletion of the lithospheric mantle. As another point of divergence they are reporting REE patterns for the IL resembling those described for the Erro-Tobbio unit by Piccardo at al. (1992).

It is not clear whether the mass-balance problem for the gabbros as found in the ZSF exists also in the ophiolites of the Ligurian Apennines. The data reported for gabbros are all showing positive Eu-anomalies. There are no analyses reported having negative Eu-anomalies, neither from highly evolved gabbros, nor from basalts, except for 2 basalt analyses from older sources quoted in Piccardo et al. (1992).

As far as can be judged without having isotope data for the ZSF rocks, the situation in the ZSF seems to be broadly similar to the EL. Here as there a genetic link between the peridotites and the basalts seems to be missing. A difference is that in the ZSF a link between gabbros and basalts has to be denied. The trace element patterns of the ZSF peridotites are similar to the EL 'continental lithospheric mantle'-type; representatives of the asthenospheric type (a potential source for the basalts) have not been found. A possible reason might be that this mantle never came close enough to the surface, because the oceanic development was arrested before the last part of the continental lithosphere slipped from under the European upper plate during extension.

On the whole, there is a close similarity between the magmatic development in the ZSF and in the development of part of the Ligurian ophiolites. Both are characterized by the lack of a genetic link between peridotites and basalts. They share this feature with a few other ophiolites, like Xigaze and Trinity (Gopel et al. 1984, Gruau et al. 1995). Other traits of this type of ophiolite are the missing of a sheeted dike complex and the occurrence of gabbros not as continuous layers but rather as discrete plutons (Rampone et al. 1998). This last point can not be assessed for the ZSF, where the original lithological setup has been thoroughly reworked tectonically, but a sheeted dike complex is indeed absent, despite Barnicoat & Fry's (1986) contention to the contrary.

The ZSF, compared to Liguria, is either a less complete record of the events, or, more likely, the oceanic development has been arrested in a very early stage. At the moment geochronological data are not as complete as for the Ligurian ophiolites. Rubatto et al. (1998) found an age of ca. 164 Ma for the gabbro intrusion (Zircon SHRIMP dating for Mellichengabbro and Allalingabbro). A magmatic age for the basalts or an age for the depletion event of the peridotites however is as yet missing.

Nevertheless it seems warranted to assume in the light of the findings presented in the present study that the ZSF has developed as a simple shear extensional rift basin of very limited extent. The small extent is also corroborated by the extremely immature character of some of the associated sediments (Rubatto et al 1998). On the other hand, could such a small ocean basin produce an accretionary prism large enough to explain the volume of the Tsaté nappe? Thus we end our work with an open question.

Acknowledgements

My gratitude goes to the following persons:

to Prof. Dr. Jürgen Hoefs for giving me the opportunity to this work, including the possibility to use the facilities of the Geochemisches Institut, but particularly for encouragement in times of unemployment and a job as Wissenschaftliche Hilfskraft, which helped me also working on the present project. He also improved this text with some criticisms;

to Prof. Dr. G .Wörner for willing to be coreferent, thus increasing his workload considerably, and for some proposals for improving of the text;

to Dr. Alfred Schneider, who introduced me to the geology of the Zermatt area and kindled my interest in its various aspects, manifesting itself in a diploma thesis conducted under his supervision; also some of the specimens (designated 'AS') were collected by him;

to Dr. Klaus Simon for all kinds of kind support, encouragement, discussions, and for reading and criticisms on the first draft of this paper;

to Prof. Dr. John Sheraton for allowing me to include our joint paper here as an appendix;

to Prof. B. Storre for pointing out a faulty argument to me;

to Prof. Dr. Oberhänsli, who kindly sent me a photocopy of the yet unpublished Swiss GK 25 of the Matterhorn area;

to Dr. C. Münker, who brought me two specimens (CM 1-98 and CM2-98) from Pfulwe pass;

to Mr. Tondok, who repaired the jaw breaker, broken by me in the course of this work;

to Ms. Dietrich and to Mr. E Schiffczyk for support with various aspects of the analytical work;

to Dr. Robert Schönhofer, alpinist-geologist, going with me to the Allalinpass for sampling and carrying rock specimens;

to Marko Gernuks, allowing me to quote his diploma thesis;

to the students of Dr. Schneider's Zermatt excursion 1996, who helped me with the tough work of rock transportation;

to Dr. J. Ganguin and Dr. T. Widmer for copies of their Ph.D.-theses;

to Dr. Michael Stipp for making photocopies in the library of the Basel university;

to Roger Nielsen, Oregon State University, Oceanic and Atmospheric sciences, who kindly sent me some D values by e-mail, in a time when the GERM homepage was down;

to Jürgen Heuer, who helped me to create a decent pdf-file;

to my parents, Dr. G. Himmelheber and Dr. I. Himmelheber for funding the printing, and generally for beeing a great support.

Appendix 1: petrographic description of the samples

The specimens are described according to the following format:

Specimen designation
    location (descriptive; Swiss national grid),
    comments,
    macroscopic petrographic description,
        minerals according to microscopy (amounts in % are only rough estimates)

Serpentinites

AS 1-96
    kleines Matterhorn, 622.6 / 87.3
    loose piece from tunnel construction
        atg 50%
        chl
        ol and augit 30%
        mgt 20%

AS 2-96
    Gobba di Rollin, 623.0 / 85.8
        atg
        ol 10%
        mgt 10%

AS 3-96
    piece from avalanche. Unequivocal relationship to outcrop.
    above road Grünsee-Riffelboden. at the western Ritzengrat. 626.45 / 94.7
        atg 80% (some aggregates forming rhomboid pseudomorphs Ø ca. 0.5 mm)
        mgt 15%
        ol 5% (rare; distributed evenly)

AS 9-96
    directly to the SW of Hotel Gornergrat. 626.7 / 92.45
        atg (partly aggregated into coarse oval shapes aligned in the foliation)
        mgt 10% (elongated oval aggregates)
        ol 7%. (elongated, partly displaying mortar texture)
        augit 3% (also diallag)

AS 12-96
    road Zermatt-Furri. 100 m N of restaurant Alm. 623 / 94.6
        atg
        mgt 15%
        ol 10%

AS 16-96
    path Riffelalp-Grünsee, "roadcuts" of the ski piste, 200 m E of station. 624.6 / 95.0
        atg
        mgt 10%
        ol 10%
        px
        fracture filled with calcite

AS 19-96
    road Grünsee-Riffelboden. 50 m E P2370. 625.4 / 94.7
        atg
        ol 15% (along shear zones)

mgt 0.5%

Hi 1-96

ca. 70 m NW Hotel Schwarzsee. 620.8 / 93.4
loose piece from constructions
contains chalcopyrite; light stripes, 2-10 mm thick.
> atg
> amphibole (idiomorphic, with kernels of px)
> mag 2% (in veins. also dispersed, rather large: Ø 1 mm

Hi 3-96

between Hirli and Schwarzsee, 620.3 / 93.5
kataclastic serpentinite
> atg (aggregated into oval domains with common extinction).
> mag 5% (around the atg domains)
> ?tit
> ol (rare)

Hi 6-96

Lichenbretter. 622.2 / 91.7
loose piece from construction (foundation of cable cabin post)
> atg
> mgt 5%

Hi 7-96

road from Zermatt to Furi. north of P. 1824. 623.2 / 93.7
> atg
> mgt 20%. (in part attached to ol)
> ol (partly very elongated)

Hi 9-96

on the way to the "glacier garden". 622.55 / 93.8
no thin section

Hi 13-5-96

Area polished by glacier on the right shore of Gornergletscher, close to path. 623.3 / 92.5
> atg
> mgt (> hem) 10% (or perhaps ti-chmt?)

Hi-14-96

Gagenhaupt. 623.5 / 92.7
wildly crenulated
> atg
> ol 35% (partly very large)
> mgt 10%
> chl?

Hi 15-96

from the creek N of ridge Gagenhaupt-Riffelhorn. 624.3 / 92.6
> atg 20%
> mgt 5%
> ol 10%
> cracks filled with ?tlc or ?calcite

Hi 21-96

south of Schwarzgrat, rock 300 m south of P. 3198. 630.2 / 96.6
loose piece.
dark green with oval lighter domains (ca. 4*2 mm)
> chl 80%

atg 5% (in the relic oval domains)
mgt > hem 2%
px. partly diallag > tlc

Hi 27-3-96
E of Spitzi Flue. 631.25 / 96.25
derivative from serpentinite. Tectonized.
chl 40% growing from fractures parallel to foliation
atg 30%
ol or perhaps amph
mgt (some small, but also large, idiomorphic individuals overgrowing the foliation)

Hi 28-96
Findelnalp, at the path. 627.2 / 95.8
light green and dark bluish to gray
atg 90%
mgt 5%
ti-chmt (or hem?)

Hi 28-2-96
same location as above
light green, cavernous; magnetite in schlieren
from outcrop
atg
Ti-chm or hem?

Hi 32-9-96
left side moraine of Allalin glacier. lower end. before Mattmark dam
loose piece from moraine
atg (partly undeformed microlithons (pseudomorphs?); partly isoclinaly folded domains)
mag 5%
ol 20% (sheared)

Metagabbros

CM 2-98

E ridge of Spitzi Flue. 631.26 / 96.26
From sheeted mafic complex
flaser texture: elongated albite in matrix (ca. 50 %) of dark green hbl + zo aggregates.
  No microscopic data.

Hi 4-96

path on little Ridge S Schwarzsee, 620.8 / 93.3
Flasergabbro: omp Ø 1-15 mm (the smaller ones probably produced by tectonic action from the larger ones),
light green. in white matrix. with grt (Ø 4 mm). Foliation and lineation are defined by the orientation of the
omp-relics.
  omp 10% (many inclusions - hbl) developing into ab-hbl-symplectite
  hbl 30%
  zo-domains (30%) containing
  ?ab (10%) in corners
  a little white mica 2%

Hi 16-96

near Fluehorn. ca. 400 m SW of P. 3225. 629.65 / 96.65
fine grained; light greenish; planar texture. Foliation planes mica-coated. with darker knots (Ø 1 mm. ep)
loose piece from scree
  ep 20%
  chl 20% (in nests and matrix)
  ab 40% (small poikiloblasts)
  hbl 20%. granoblastic. idiomorphic
  tit
  little Carb

Hi 18-96

S Schwarzgrat. ca. 100 m S of P. 3225. 629.8 / 96.85
A parallel texture is defined by the green, fine-grained matrix flowing in a flasery way around white eyes (Ø ~ 5
mm, elongated to different degrees - 1:2 to 1:20).
loose piece from scree
    33% aggregates of hbl and hbl-ab-symplectite (flasery matrix)
    66% aggregates of
              zo 40%
              chl 20%
              ab 20%
              tit
              white mica
    carb on transverse fractures (small amounts)

Hi 19-96

S Schwarzgrat. 629.9 / 96.95
large omp. dark green. Ø 5-10 mm. slightly elongated. subhedral. ca. 70 %. in white matrix.
loose piece from scree
40% large omp with inclusions(idiom. laws, ?zircon); tectonically almost unstrained (undulous extinction).
  fractures filled with Ab
  22% zo + 12% ab
  15% green-blue hbl (blasts) in 15% ab; also symplectite
  tit
  very little bt and chl

Hi 24-96

Rib N of above location. 631.9 / 98.0
This specimen is from the area of the purported "sheeted dike complex" of Fry & Barnicoat (1986). It is a
metagabbro containing a boudinaged metabasaltic dike. Here the metagabbro is described.

Green sigma clasts. Ø 2-5 mm. in white matrix. Flasery planar texture.
　　?hbl 15 %: textural omp-relics?
　　blue-green hbl 20%
　　zo 10%
　　ab (poikiloblasts) 20%
　　little white mica
　　chl
　　tit

Hi 30-1-96
　at the Allalinpass. 634.8 / 98.6
　loose piece at the foot of the wall.
　green and white cm-size domains intermingling in an unordered texture
　　　zo (2 generations) 70%
　　　hbl (?glc; partly idiomorphic) 20%
　　　white mica
　　　tit (< rt)
　　　ab-hbl symplectite 10%

Hi 30-2-96
　at the Allalinpass. 634.8 / 98.6
　loose piece at the foot of the wall.
　similar to Hi 30-1-96. but schistose
　　　ab as porphyroblasts and in symplectitic intergrowth 30%
　　　chl 20%
　　　zo 15% (sometimes displaying thin rims of ep)
　　　hbl 30% (first generation: large, tectonized; 2nd generation: idiomorphic)
　　　tit (< rt) 7%

Hi 30-3-96
　at the Allalinpass. 634.8 / 98.6
　loose piece at the foot of the wall.
　green matrix; thin strips of ep form a flasery planar texture. Blackish blasts (mgt?)
　　　ep (fine-grained plaster) 30%
　　　hbl 30%
　　　ab (poikiloblastic) 10%
　　　chl 10%
　　　tit

Hi 30-6-96
　at the Allalinpass. 634.8 / 98.6
　loose piece at the foot of the wall.
　Flasergabbro texture: green, intermediate sized, elongated domains, between white stripes. Some grt, Ø 2 mm
　　　hbl-ab-symplectite
　　　blue-green hbl (hbl total: 40%)
　　　ep-strings. These 3 elements show an anastomosing shape preferred orientation
　　　ab (total 20%)
　　　white mica
　　　rare omp-relics
　　　grt
　　　tit

Hi 32-1-96
　left side moraine of the Allalin glacier. lower end, before the Mattmark dam.
　loose piece from moraine.
　Coronite. unordered texture of 3 different domains each having Ø 10-20 mm: light green to whitish domains;
　　　brownish domains; blue domains with grt-coronae. The interior of the coronae is filled with a white
　　　substance, wherein needle-shaped hbl has developed.
　　　domain 1. very fine-grained (felty) saussurite

domain 2. ?omp + tlc + ?cpx + hbl. This is sometimes surrounded by:
1. green-blue hbl
2. omp
3. grt
domain 3: grt-garlands. filled with intergrown ep + glc + tlc

Hi 32-3-96

left side moraine of the Allalin glacier. lower end, before the Mattmark dam.
loose piece from moraine.
saussurite-omp-type: cpx, sturdy, anhedral, blackish, Ø 2-20 mm, in white matrix (saussurite).
omp: large, sometimes surrounded by concentric shells of
1.?omp + opaque + tlc.
2. grt
ol: small, surrounded by concentric shells of
1. chl? anth?
2. px
3. grt
saussurite

Hi 32-8-96

left side moraine of the Allalin glacier. lower end, before the Mattmark dam.
loose piece from moraine.
Saussurite-omp-type: light green smaragdite with diffuse outline, and dark-green aggregates containing white mica (also diffuse), in white matrix. A small vein with needle-shaped white minerals. A part of the specimen is more bluish.
greenish part:
omp: large relics; beginning transformation into smaragdite
grt (a little)
rt
zo (small in matrix; sometimes larger)
white mica (a little)
?glc (idiomatic, corroded)
blueish part:
much matrix: zo + hbl
cld (nests)
omp
tlc
white mica
amph (glc? > blue-green hbl)
grt
vein:
tlc-ky-glc

Hi 32-10-96

left side moraine of the Allalin glacier, lower end, before the Mattmark dam.
loose piece from moraine.
Saussurite-omp-type: light green diffuse domains. partly smaragdite, and dark green diffuse domains with lighter core (cld+tlc). Some py. Specimen contains a vein.
omp (plaster and small grains transformed into smaragdite) 80%
cld + tlc 15%
white mica
glc?
vein:
cld, tlc; occasionally rt. omp. white mica. chl

Hi 32-11-96

left side moraine of the Allalin glacier. lower end, before the Mattmark dam.
loose piece from moraine.

Dark green (hbl+chl) matrix with grt-coronae around former ol. Peculiar white "nests" surrounded by recrystallized chl

concentric corona-structures:

	core:	zo + omp or zo + ab?
	intermed. layer:	grt
	rim:	chl
rest:	hbl	
	ep / zo (occasionally)	
	ap. rt	

Hi 32-13-96

left side moraine of the Allalin glacier. lower end, before the Mattmark dam.

loose piece from moraine.

Similar to Hi 32-10-96. but smaller domains; dark green domains (cld?) without light core

cld > tlc (> sometimes: opaque)

"matrix" with omp

glc > blue-green hbl very large crystals (post-matrix?)

zo

rt

Fe-Ti gabbros

Hi 27-1-96

E of Spitzi Flue. 631.2 / 96.3

elongated white aggregates (2*4 mm and smaller) in fine grained green matrix; some grt (Ø 3 mm)

    ep/zo fine grained 30 % (in the white flasery domains and elsewhere)

    blue-green hbl-Ab-symplectite 60 % (partly pseudomorphic after glc)

    white mica 10% (nests)

    rt. tit

    grt > chl (usually poor in inclusions)

    qtz

    the white flasery domains are made up of white mica (very large crystals) and zo.

Hi 27-2-96

E of Hi 27-1-96. 631.2 / 96.3

Strange variety, containing oval aggregates of hbl + ep. Fine grained dark green matrix. Hexagonal ?grt-pseudomorphs (darker at their rim, lighter at their core)

    ab (poikiloblastic) 60%

    blue-green hbl (small, contained in ab; partly larger round aggregates (20%)

    ep dispersed and in the round aggregates

    chl

    white mica

    rt > tit

    hem (rare)

Hi 30-4-96

at the Allalinpass. 634.8 / 98.6

loose piece at the foot of the wall.

green matrix; light stripes of ep (Ø ca 2 mm. elongated and sometimes fish-shaped) are forming a flasery planar texture; grt Ø ca. 3 mm

    hbl-ab-symplectite very fine-grained; (Σ ab 35%. Σ hbl 35%)

    grt (surrounded by ab + chl)

    ep 20%

    chl

    tit < rt

    ab (sometimes larger poikiloblastic grains)

Hi 30-5-96

at the Allalinpass. 634.8 / 98.6

loose piece at the foot of the wall.

similar to Hi 30-4-96; grt: beginning transformation into chl; white and yellowish stripes. Zones enriched in ep.

    hbl-ab-symplectite 45% + 25%

    glc-relics (occasionally)

    ilm > tit

    ep 15%

    grt 5%. Inclusions: tit. zo; rim occasionally transformed into chl

Hi 32-12-96

left side moraine of the Allalin glacier. lower end, before the Mattmark dam.

loose piece from moraine.

Relatively fine-grained. Green matrix (light hbl, dark chl). containing white aggregates (slightly elongated; ca. 2*4 mm) and also grt (Ø 3 mm). Carbonate, weathering brown.

    ep 20 %

    blue-green hbl-ab-sympl (40 %)

    omp 15%

    grt 10%

    white mica 10 %

    glc

    rt > tit

    carb

    qtz

Hi 32-2-96

left side moraine of the Allalin glacier. lower end, before the Mattmark dam.

loose piece from moraine.

dark, coronitic texture. Omp-relics. Ø 5-15 mm. rims transformed into chl. Reddish (grt)-aggregates. Some Pyrite.

> omp-relics 30%
> grt 30%
> hbl-ab-symplectite 30%
> ep
> glc > blue-green hbl
> rt > tit

Hi 32-4-96

left side moraine of the Allalin glacier. lower end, before the Mattmark dam.

loose piece from moraine. Contains a vein.

Similar to Hi 32-2-96. This specimen is divided by a vein into two different parts:

Part 1: coarse omp-relics (Ø up to 20 mm) and grt-aggregates. occasionally py.

Part 2: more fine-grained; dark green matrix. containing small grt(Ø 1-2 mm). Occasionally Py.

> Part 1:
>> omp 30%, rims: beginning transformation into ?anthophyllite
>> grt 30% (inclusions of glc)
>> green-blue hbl 12%
>> ab 8%
>> ?cld
>> rt > ilm > tit
> Part 2:
>> omp; rims: beginning transformation into ?anthophyllite
>> ab-hbl-symplectite 30%
>> ep 10%
>> grt 10% (inclusions of glc)
>> rt > tit

Hi 32-5-96

left side moraine of the Allalin glacier. lower end, before the Mattmark dam.

loose piece from moraine.

Similar to Hi 32-4-96

> omp-relics. Ø 5-40 mm. subhedral; grt and hbl. some py. Many late white veinlets.
> omp tectonized 35%
> grt 35 %
> blue-green hbl 30%
> rt > tit (very large)
> strange symplectite

Hi 32-6-96

left side moraine of the Allalin glacier. lower end, before the Mattmark dam.

loose piece from moraine.

Similar to Hi 32-5-96. Omp anhedral. Much grt. The larger part is reddish; the rest is more greenish.

> reddish part:
>> grt 40% > hbl + chl; inclusions: tit, zrc, ep
>> omp 20% > ab-hbl-symplectite (little ab)
>> rt 10% (macroscopically visible)
>> green-blue hbl 15% postdeformational
>> partly glc
> greenish part
>> omp-relics. more and larger 20%
>> glc 10% (later than omp, as less tectonized; sometimes as core of green-blue hbl (20%))
>> ep
> grt 20%
> rt 10%

Metabasalts

CM 1-98

E ridge of Spitzi Flue. 631.26 / 96.26
From sheeted mafic complex
Dark metamafite, similar to Hi 27-1-96. Dark green matrix (hbl). Strongly elongated ab is marking a planar texture; grt (Ø 1-2 mm), rims transformed into chl; some white mica
No microscopic data.

Hi 2-96

Between Hirli and Schwarzsee. Rock step in track of skiing lift. 620.3 / 93.5
Fine-grained ab-amphibolite. A planar texture is defined by light stripes (elongated blasts. 0.2*2 mm and longer) and dark stripes (hbl). Grt (Ø 1 mm)

ab 30%
grt 5 %, partly almost completely replaced by chl
hbl 30% idiomorphic. Parallel, slightly undulatory texture
zo 20%
tit 10%
ab-hbl-symplectite
hem
chl

Hi 22-96

S of Schwarzgrat, 200 m S of P. 3198. 630.4 / 96.5
Loose piece, found right below a larger serpentinite cliff
Fine-grained. dark blue-green. A few small white flasers.

hbl defining planar strongly undulating texture
ep (partly large relics) 15%
ab. poikiloblastic 40%
blue-green hbl, corroded by ab 20%
very much tit (partly with rt-core) 10%
chl 15%
bt interstitial

Hi 24-96

Rib N of above location. 631.9 / 98.0
This specimen is from the area of the purported "sheeted dike complex" of Fry & Barnicoat (1986). It is a metagabbro containing a metabasaltic dike. Here the metabasalt is described.
Fine-grained. Planar texture defined by fine, elongated, white strips in the bluish matrix. In some strips ep is enriched. Grt Ø 1-4 mm

grt 15%
blue-green hbl 20%
ab (poikiloblastic) 20%
tit (< ilm) very much 10%
ep 10%
chl. bt?
py? (ilm?)

Hi 25-96

Mellichen, 300 m NE P. 2870. 631.6 / 97.8
Metapillow. White domains, roundish to flasery. in fine-grained green matrix; grt and pseudomorphs after grt, Ø ca 2 mm
Pillow core:

hbl-ab-symplectite 60%, partly recrystallized. partly hexagonal pseudomorphs (after glc?)
ep 20%
grt 5%
rt > tit
opaque

Pillow rim (very fine-grained)

hbl-ab-symplectite 70%
grt 15%. rims > hbl

ep 20%
hem
qtz
rt > tit

Hi 29-2-96
    End of the Grand-Dixence road in the Täsch valley. 630.9 / 98.3
    loose piece from moraine
    Metapillow
    Ep-rich stripes and ab-blasts.
        Pillow core:
            grt 5-10% (rim > bt) Inclusions: ep, white mica, tit. Rim free from inclusions.
            hbl-ab-symplectite (patches) 30%
            ab also as larger individuals 10%
            ep 10%
            chl and bt; together 15%
            rt > ilm; tit
            carb 10%
        pillow rim:
            ab 30% (recrystallized)
            ep 10%
            carb 20%
            grt 15% (idiomorphic)
            occasionally glc-ab-symplectite, idiomorphic
            rt > ilm > tit

Hi 29-4-96
    End of the Grand –Dixence road in the Täsch valley. 630.9 / 98.3
    loose piece from moraine
    Metapillow
    Green, fine-grained matrix with white stripelets defining a flasery planar texture; grt Ø 3 mm, largely
    transformed into chl.
        Pillow core:
            white mica
            hbl-ab-symplectite (20 + 40 %)
            zo (10%) forming oval aggregates with white mica and ab
            tit (< rt)
            carb 10%
            chl 10%
            grt relatively rare, largely replaced by chl
        pillow rim:
            ab 30%
             carb 20%
            chl 15%
            hbl 15%
            ep 15%

Hi 31-96
    Vor der Wand. 633.2 / 99.6
    Fine-grained, dark green matrix with white mica, containing ab-blasts. partly round, Ø 1-2 mm, partly
    elongated (planar texture). small grt (Ø 1 mm). Surface is cavernous (weathering carbonate)
        green-blue hbl (very small) 40%
        ab 30%
        zo 17%
        tit (<rt)
        white mica
        grt 10%
        carb
        chl

Metarodingites

Hi 8-96

„Gletschergarten". Old talcum mine. 622.7 / 93.8
Dark green, fine-grained, with light blasts. Strongly foliated (sheet silicates). S-planes are having a light green luster. Specimen feels soapy.
> chl 80%
> opaques 5%
> ol and di 20%:
> carbonate (vein)

Hi 10-96

Glacially polished area at right. shore of Gornergletscher, below P. 2161. 623.0 / 93.0
Light green (ep). Fine-grained. With rusty spots, Ø 0.2 - 1 mm. Weak foliation.
> domains 1: vesuvianite or perhaps (?)brucit
> domains 2: chl + vesuvianite or perhaps (?)brucit

Hi-12-1-96

About 200 m south of location of specimen Hi 10-96, 50 m above footpath. 623.1 / 92.8
Rodingite with diffuse contact to serpentinite
Dark green, fine-grained, with infolded light bands.
> chl 60%
> ?vesuvianite 40% (amount varies with domain)

Hi 12-2-96

Same location as above.
Light flasery patches (ca. 2*15 mm. Defining a planar texture) in dark green matrix. The proportion of the patches decreases away from contact to serpentinite. Specimen contains some serpentinite
> Part 1 (fine-grained, light and greenish):
> chl 30%
> mgt 5%
> hbl 50%
> uvarovite with core of spinel (rare)
> Part 2 (medium grained, light to dark green):
> patchy domains
> hbl 30% (hypidiomorphic) + ?px
> chl 15% + atg 40%
> some opaques

Hi 13-3-96

Still further south from location of specimen Hi 10-96, close to footpath
Similar to Hi 12-2-96, but lighter.
Light green domains: medium grained (textural relic) in bluish-gray matrix
> chl 40%
> vesuvianite 60% (partly large round ovals)

Chloritoschist

Hi 17-96

Close to Fluehorn, E of location of specimen Hi 16-96. 629.6 / 96.8
Dark green (chl). Fine-grained. Feels soapy.
The location is interpreted as part of a ca. 2 m wide thrust horizon.
It belongs to "Zone 6" as defined in Bearth (1973)
> chl
> ?qtz 2%
> ore 5%

Veins

Hi 32-4-96

Left side moraine of the Allalin glacier. Lower end, before the Mattmark dam.
Loose piece from moraine.
Omphacite vein.

Hi 32-10-96

Left side moraine of the Allalin glacier. Lower end, before the Mattmark dam.
Loose piece from moraine.
> cld
> tlc (some)
> rt, omp, chl, white mica (occasionally)

182

Appendix 2: chemical data

Normal Gabbros

Specimen	CM 2-98	Hi 4-96	Hi 16-96	Hi 18-96	Hi 19-96	Hi 24-96	Hi 30-1-96	Hi 30-2-96	Hi 30-3-96	Hi 30-6-96	Hi 32-1-96
SiO2	53.6	49.3	47.1	46.4	51.1	51.2	45.6	47.5	44.2	44.0	
TiO2	0.348	0.318	0.134	0.146	0.629	0.378	0.307	0.856	0.267	0.433	
Al2O3	17.7	17.0	19.3	22.4	16.6	17.0	23.3	16.1	20.3	14.2	
Fe2O3(t)	3.51	5.86	5.49	4.67	6.27	4.96	3.94	8.58	5.77	6.29	
Fe2O3	1.03	1.45	1.82	0.96	1.20	1.84	2.31	1.97	3.08	1.13	
FeO	2.2	4.0	3.3	3.3	4.6	2.8	1.47	6.0	2.4	4.6	
MnO	0.070	0.104	0.084	0.075	0.121	0.096	0.069	0.120	0.101	0.127	0.145
MgO	7.64	11.19	11.93	8.87	7.53	8.86	5.57	11.52	10.1	8.01	8.38
CaO	11.69	11.95	6.84	11.76	12.31	11.02	18.89	8.12	15.51	10.69	11.95
Na2O	4.1	2.42	3.88	2.89	3.64	4.16	1.76	3.33	1.12	3.13	3.70
K2O	0.32	0.09	b.d.	0.09	0.08	0.09	b.d.	0.10	0.07	0.08	b.d.
P2O5	b.d.	b.d.	b.d.	b.d.	0.058	0.022	0.062	0.070	0.022	0.026	0.027
S	0.0	0.0	0.1	b.d.	0.0	0.0	0.0	0.1	0.0	0.1	0.0
CO2	0.028	0.061	0.37	0.106	0.05	0.03	0.026	0.05	0.029	0.13	0.05
H2O	2.58	3.72	5.20	4.31	1.93	2.53	2.46	4.53	3.74	2.64	1.85
SUM	101.4	101.6	100.0	101.3	99.9	100.1	101.8	100.3	101.0	89.3	99.9
Mg#	84	82	84	82	74	81	77	76	80	75	74
Li	13.62	10.36		12.80	1.91		4.35		10.95	7.66	1.78
Sc	42	28	10	6	41	44	16	38	21	43	47
V	121	108	32	28	179	131	53	182	59	144	201
Cr	479	914	269	278	66	728	82	219	743	236	54
Co	23	41	40	33	24	33	20	53	61	38	34
Ni	106	245	330	252	78	132	172	188	294	91	86
Cu	17	150		34			16		38		
Zn	b.d.	19	29	13	43	43	b.d.	54	13	80	36
Ga	11	11	12	14	16	14	18	10	13	12	13
Rb	7.52	1.19		0.97			b.d.		0.85		
Sr	333	130	196	244	343	348	384	130	694	355	357
Y	7.10	7.13	1.92	2.21		7.02	4.87		5.98	11.02	
Zr	18	19	13	15	65	33	33	48	25	42	33
Nb	0.37	0.34	0.06	0.60	4.22	2.31	1.06		1.16	1.56	1.23
Mo	0.14	0.06	0.07	0.12	0.13	0.17	0.27		0.24	0.78	4.08
Sn	0.62	0.34	0.28	0.53	3.86		1.27		0.59	1.10	1.56
Ba	25.39	7.31	2.61	9.49	8.21	8.31	2.71		8.46	5.39	4.75
La	1.89	0.73	0.50	1.46	2.70	0.99	2.34		2.26	1.87	1.17
Ce	3.86	1.80	1.10	2.70	7.34	2.61	5.47		4.86	4.86	3.22
Pr	0.88	0.44	0.19	0.47	1.16	0.46	0.98		0.91	0.85	0.63
Nd	3.30	2.19	0.92	1.61	5.97	2.55	3.80		3.14	4.31	3.52
Sm	1.03	0.77	0.26	0.45	1.90	0.92	0.81		0.92	1.34	1.32
Eu	0.56	0.42	0.34	0.61	0.82		0.51		0.32	0.71	0.65
Gd	1.01	0.68	0.30	0.40	2.06		0.92		1.00	1.55	1.49
Tb	0.18	0.13	0.05	0.04	0.38	0.18	0.12		0.14	0.29	0.26
Dy	1.30	1.11	0.38	0.41	2.60	1.17	0.88		0.95	1.90	1.83
Ho	0.26	0.33	0.08	0.09	0.55	0.24	0.20		0.20	0.39	0.36
Er	0.61	0.57	0.19	0.23		0.66	0.40		0.47	1.08	1.00
Tm	0.12	0.09	0.03	0.04	0.25		0.07		0.08	0.16	0.14
Yb	0.59	0.56	0.18	0.21	1.73	0.67	0.42		0.48	0.98	0.89
Lu	0.10	0.11	0.03	0.05	0.26	0.10	0.08		0.09	0.92	0.14
Th	b.d.	b.d.	b.d.	b.d.		0.01	b.d.		b.d.	0.02	
U	0.09	b.d.	b.d.	b.d.	0.08	0.02	0.08		0.44	0.02	0.01

Specimen	Hi 32-3-96	Hi 32-8-96	Hi 32-10-96	Hi 32-11-96	Hi 32-13-96
SiO2		49.6	49.3	44.2	
TiO2		0.467	0.225	0.273	
Al2O3		17.9	18.2	13.0	
Fe2O3(t)		2.99	5.31	11.99	
Fe2O3		0.77	0.95	2.77	
FeO	4.0	2.0	3.9	8.3	4.6
MnO		0.069	0.075	0.230	
MgO		7.56	10.67	17.68	
CaO		15.49	9.91	8.29	
Na2O		3.56	4.04	1.5	
K2O		0.10	b.d.	0.07	
P2O5		0.021	b.d.	0.027	
S	0.0	0.0	0.1	0.1	0.1
CO2	0.12	0.07	0.06	0.070	0.03
H2O	1.56	1.87	2.86	4.74	3.68
SUM		99.5	100.3	101.2	
Mg#		85	82	77	
Li	0.53	1.13	1.80	2.14	1.52
Sc		47	15	24	
V		158	51	71	
Cr		2599	566	576	
Co		22	41	80	
Ni		165	354	604	
Cu				57	
Zn		18	46	28	
Ga		13	10	10	
Rb				0.84	
Sr		410	187	144	
Y	8.02		5.06	5.50	
Zr		28	28	17	
Nb	0.34	0.34	1.74	0.61	0.40
Mo	0.86	0.23	0.59	0.13	0.47
Sn	0.52		1.07	0.83	0.42
Ba	5.75	12.07	1.80	3.22	2.51
La	0.71	0.96	3.63	1.78	0.73
Ce	2.07	2.70		3.64	1.78
Pr	0.43	0.53	1.12	0.59	0.31
Nd	2.40	2.83	4.97	2.41	1.62
Sm	0.85	1.08	1.49	0.76	0.60
Eu	0.45	0.48	0.57	0.49	0.42
Gd	1.07	1.04	1.79	0.73	0.59
Tb	0.21	0.22	0.24	0.13	0.11
Dy	1.43	1.42	1.34	1.04	0.71
Ho	0.29	0.29	0.24	0.23	0.14
Er	0.72	0.71	0.54	0.46	0.37
Tm	0.10	0.11	0.07	0.10	0.06
Yb	0.68	0.70	0.44	0.47	0.35
Lu	0.10	0.10	0.06	0.08	0.05
Th	b.d.		0.14	b.d.	b.d.
U	0.01	0.01	0.02	0.33	0.00

Fe-Ti gabbros

Specimen	Hi 27-1-96	Hi 27-2-96	Hi 30-4-96	Hi 30-5-96	Hi 32-12-96	Hi 32-2-96	Hi 32-4-96	Hi 32-5-96	Hi 32-6-96
SiO_2	50.4	50.2	46.7	49.5	48.9	43.4	45.2	41.7	43.0
TiO_2	1.155	2.667	1.242	2.147	1.648	6.929	3.932	8.010	6.754
Al_2O_3	19.6	16.9	18.5	15.5	14.7	13.6	13.6	13.1	14.5
$Fe_2O_3(t)$	6.55	9.51	10.25	10.49	8.49	16.24	16.16	16.13	14.70
Fe_2O_3	2.73	3.95	2.41	2.75	2.13	4.04	3.01	2.49	2.30
FeO	3.4	5.0	7.05	7.0	5.72	11.0	11.8	12.3	11.2
MnO	0.115	0.148	0.191	0.156	0.139	0.331	0.219	0.271	0.237
MgO	6.21	5.86	8.52	7.23	5.39	5.54	6.34	6.18	6.72
CaO	10.12	6.89	8.66	10.56	13.25	11.54	10.74	11.26	10.87
Na_2O	4.46	6.28	3.33	3.82	3.88	2.87	3.15	2.98	2.70
K_2O	0.16	0.26	0.41	0.07	0.08	b.d.	b.d.	b.d.	b.d.
P_2O_5	0.184	0.481	0.144	0.276	0.256	0.107	0.047	0.435	0.027
S	0.0	0.0	b.d.	0.0	0.0	0.4	0.2	0.2	0.1
CO_2	0.02	b.d.	0.007	0.014	3.389	0.020	0.06	1.010	0.01
H_2O	3.06	2.31	4.04	2.30	1.46	0.99	1.41	1.02	1.51
SUM	101.7	101.0	101.2	101.3	101.0	100.8	99.7	101.0	99.9
Mg#	1	59	66	62	60	44	48	47	52
Li	10.89	11.39	29.25	8.96	20.23	5.61	2.94	3.02	2.38
Sc	23	30	35	38	31	52	48	46	48
V	142	247	139	272	201	824	848	813	732
Cr	168	26	186	163	163	26	24	33	55
Co	30	17	48	35	31	58	52	43	38
Ni	93	21	145	83	87	25	76	31	83
Cu	85	6	63		37	35		11	
Zn	32	34	44	30	42	25	70	23	34
Ga	15	15	14	15	15	15	20	17	15
Rb	2.97	2.01	5.18	1.33	2.00	0.70	0.80	1.05	
Sr	243	550	371	362	237	45	65	14	209
Y	17.05	33.99	20.70	36.47	26.51	9.97	17.36	26.98	
Zr	116	302	105	197	155	105	106	160	103
Nb	6.10	13.24	5.04	6.50	6.61	1.55	5.69	19.42	7.96
Mo	0.14	0.18	0.14	7.12	0.51	0.09	0.69	4.47	0.37
Sn	4.84	10.22	2.93	0.16	5.14	0.57	2.16	7.04	1.99
Ba	10.76	36.68	64.99	5.81	3.33	11.36	2.00	2.24	6.11
La	7.49	17.05	6.85	10.09	9.46	4.50	5.22		2.52
Ce	19.49	47.48	19.30	35.41	27.00	13.50			7.94
Pr	2.90	7.05	2.67	5.39	4.45	1.32	1.74		1.49
Nd	12.61	27.65	11.32	24.56	17.81	6.33	9.12		
Sm	2.84	6.94	3.08	6.70	4.56	2.00	3.06		3.23
Eu	0.89	1.76	1.08	1.96	1.47	0.89	1.35		
Gd	3.52	8.09	3.35	7.03	5.25	2.62	4.38		3.14
Tb	0.49	1.14	0.55	1.02	0.83	0.38	0.73		0.59
Dy	3.08	6.79	3.65	7.09	5.03	2.42	4.86	4.98	3.81
Ho	0.68	1.51	0.83	1.42	1.05	0.52	1.02	1.16	0.75
Er	1.74	3.63	1.85	3.26	2.52	1.33	2.59	2.40	2.03
Tm	0.25	0.60	0.33	0.67	0.42	0.20	0.44	0.45	0.30
Yb	1.47	3.27	2.00	3.06	2.43	1.14	2.36	2.53	1.91
Lu	0.27	0.50	0.33	0.50	0.38	0.18	0.39	0.45	0.28
Th	0.25	0.53	0.11	0.21	0.25		0.17	b.d.	0.02
U	1.82	2.88	1.49	1.13	4.35	0.03	1.08	b.d.	

Basalts

Specimen	CM 1-98	Hi 2-96	Hi 22-96	Hi 24-96	Hi 25-96r	Hi 25-96c	Hi 29-2-96c	Hi 29-2-96r	Hi 29-4-96c	Hi 29-4-96r	Hi 29-4-96i	Hi 31-96
SiO_2	51.0	49.0	50.7	42.7	49.9	50.1	49.7	51.1	48.3	44.2	46.5	48.9
TiO_2	1.821	1.545	1.180	3.803	1.967	2.088	1.903	1.534	2.044	1.577	1.020	1.99
Al_2O_3	16.9	16.0	15.2	14.7	15.2	15.0	15.2	13.8	15.2	14.1	18.1	15.5
$Fe_2O_3(t)$	9.38	8.89	9.49	13.48	10.33	10.91	9.90	8.54	10.15	7.32	7.92	9.95
Fe_2O_3	3.09	2.53	2.22	3.29	2.89	2.38	2.27	2.87	2.61	2.29	2.26	2.65
FeO	5.7	5.7	6.5	9.2	6.7	7.7	6.9	5.1	6.8	4.5	5.1	6.6
MnO	0.175	0.153	0.154	0.208	0.162	0.171	0.236	0.200	0.148	0.120	0.137	0.15
MgO	6.41	7.40	8.50	5.72	5.31	6.55	5.19	3.59	5.31	4.11	3.27	5.2
CaO	9.27	10.72	6.21	11.29	11.65	9.00	11.23	14.71	9.82	14.08	12.44	11.09
Na_2O	4.28	3.41	4.89	3.44	3.27	4.09	3.41	1.77	4.57	3.84	3.14	4.57
K_2O	0.07	b.d.	0.12	0.10	b.d.	0.20	0.29	0.09	0.28	1.53	0.93	0.26
P_2O_5	0.294	0.179	0.183	1.891	0.330	0.271	0.259	0.269	0.233	0.337	0.335	0.26
S	0.0	0.1	0.1	0.2	0.1	0.0	0.0	0.1	0.0	0.0	0.1	0.1
CO_2	0.022	0.04	0.13	0.02	0.03	0.04	1.22	3.33	2.16	0.96	5.94	1.84
H_2O	2.14	2.96	3.87	2.32	1.87	2.34	3.05	5.42	3.60	3.42	3.13	1.83
SUM	101.1	99.7	100.0	98.9	99.4	99.9	100.9	103.8	101.1	95.1	99.2	100.9
Mg#	61	66	68	50	55	58	55	49	55	57	49	55
Li	7.92				3.54						11.67	9.78
Sc	31	35	21	28	28	30	30	33	32	31	31	34
V	220	232	170	356	235	229	233	200	246	189	189	249
Cr	134	254	309	36	144	155	172	140	145	115	115	164
Co	34	37	42	39	34	35	45	30	34	27	27	35
Ni	87	102	336	19	107	110	97	66	101	71	71	100
Cu	35											59
Zn	40	50	81	93	75	94	75	45	87	54	54	58
Ga	16	16	16	17	17	15	19	23	14	16	16	15
Rb	0.84	b.d.									21.99	3.33
Sr	183	181	169	349	181	140	218	604	419	738	738	280
Y	26.51	29.29	26.79	95.98	34.31		36.93		29.95		27.74	32.56
Zr	184	129	108	127	193	205	190	155	203	157	157	188
Nb	7.86	2.65	4.07	7.19	6.82		5.08	4.37	7.60		6.52	7.59
Mo	0.47	0.70	0.13	0.77	0.26	0.36	0.30	0.29	0.22		0.32	0.88
Sn	5.33	2.85	2.48	1.53	3.47	4.15	3.52	3.37		3.69	6.43	6.08
Ba	5.60	5.38	12.64	6.44	2.89	12.00	15.15	6.05	13.35	86.14	82.37	26.64
La	11.29	4.43	4.99		4.46	8.27	6.92	6.81	5.20	10.28	11.00	10.59
Ce	34.04	14.08	14.29	85.47	13.25		20.70	18.70			30.37	31.57
Pr	4.92	2.47	2.24	12.63	2.11	3.80			2.26	3.40	4.20	4.70
Nd	20.29	13.04	11.03	66.90	10.68	19.06	16.94	15.73	11.51	15.35	22.17	23.73
Sm	4.91	4.28	3.09	17.33	3.10	5.34	4.71	4.77	3.37	4.26	5.44	5.71
Eu	1.60	1.39	1.03	4.55	1.00	1.62	1.58	1.52	0.95	1.58	1.52	1.77
Gd	5.78	4.33	3.54	18.40	3.56	5.82	5.24	5.14	3.74	5.04	5.52	6.63
Tb	0.75	0.73	0.62	2.69	0.65	0.96	0.89	0.82	0.62	0.78	0.86	0.91
Dy	5.30	4.86	4.37		5.15		5.67	5.18	4.49		6.17	6.72
Ho	1.20	1.01	0.94	3.04	1.15	1.34	1.16	1.04	0.99	0.97	1.15	1.37
Er	2.62	2.71	2.57	7.82	3.34	3.72		2.83	2.70	2.79	2.72	3.09
Tm	0.47	0.42	0.39	1.04	0.53	0.56	0.50		0.42	0.40	0.40	0.59
Yb	2.45	2.80	2.51	6.22	3.35	3.59	3.22	2.86	2.81	2.67	2.37	2.95
Lu	0.39	0.39	0.37	0.83	0.47	0.54	0.47	0.41	0.37	0.38	0.42	0.50
Th	0.20	0.10	0.16	0.19	0.24	0.35	0.23	0.18	0.25	1.47	0.41	0.29
U	2.42	0.05	0.11	0.11	0.09	1.50	0.12	0.14	0.28	0.27	8.45	4.20

Serpentinites

	AS 1-96	AS 2-96	AS 3-96	AS 9-96	AS 12-96	AS 16-96	AS 19-96	Hi 1-96	Hi 3-96	Hi 6-96
SiO_2	38.5	39.5	39.5	39.4	40.0	38.8	41.0	44.5	39.9	41.2
TiO_2	0.034	0.025	0.024	0.111	0.029	0.052	0.043	0.030	0.032	0.027
Al_2O_3	1.2	1.2	1.8	1.6	1.4	1.9	1.7	1.7	1.3	1.8
$Fe_2O_3(t)$	9.63	8.90	7.86	8.37	6.56	8.73	6.22	8.48	7.35	5.87
Fe_2O_3	6.69	6.34	6.01	6.05	4.84	7.22	3.05	4.53	5.44	3.15
FeO	2.65	2.31	1.67	2.09	1.55	1.36	2.85	3.56	1.72	2.45
MnO	0.176	0.098	0.092	0.113	0.107	0.134	0.076	0.101	0.132	0.098
MgO	41.37	37.84	38.05	38.10	39.61	38.18	38.48	29.46	38.40	38.59
CaO	0.32	b.d.	0.21	0.58	b.d.	b.d.	0.95	7.46	0.27	b.d.
Na_2O	b.d.	b.d.	b.d.	b.d.	b.d.	b.d.	b.d.	b.d.	b.d.	b.d.
K_2O	b.d.	b.d.	b.d.	b.d.	b.d.	b.d.	b.d.	b.d.	b.d.	b.d.
P_2O_5	b.d.	b.d.	b.d.	b.d.	b.d.	b.d.	0.142	b.d.	b.d.	b.d.
S	0.2	0.2	0.1	0.1	0.2	0.1	0.1	0.2	0.1	0.1
CO_2	0.07	0.10	0.02	0.10	0.08	0.11	0.04	0.03	0.70	0.16
H_2O	8.19	11.51	12.17	10.68	11.83	10.99	11.37	7.74	11.17	12.21
SUM	99.4	99.1	99.7	98.9	99.6	98.8	99.8	99.3	99.2	99.8
Mg#	91	91	92	91	93	91	94	89	92	94
Li	0.92	0.31	0.99	0.42	0.52	0.46	0.24	2.79	0.24	0.34
Sc	b.d.	b.d.	9	9	b.d.	9	b.d.	7	9	7
V	43	40	48	60	45	59	83	26	42	41
Cr	2463	2906	2411	2418	2566	2567	2234	2590	1330	1700
Co	123	108	98	103	121	89	98	193	89	87
Ni	2486	2389	2106	2157	2520	1716	2498	2167	1542	1438
Cu	12.10	3.02	6.14	2.84	11.70	6.93	7.98	8.62	5.35	12.62
Zn	52	42	39	35	39	42	37	41	34	37
Ga	b.d.	b.d.	b.d.	b.d.	b.d.	b.d.	b.d.	b.d.	b.d.	b.d.
Rb										
Sr	0.31	0.18	0.47	1.51	b.d.	0.32	1.57	1.48	4.26	b.d.
Y	0.50	0.44	0.59	0.72	0.81	0.78	1.80	0.53	0.28	0.44
Zr	2.67	0.63	2.96	1.22	0.52	0.94	1.05	2.30	0.63	0.46
Nb	b.d.	b.d.	0.16	0.07		0.10	0.08	0.12		b.d.
Mo										
Sn										
Ba	0.68	0.44		0.57			0.26	0.30	1.26	
La	b.d.	b.d.	0.04	b.d.	0.07	0.04	0.25	b.d.	0.05	b.d.
Ce	0.05	0.12	0.12	0.10	0.18	0.12	0.93	b.d.	0.10	0.08
Pr	0.02		0.02	0.02	0.04		0.14	0.02		
Nd	0.10	0.11	0.11	0.10	0.20	0.11	0.94	0.14	0.04	0.11
Sm					0.09		0.28	0.17		
Eu										
Gd	0.06	0.07	0.06	0.06	0.13	0.08	0.43	0.13		0.05
Tb	0.02	0.01	0.01	0.02	0.02	0.02	0.06	0.03		0.02
Dy	0.10	0.07	0.11	0.11	0.12	0.12	0.35	0.17	0.03	0.06
Ho	0.02	0.02	0.02	0.02	0.03	0.03	0.10	0.04		0.03
Er	0.08	0.06	0.09	0.09	0.11	0.10	0.25	0.09	0.03	0.07
Tm	0.02	0.01	0.02	0.02	0.03	0.02	0.04	0.03	0.01	0.01
Yb	0.08	0.06	0.09	0.11	0.09	0.12	0.23	0.10	0.07	0.07
Lu	0.02	0.01	0.01	0.02	0.02	0.02	0.03	0.02	0.02	0.01
Th										
U										

187

	Hi 7-96	Hi 13-5-96	Hi 14-96	Hi 15-96	Hi 21-96	Hi 27-3-96	Hi 28-96	Hi 28-2-96	Hi 32-9-96
SiO_2	39.5	40.0	39.2	40.8	40.3	40.9	39.5		39.7
TiO_2	0.038	0.028	0.027	0.061	0.103	0.223	0.040		0.048
Al_2O_3	1.9	1.4	1.2	1.1	3.8	3.5	1.4		2.2
$Fe_2O_3(t)$	6.66	8.15	8.35	8.92	8.22	7.95	9.61		7.64
Fe_2O_3	4.82	6.05	6.05	5.41	3.29	2.09	6.77		3.74
FeO	1.66	1.89	2.07	3.16	4.44	5.27	2.55	3.02	3.51
MnO	0.105	0.114	0.143	0.104	0.115	0.126	0.111		0.105
MgO	39.22	37.83	39.78	35.85	34.14	34.63	37.45		38.76
CaO	b.d.	b.d.	b.d.	1.16	1.65	0.73	b.d.		0.50
Na_2O	0.23	b.d.	b.d.	b.d.	b.d.	b.d.	b.d.		b.d.
K_2O	b.d.	b.d.	b.d.	b.d.	b.d.	b.d.	b.d.		b.d.
P_2O_5	b.d.	b.d.	b.d.	b.d.	b.d.	b.d.	b.d.		b.d.
S	0.1	0.1	0.1	0.1	0.1	0.1	0.1	0.1	0.1
CO_2	0.14	0.02	0.17	0.22	0.06	0.06	0.10	0.14	0.19
H_2O	11.93	12.93	9.88	11.10	11.75	11.63	11.82	11.50	10.33
SUM	99.6	100.3	98.6	99.0	99.7	99.3	99.8		99.2
Mg#	93	92	92	90	91	91	90		92
Li	0.21	1.50	0.77	2.56	0.84	0.47	0.28	1.54	4.07
Sc	b.d.	b.d.	b.d.	b.d.	b.d.	9	8		b.d.
V	59	36	34	43	65	73	54		51
Cr	2846	3144	2539	4364	2195	2346	2788		2594
Co	106	79	107	112	100	96	72		99
Ni	2281	1611	2244	2261	1847	1933	1251		1934
Cu	10.33	14	9.48		22.64	18.07	14.51	16.80	29.97
Zn	69	49	49	98	51	48	37		45
Ga	b.d.	b.d.	b.d.	b.d.	b.d.	b.d.	b.d.		b.d.
Rb									
Sr	0.17	0.14	0.19	2.19	1.67	1.71	0.15	0.33	1.34
Y	1.83	0.19	0.25	0.19	1.45	2.81	0.33	1.05	0.38
Zr	5.95	0.67	0.91	0.69	3.16	1.82	0.78	2.71	3.29
Nb	0.21	b.d.	b.d.	0.20	0.30	0.63		0.11	0.19
Mo		0.07							
Sn									
Ba		0.73	1.21	0.39	0.71	2.14	0.37	0.58	1.13
La	0.13	b.d.	b.d.	0.06	0.10	0.18	0.05	0.05	b.d.
Ce	0.40	b.d.	0.05	0.11	0.32	0.68	0.18	0.15	0.07
Pr	0.07		0.02		0.06	0.12	0.02	0.03	
Nd	0.37	0.05	0.08	0.08	0.30	0.61	0.10	0.27	0.10
Sm	0.19				0.14	0.24		0.16	0.07
Eu	b.d.					b.d.			
Gd	0.20	0.06	0.08		0.22	0.31	0.06	0.14	0.09
Tb	0.04	0.01	0.01	0.01	0.03	0.07	0.01	0.03	0.01
Dy	0.33	0.05	0.07	0.03	0.28	0.38	0.05	0.20	0.08
Ho	0.07	0.01	0.02		0.06	0.10	0.02	0.05	0.02
Er	0.24	0.04	0.05	0.02	0.19	0.33	0.05	0.11	0.07
Tm	0.04	0.01	0.01		0.03	0.05	0.01	0.02	0.02
Yb	0.25	0.06	0.06	0.02	0.20	0.34	0.06	0.11	0.09
Lu	0.04	0.01	0.02		0.04	0.06	0.01	0.02	0.03
Th									
U									

	Metarodingites					Chloritoschists		Veins	
	Hi 8	Hi 10-96	Hi 12-1-96	Hi 12-2-96	Hi 13-3-96	Hi 11	Hi 17-96	Hi 32-4-96	Hi 32-10-96
SiO2	32.3	34.8	36.6	43.7	34.5	34.0	29.7	55.3	30.8
TiO2	1.146	0.316	0.078	0.025	0.044	0.146	0.633	0.082	0.055
Al2O3	15.6	15.6	13.8	3.9	18.6	10.5	17.9	9.9	35.2
Fe2O3(t)	10.44	8.04	4.66	6.74	3.69	6.03	10.74	6.95	11.27
Fe2O3	0.64	7.01	2.66	4.65	2.38	4.14	1.33	2.57	1.94
FeO	8.8	0.9	1.8	1.9	1.2	1.7	8.5	3.9	8.4
MnO	0.398	0.209	0.063	0.058	0.141	0.054	0.133	0.037	0.060
MgO	26.13	13.05	26.1	27.4	14.67	36.22	28.74	8.80	11.47
CaO	4.23	23.8	10	11.58	23.01	0.23	0.44	12.15	1.74
Na2O	b.d.	b.d.	b.d.	b.d.	b.d.	b.d.	b.d.	5.76	1.58
K2O	b.d.	b.d.	b.d.	b.d.	b.d.	b.d.	b.d.	b.d.	b.d.
P2O5	0.115	b.d.	b.d.	b.d.	b.d.	b.d.	0.213	0.029	0.027
S	0.1	0.0	0.0	0.1	b.d.	0.1	0.1	0.0	0.1
CO2	0.38	0.030	0.033	0.038	0.024	0.16	0.05	0.20	0.01
H2O	11.53	5.99	9.19	6.16	7.25	12.97	12.86	1.23	7.14
SUM	101.4	101.7	100.3	99.5	101.8	100.2	100.6	100.0	98.5
Mg#	85	79	93	90	90	93	86	75	70
Li	7.09	2.17	0.76	0.54	3.05	0.11	8.10		0.37
Sc	28	41	11	11	14	12	11	30	b.d.
V	144	137	43	27	b.d.	52	55	832	45
Cr	682	364	1637	2572	257	677	451	52	297
Co	68	46	62	97	38	41	85	28	64
Ni	576	287	1304	2140	620	566	487	86	500
Cu		16	6	24	31				
Zn	143	23	16	17	8	30	100	87	102
Ga	11	b.d.	8	7	b.d.	b.d.	12	24	21
Rb		b.d.	1.21	b.d.	b.d.		1.31		
Sr	7.90	18.47	5.58	5.63	36	b.d.	b.d.	17	16
Y	12.56	8.94	2.35	0.19	0.34	2.92	8.76	0.84	0.10
Zr	94	19	5.73	2.96	3.65	13	48	2.39	4.02
Nb	0.93	0.47	0.21	0.13	0.22	0.09	2.53		0.93
Mo	0.07	0.13	0.07	b.d.	0.11		0.13	0.15	0.10
Sn	0.29	0.87	0.39	0.25	0.08	0.27	0.69	1.07	1.56
Ba	0.12	0.28	0.69	1.68	0.43	b.d.	0.28	1.09	3.69
La	6.03	2.53	0.40	0.45	0.99	0.39	2.09	0.11	b.d.
Ce	12.48	5.57	0.27	0.24	0.45	1.24	7.99	0.05	b.d.
Pr	1.96	0.98	0.20	0.06	0.25	0.26	1.49	0.03	0.01
Nd	9.52	3.79	0.29	0.16	0.39	1.32	9.49	0.05	b.d.
Sm	2.77	1.28	0.13	b.d.	0.26	0.47	1.97	0.07	b.d.
Eu	0.66	0.60	b.d.	b.d.	0.17	0.25	0.27	b.d.	b.d.
Gd	3.58	1.18	0.12	0.03	0.10	0.47	1.82	0.06	0.01
Tb	0.53	0.21	0.04	0.01	0.02	b.d.	0.29	0.02	b.d.
Dy	3.11	1.63	0.45	0.04	0.07	0.58	1.84	0.17	0.02
Ho	0.64	0.37	0.07	0.01	0.04	0.12	0.37	0.02	b.d.
Er	1.61	0.74	0.19	0.05	0.02	0.27	0.85	0.08	0.01
Tm	0.24	0.15	0.05	0.00	0.01		0.14	0.01	0.00
Yb	1.33	0.74	0.21	0.02	0.05	0.20	0.74	0.07	0.01
Lu	0.21	0.10	0.04	0.02	0.01	0.03	0.12	0.01	b.d.
Th	0.10	b.d.	0.04	b.d.	b.d.		b.d.	b.d.	
U	0.09	0.48	0.44	0.68	1.19	0.02	b.d.	b.d.	b.d.

Appendix 3: paper of W. Himmelheber and J.W. Sheraton (in preparation)

A statistical method for studying element mobility revised

Wendelin Himmelheber and John W. Sheraton

Abstract

Conventional methods of calculating inter-element correlation coefficients using percentage data can lead to erroneous results because the constant-sum constraint forces correlations upon the data. This problem can be avoided by the use of logarithms of element ratios, this method being due to J. Aitchison. The new method was applied to analyses of granulite-facies Archaean gneisses from northwest Scotland and their retrograde amphibolite-facies equivalents (both undeformed and sheared), and compared to the results obtained by application of standard statistics, taken from the literature, in order to study the chemical changes associated with retrogression. The results from the two methods are similar in some respects, but there are many significant differences, notably the much reduced correlations of Si with other elements when log-ratios are used, and widely differing results as concerns correlations with minor and trace elements. The new method also appears to give more easily interpretable results. Most of the main features of the log-ratio correlation coefficient matrix for the granulite-facies gneisses can be explained in terms of igneous processes, whereas many of the significant changes of element correlations in the retrograde gneisses can be explained by metamorphism in the presence of hydrous fluids, with formation of a relatively simple mineralogy (biotite–hornblende–quartz–plagioclase).

Introduction

In a frequently quoted paper, Beach and Tarney (1978) proposed a statistical method for the investigation of changes of bulk-rock chemistry associated with metamorphism. Two groups of samples are necessary, only one of which displays

191

the effects of a particular metamorphic event. Correlation coefficients between every pair of elements are calculated for each group and the two matrices are compared. Usually it is found that some correlations increase, whereas others decrease. The significance of these changes can be assessed using Fisher transforms, as proposed by Beach and Tarney (1978). Ideally, the changes can be related to mineralogical changes associated with the metamorphic event, and the mobility or immobility of elements can be assessed.

A particular virtue of this method is that it can be applied in cases when the isocon method of Grant (1986) cannot be used because of uncertainties about the relationships between the protoliths and altered equivalents. Beach and Tarney (1978) applied their method to Sheraton's (1969) data set of nearly 250 X-ray fluorescence analyses of Archaean Lewisian high-grade gneisses of northwest Scotland (Sheraton, 1970; Tarney et al., 1972; Sheraton et al., 1973). These gneisses were originally metamorphosed under granulite-facies conditions, but most were later retrogressed to amphibolite facies. The retrograde samples were subdivided into undeformed gneisses and those from shear zones. As a particular effort was made to collect both compositionally and geographically representative samples, they should be relatively unbiased and thus amenable to statistical analysis. However, the granulite-facies group has a more mafic average composition than either of the retrograde groups, because felsic rocks are more susceptible to retrogression. The question of whether this introduces a significant bias into the results is discussed below. Using this method, Beach and Tarney (1978) arrived at mostly quite plausible conclusions. Changes in major element correlations were apparently related to mineralogical changes involving the production of biotite–hornblende–quartz–plagioclase assemblages with relatively constant Fe/Mg ratios in the amphibolite-facies gneisses. Changes in minor and trace element chemistry were discussed in much less detail by Beach and Tarney (1978) and seem to be less well understood.

When Beach and Tarney (1978) wrote their paper, the log-ratio method of Aitchison (1986; also summarized in Rollinson, 1992, 1993) had not been published. As already Pearson (1897) pointed out, standard statistical analysis using percentage data can lead to spurious results, because the constant-sum constraint forces correlations upon any set of percentage data. Therefore it seemed advantageous to try the log-ratio method, which avoids this problem, on the same data set and to compare the results. Indeed, we think that with Aitchison's method much better and more easily interpretable results can be obtained. Possibly the confusing results that are often obtained when normal correlation coefficients are used on geochemical data sets were the reason that the method of correlation matrices has so rarely been applied. Although the paper by Beach and Tarney has frequently been quoted as testimony to the importance of large fluid fluxes for element mobility and deformation (development of hydrated shear zones), we could find only three other papers which actually use such correlation matrices (some even reproducing the printing error – missing square root – near the top of page 330 of Beach and Tarney, 1978): Weaver et al. (1981), Watkins (1983), and Schnetger (1994).

Remarks on the method

Aitchison's method hinges on the observation that the ratio of two components of a composition remains unaffected by the effect of dilution by any other component, which enforces correlation on the whole system. Therefore it is proposed that element ratios are used as the basic parameter. Because these can vary over a range of more than 12 orders of magnitude (element abundances range between 100 % and <1 ppm), logarithms of the ratios (log-ratios) are used. An obstacle to the widespread use of the Aitchison method is probably the fact that it is not contained in standard statistics packages for computers. Yet, to our knowledge there exist at the moment three computer programs for the Aitchison method. One is CODA, written by Aitchison himself in Basic, obtainable from Chapmann & Hall. Another was written by Marcus (1993) in Matlab. A third one was written by one of us (Himmelheber) in Turbopascal. The last two are shareware.

Of the several mathematically equivalent methods of data treatment, the conceptually most simple one is the matrix of log-ratio variances. In this matrix, for any pair of elements the value of

$$\text{var}(\log(c_i/c_j)) = \sum_{k=1}^{N} \left(\ln(\frac{c_{ik}}{c_{jk}}) - \sum_{k=1}^{N} \ln(\frac{c_{ik}}{c_{jk}})/N \right)^2 / (N-1)$$

[c_{ik}: concentration of component i in specimen k; N: number of specimens]
is given. The log-ratio variance of two elements may be visualized as a measure of the angular spread of the data in a plot of one element against the other. When the data points lie on a straight line through the origin of the coordinate system, their

ratios will be identical and their angular spread will be zero. Thus, a variance close to zero indicates a strong linear relationship (i.e., correlation) between two elements.

Another, intuitively less appealing, way of describing the distribution of the data is the centered log-ratio covariance matrix. This is a covariance matrix, computed from centered element contents, that is, the logarithms of the element contents of a specimen are divided by the logarithm of the geometric mean of all the element contents of that specimen. This covariance matrix and the log-ratio variance matrix can be transformed into each other and thus are mathematically equivalent (Aitchison, 1986):

$$cov(c_i, c_j) = 1/2 * \left(\sum_{k=1}^{m}(var(c_i, c_k))/m + \sum_{k=1}^{m}(var(c_k, c_j))/m - \sum_{k=1}^{m}\sum_{l=1}^{m}(var(c_k, c_l))/m^2 - var(c_i, c_j) \right)$$

[m: number of elements/oxides]

Although normalizing to the geometric mean of a specimen's contents of all elements doesn't seem to make any geochemical sense, the two forms of describing the correlation structure of a sample are mathematically equivalent, and the covariances are the basis for all methods of multivariate analysis.

From the centered log-ratio covariance matrix, in turn, the definition of a log-ratio correlation coefficient can be obtained: $r(c_i, c_j) = cov(c_i, c_j)/sqrt(cov(c_i, c_i)*cov(c_j, c_j))$. In comparison to the log-ratio variance, this correlation coefficient has several advantages as a measure of correlation: it is insensitive to degenerate cases (e.g. a very small circular cloud of data points would display a small log-ratio variance); it is an absolute, not a relative, measure; and standard testing procedures are applicable. It seems to have been introduced into the literature by Marcus (1993).

A word of caution is, however, due here. Contrary to the log-ratio variance, but similar to the "crude" correlation coefficient, the log-ratio correlation coefficient is dependent on the elements selected for analysis and on their behavior (Aitchison's "subcomposition difficulty") and can even change sign when the subset of elements considered is changed. As can be gained from the formulas given, this dependence is inherited from the centered logratio covariance matrix. Also it is subject to the other difficulties of the crude correlation coefficient, as the "negative bias difficulty" (there must be at least one negative entry in every row and column of the matrix) and the "null correlation difficulty" (independence of variables is not necessarily expressed as a zero value for the correlation coefficient). These problems are perhaps the reason that Aitchison (1986) did not define the log-ratio correlation coefficient. There is however a difference to the situation of the crude correlation coefficient: The behavior of the log-ratio correlation coefficient is mathematically predictable. In practical work, it seems that a careful selection of elements, adapted to the problem studied, leads to sensible results. As a measure of caution Table 1, in addition to the results of the application of the log-ratio correlation coefficient to the data of Sheraton (1969), gives also the values of the log-ratio variance.

It must be stressed here that both the percentage data and their centered log-ratios are very far from normal distributions (Fig. 1). This means that student t-testing for significance of correlation is not valid, and neither are the criteria of significance for the differences of Fisher transforms (see below). Although this problem, of which we have found no treatment in the literature, is beyond our mathematical expertise, it is hoped that the thresholds of 99% significance indicated in Tables 1, 2, and 3 at least give some indication of the strength of the correlations.

In order to obtain a clearer picture of the groupings of elements which are correlated, the ordering of elements has been changed from the standard one. As it was found that correlations involving H, C, S, and Cl were of no clear petrological significance, these elements were omitted in the analysis presented here. For ease of comparison, Table 2 gives the conventional correlation coefficients (termed 'crude' by Aitchison, 1986) with the same element ordering. The results are not exactly identical to those presented by Beach and Tarney (1978). Some small and not significant deviations are due to different rounding errors of the programs used, whereas some larger discrepancies are probably due to errors in the manual data transfer. Also, Beach and Tarney (1978) set all values below the detection limit to the value of the particular detection limit, whereas we ignored these values in the computation. Table 2 also differs from Beach and Tarney's Table 1 in not treating Fe^{2+} and Fe^{3+} as separate components. It would be interesting to try and relate oxidation state to abundances of other elements, but their treatment as separate components is statistically unjustified. However, sheared gneisses appear to be more oxidized than undeformed rocks (Beach and Tarney, 1978).

It is found that the results of the conventional and log-ratio methods are significantly different. As will be shown, the latter method gives a much clearer and more easily interpretable picture of the inter-element relationships and their changes during metamorphism.

The correlation matrix and its interpretation

The correlation matrix for the granulite-facies gneisses (Table 1) includes two large groups of elements, which have positive correlation coefficients within groups, but negative ones between groups. The first of these groups comprises the compatible ferromagnesian elements Ca, Fe, Mn, Mg, Cr, and Ni. In contrast, the second group comprises elements which are generally incompatible during mantle melting events (Na, Sr, Pb, P, Zr, Ce, and Ba) and which are present mainly in feldspar or accessory minerals. As the Lewisian gneisses are thought to have been derived from calc-alkaline igneous protoliths (Tarney et al., 1979; Weaver and Tarney, 1980), these groupings may be at least partly explained by igneous processes, such as fractional crystallization of mafic phases (e.g., hornblende or pyroxene) or, more likely, partial melting processes.

Al is most strongly correlated with Si, and also shows some correlation with the compatible elements. In marked contrast to the older method, the only other significant correlation of Si is with Na, and that is only weak. K and Rb appear to form a distinct 'phyllosilicate' group, probably largely reflecting the distribution of primary biotite, although high-temperature plagioclase can contain significant amounts of the orthoclase component. Ti also stands apart, except for low positive correlations with Fe and Mn, suggesting only poor correlations between the abundances of ilmenite and ferromagnesian silicates, possibly resulting from the moderately to strongly incompatible behavior of Ti in mafic magmatic systems.

The above picture is by and large affirmed by the matrix of log-ratio variances. One difference however is to be observed in the behavior of the elements Al, Na, Sr, Pb and to a lesser extent also Si and K: besides displaying their respective affiliations to the compatible ferromagnesians and the incompatible elements, they show also some correlation between each other, thus pointing to the importance of feldspar in the mineralogy of the specimens.

The next step is to compare the correlation matrices for the granulite-facies and retrograde samples. Beach and Tarney (1978) proposed a method for judging the significance and amount of change of correlation coefficients: the matrix of differences of Fisher transforms. As well as being strictly dependent on the normality of the data distribution, this statistical measure also has to be treated with care, because it combines cases that should ideally be kept apart. For example, a zero correlation becoming negative and a positive correlation being reduced in size both have a Fisher transform difference with the same sign. However, with this proviso in mind, the matrix of Fisher transform differences (Table 3) can be used to compare the magnitudes of changes in correlation coefficients and to give an indication of their significance levels. As a parallel to Table 1, Table 3 contains also a tentative measure of the change of log-ratio variances, defined as $\log(\text{var}_{\text{group1}}/\text{var}_{\text{group2}})$.

When the granulite-facies and undeformed amphibolite-facies gneisses are compared, several points emerge. The most pronounced change is the formation of a new group of elements: the (sodic) 'plagioclase' group (Si, Al, Na, and Sr). The correlation Si–Al is much increased, and the increased correlation between Al and Na, as measured by Fisher transform difference, is the highest in the whole matrix. The originally positive correlation of Ca with Al is lost, so that Ca clearly does not belong to this group, and negative Ca–Na and Ca–Sr correlations are also largely lost. There are significant negative correlations of Si with the ferromagnesian elements, which are conspicuously absent in the granulite-facies gneisses. Pb loses some of its correlation with Sr and Na, and, together with Ba, appears to join K and Rb in the phyllosilicate group. However, the latter elements also show increased correlations with Si, Al, and Na, as well as negative correlations with ferromagnesian elements, and thus appear to develop an affinity with the plagioclase group, possibly reflecting the presence of much of the K and Rb in plagioclase in addition to phyllosilicate minerals (biotite and minor muscovite). This association may also reflect the higher abundance of secondary biotite in felsic layers, as biotite is the most common alteration product of pyroxene in felsic gneisses, hornblende only predominating in mafic to intermediate compositions (Sheraton, 1969). However, the same association appears when only the more mafic compositions are investigated (see below).

There are several noteworthy changes affecting the ferromagnesian elements. Strong negative correlations are developed with both the plagioclase and phyllosilicate groups. Al, in particular, changes from a weak positive to a strong negative correlation as it joins the plagioclase group. Ca loses much of its correlation with the ferromagnesians, but the coefficients remain positive. Ti acquires significant positive correlations with the ferromagnesian group, and negative correlations with many elements from the other groups. As stated by Beach and Tarney (1978), the common factor for the ferromagnesian element group appears to be that they are all hosted by hornblende, particularly in the more mafic rocks.

The remaining elements (P, Ce, and Zr) are those mainly contained in the accessory minerals apatite and zircon. Their original mutually positive correlations are largely lost in the retrograde gneisses. Negative correlations with the ferromagnesian group are also lost or much reduced, as are negative correlations with Al and positive ones with Na and Sr.

At this point, the question of whether or not the generally more felsic character of the retrograde gneisses introduces a systematic bias into the results needs to be considered. In order to test this, we selected a subset of the retrograde gneisses with a distribution of SiO_2 contents (<50 %, 50–60%, and >60% SiO_2) to mirror the respective numbers of specimens in the granulite group. Several retrograde subsets were arbitrarily selected in this way, and the resulting correlation matrices were quite similar to the original matrix, except that Ti does not appear to join the ferromagnesian group.

The picture is very similar when the granulite-facies rocks are compared to the sheared gneisses, but there are a few minor differences. For example, correlations of Ca with the ferromagnesian elements are not reduced to the same degree, and Ti's gain of correlation with this group is larger. Pb loses all significant correlations. The coherence between P, Ce, and Zr is somewhat greater than in the undeformed retrograde gneisses. Zr develops a significant positive correlation with Al and retains its correlations with Na and Sr, suggesting a possible affiliation to the plagioclase group of elements.

The changes in log-ratio variance point in the same direction. Only two points are mentioned here: The plagioclase group of elements, already present in the granulite's log-ratio variance matrix, is much enhanced upon retrogression. The somewhat enigmatic affiliation of Zr with this groups is also observed.

In some respects, the interpretation of these changes is fairly straightforward, as has already been implied by use of the terms 'plagioclase' and 'ferromagnesian' groups of elements. As pointed out by Beach and Tarney (1978), large amounts of hydrous fluid which passed through the rocks during the retrogression would tend to smooth out some of the geochemical differences, especially in the shear zones. At the same time, a relatively simple mineralogy (biotite–hornblende–quartz–plagioclase, plus minor phases) was developed.

The ferromagnesian elements were apparently least affected by the retrogression, consistent with them being largely taken up by amphibole upon alteration of pyroxene. The changes that have occurred can be understood in conjunction with the formation of the 'plagioclase' group and the influence of hydrous fluids. Some addition of Na from the fluid phase may have occurred (Beach and Tarney, 1978), although the hornblende-forming reactions require breakdown of plagioclase, which could provide Na, as well as Al. However, retrograde plagioclase itself tends to be more sodic (An20–30) than granulite-facies plagioclase (An30–50), although this is dependent on the degree of recrystallization, as even sheared gneisses contain relic grains of more calcic plagioclase (Sheraton, 1969). This tendency of plagioclase to assume a more restricted compositional range could possibly account for the enhanced correlations between plagioclase group elements, particularly Al, Na, and Sr, the contents of which would be much higher in plagioclase than in hornblende. Formation of more sodic plagioclase, as well as replacement of clinopyroxene by hornblende in the more mafic gneisses, liberated excess Ca, which entered secondary epidote/clinozoisite or calcite (Sheraton, 1969; Sills, 1983). Part may also have been lost from the gneisses to offset any gain of Na from metasomatic fluids; amphibolitization of ultramafic pods could use up excess Ca (Beach and Tarney, 1978). Some loss of Ca was also deduced from a study of retrograde mineral reactions in shear zones (Beach, 1980), which commonly contain small chlorite–epidote–carbonate–prehnite veins (Sheraton 1969). Such partial redistribution of Ca could explain much of its reduced correlation with the ferromagnesian elements in retrograde gneisses, in which it is hosted by hornblende, plagioclase, epidote group minerals, and minor titanite and calcite.

The increased correlation of Ti with the ferromagnesian elements in retrograde gneisses is difficult to explain, because, although some might enter the hornblende structure, much of the original ilmenite survives or is partly replaced (i.e., rimmed) by titanite. Retrograde amphiboles analyzed by Sheraton (1969) have only moderate Ti contents (0.46–0.64 % TiO_2), much lower than those from granulite-facies rocks (2.09 %). However, the increased correlation may merely reflect the more felsic average composition of the retrograde rocks, as Ti is generally incompatible only in mafic igneous systems.

The reduced correlations between the 'accessory mineral' trace elements P, Ce, and Zr are also difficult to explain. They suggest some mobility, but there is no obvious petrographic evidence for recrystallization of apatite or zircon. Also puzzling are the positive correlations which Zr acquires with the plagioclase group elements.

The ubiquitous presence of a fluid phase implied by the development of hydrous retrograde minerals was presumably responsible for some redistribution of the more soluble incompatible elements. Small amounts of sericite were commonly formed during retrogression, although some of this undoubtedly reflects the presence of significant amounts of K in the original plagioclase. This may be why the formation of the plagioclase group of positively correlated elements was accompanied by increased correlations of that group with K, Rb, and Pb. Chloritization of biotite would also lead to mobility of the latter elements. Nevertheless, there seems to be little evidence for any overall addition of mobile elements, such as K and Rb, during the retrogression.

This general description holds for both the undeformed retrograde gneisses and the gneisses from shear zones. Thus, the changes from granulite-facies to undeformed amphibolite-facies gneisses are very similar to those from granulite-facies to shear-zone gneisses, as is shown by the few significant entries in the Fisher transform difference matrix of the two retrograde groups in Table 3. However, the shear zones are generally supposed to have been conduits for the introduction of fluids, which then diffused into the neighboring rocks. It might be expected, therefore, that the sheared gneisses would display stronger retrograde effects. The results are somewhat equivocal in this respect, although at least some of the significant differences between the retrograde groups may be explicable in terms of enhanced metamorphic effects.

It is possible that the significantly higher correlations of Ca with the ferromagnesian elements in the sheared gneisses is because the plagioclase is even more sodic, so that proportionally more of the Ca is hosted by hornblende. However, most of the excess Ca is contained in epidote group minerals, so that a closer spatial association of these with the hornblende is more likely. Such an association may be enhanced in the more strongly deformed rocks, as much of the secondary epidote tends to occur as small inclusions in plagioclase grains in the undeformed retrograde gneisses. The increased correlations of Ti with the ferromagnesian elements in the sheared gneisses may have a similar explanation, as both retrograde and sheared gneisses have a very similar range of SiO_2 contents. The changes involving Pb, P, Ce, and Zr are particularly difficult to interpret. However, the stronger negative correlations of Zr with the ferromagnesian elements and apparently higher correlations with the plagioclase group in the shear-zone gneisses suggest an enhanced association of zircon with felsic layers in the gneiss. It may be that extreme deformation resulted in an enhanced metamorphic differentiation into mafic (hornblende+epidote+ilmenite/titanite) and felsic (plagioclase+quartz+zircon) layers; there is little evidence for any change in the preferred associations of biotite or apatite. Petrographic data appear to be broadly consistent with this suggestion, but detailed quantitative studies will be necessary to confirm it.

Conclusions

Most of the main features of the log-ratio correlation coefficient matrix for the granulite-facies gneisses can be explained in terms of igneous processes, with two groups of strongly correlated elements, the 'compatible' ferromagnesian elements (Ca, Fe, Mn, Mg, Cr, and Ni) and the 'incompatible' elements (Na, Sr, Pb, P, Zr, Ce, and Ba, present in feldspar or accessory minerals), and negative correlations between groups. Many of the significant changes of element correlations in the retrograde and sheared gneisses can be explained by metamorphism associated with the introduction of hydrous fluids, which, minor phases apart, resulted in the formation of a relatively simple mineralogy (biotite–hornblende–quartz–plagio-clase), compared to the original granulite-facies gneisses (biotite–hornblende–ortho-pyroxene–clinopyroxene–quartz–plagioclase). For example, there are marked increases in correlations between 'plagioclase' group elements (Si, Al, Na, and Sr), probably because plagioclase is higher in these elements than hornblende and tends to become more sodic in the retrograde rocks. In addition, hornblende contains significantly less Si than the pyroxene it replaced. The 'phyllosilicate' elements (K and Rb), together with Ba and Pb, also develop greater affinities with the plagioclase group, reflecting sericitization of plagioclase and/or formation of secondary biotite in the more felsic layers. Ca is much less strongly correlated with the ferromagnesian elements, being present in several minor secondary minerals (epidote, titanite, and calcite), as well as hornblende. An increased correlation of Ti with the ferromagnesian elements may just reflect the more felsic average composition of the retrograde rocks, as Ti is generally incompatible only in mafic magmas. At least some of these changes appear to have been more intense in the shear zones, which served as conduits for the fluids, although the results are somewhat equivocal in this respect.

The results from the two methods of calculating correlation coefficients (using straight concentration data and log-ratios) are similar in some respects (e.g., strong positive correlations between ferromagnesian elements, Na–Sr, K–Rb, and Ce–P), and thus Beach and Tarney (1978) arrived at basically the same interpretation, concerning the influence of a hydrous fluid and the simplified mineralogy, but there are many significant differences:

1. With Beach and Tarney's treatment, Si is positively correlated with Na, Sr, Pb, Zr, and Ba in the granulite-facies gneisses, but the correlations are much reduced upon retrogression and there is less evidence for development of a distinct plagioclase group of elements. There are strong negative correlations between Si and the ferromagnesian elements. With our log-ratio method, in marked contrast, Si shows few significant correlations in the granulites: only Si–Na and Si–Al, both of which are positive and increased upon retrogression. Si is not particularly associated with elements in either mafic or felsic minerals, which seems reasonable from a petrological viewpoint. However, negative correlations with ferromagnesian elements are present in the compositionally more varied retrograde rocks.

2. According to Beach and Tarney, the correlations of Ca with the ferromagnesian elements are higher in both undeformed and sheared retrograde gneisses, and particular attention was drawn to the better Ca–Ni correlation in the sheared, but not the undeformed, rocks. This was attributed to both elements (and Cr) entering hornblende, although it is not clear why this should apply only to the sheared gneisses. The explanation also suffers from the problem that hornblende is not the most important host to Ca. With our treatment, there are no such increased correlations, and those for the undeformed retrograde rocks are reduced.

3. Beach and Tarney's results show that correlations between ferromagnesian elements are increased upon retrogression. In contrast, our data show little change, suggesting that the ferromagnesian elements in orthopyroxene and clinopyroxene were incorporated more or less quantitatively into hornblende (±biotite), which is consistent with petrographic observations.

4. With our treatment, Al is negatively correlated with P, Ce, and Zr and weakly affiliated with the ferromagnesian elements in the granulite-facies gneisses, in contrast to Beach and Tarney's results. This is due to the somewhat compatible character of Al. The much stronger negative correlations of P, Ce, and Zr with the

ferromagnesian elements in our data suggest a tendency for apatite and zircon to occur in the more felsic granulite-facies rocks.

5. According to Beach and Tarney, many correlations of minor and trace elements with major elements are lowered upon retrogression. No detailed explanations were offered, but some elements (P, Ce, and Zr) were thought to enter accessory phases which were newly formed during retrogression. With our results, there is no such general trend towards reduced correlations, and most trace element correlations are raised or lowered according to their geochemical affiliations with the major elements. However, in agreement with Beach and Tarney, our results do show that large negative correlations of P, Ce, and Zr with the ferromagnesian elements in the granulites are much reduced during retrogression.

6. Our method gives larger negative correlations between K, Rb, and Ba and the ferromagnesian elements (and positive ones with Si) in both retrograde groups, possibly reflecting the presence of biotite as the main secondary phase in the more felsic rocks which tend to be preferentially retrogressed.

Clearly, the revised statistical technique proposed here produces significantly different results from the conventional method, notably the much reduced correlations of Si with other elements. Application of the log-ratio method to different data sets, and relating the results to petrological and geochemical changes in the rocks being studied, will be the best test of its usefulness.

Acknowledgement

We thank John Tarney for invaluable comments on the manuscript.

Fig. 1

Histograms: Exemplary distributions of centered logratios of granulite-facies gneisses. Figures on the x-axis give the upper bound of the respective quantiles.

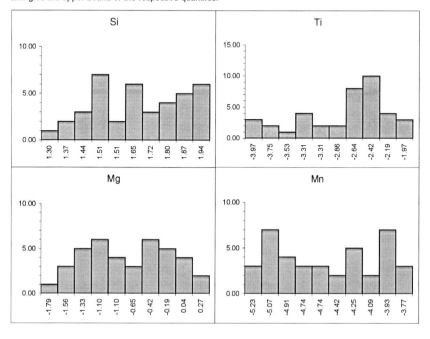

Table 1: correlation coefficients (logratio)
Figures in large print are significant at the 99 % level

granulite facies gneisses (N = 39)

	Ca	Fe(t)	Mn	Mg	Cr	Ni	Ti	Si	Al	Na	Sr	Pb	P	Zr	Ce	Ba	Rb
Ca																	
Fe(t)	0.79																
Mn	0.82	0.95															
Mg	0.89	0.78	0.80														
Cr	0.73	0.57	0.62	0.79													
Ni	0.60	0.46	0.50	0.72	0.82												
Ti	0.11	0.48	0.39	-0.02	-0.19	-0.33											
Si	0.20	-0.11	-0.10	-0.07	-0.04	0.03	-0.17										
Al	0.64	0.34	0.38	0.48	0.33	0.34	-0.19	0.66									
Na	-0.48	-0.51	-0.50	-0.67	-0.57	-0.50	-0.01	0.42	-0.06								
Sr	-0.68	-0.76	-0.79	-0.72	-0.64	-0.50	-0.34	0.10	-0.28	0.59							
Pb	-0.55	-0.40	-0.46	-0.57	-0.62	-0.58	-0.02	0.06	-0.18	0.64	0.49						
P	-0.73	-0.50	-0.56	-0.81	-0.80	-0.76	0.29	-0.24	-0.62	0.45	0.56	0.47					
Zr	-0.79	-0.78	-0.81	-0.85	-0.67	-0.55	-0.13	0.12	-0.47	0.60	0.75	0.43	0.70				
Ce	-0.70	-0.61	-0.61	-0.71	-0.70	-0.74	0.04	-0.29	-0.59	0.44	0.60	0.44	0.75	0.60			
Ba	-0.81	-0.80	-0.83	-0.71	-0.54	-0.44	-0.30	-0.12	-0.51	0.26	0.67	0.32	0.46	0.68	0.50		
Rb	0.06	-0.09	-0.05	0.23	0.18	0.21	-0.41	0.04	0.27	-0.47	-0.27	-0.24	-0.38	-0.35	-0.29	0.09	
K	-0.30	-0.44	-0.37	-0.13	-0.04	0.02	-0.44	-0.09	-0.03	-0.25	0.02	-0.09	-0.14	0.01	-0.02	0.49	0.75

Undeformed retrograde gneisses in amphibolite facies (N = 145)

	Ca	Fe(t)	Mn	Mg	Cr	Ni	Ti	Si	Al	Na	Sr	Pb	P	Zr	Ce	Ba	Rb
Ca																	
Fe(t)	0.52																
Mn	0.53	0.92															
Mg	0.32	0.79	0.77														
Cr	0.31	0.64	0.68	0.66													
Ni	0.25	0.48	0.50	0.58	0.71												
Ti	0.11	0.58	0.47	0.36	0.19	-0.03											
Si	-0.11	-0.65	-0.62	-0.71	-0.62	-0.48	-0.50										
Al	-0.04	-0.62	-0.61	-0.70	-0.69	-0.54	-0.41	0.94									
Na	-0.25	-0.67	-0.68	-0.72	-0.72	-0.57	-0.39	0.83	0.89								
Sr	-0.22	-0.62	-0.66	-0.62	-0.66	-0.54	-0.32	0.51	0.63	0.77							
Pb	-0.10	-0.52	-0.43	-0.48	-0.30	-0.29	-0.35	0.50	0.45	0.36	0.20						
P	-0.04	0.02	-0.04	0.02	-0.24	-0.10	0.16	-0.38	-0.26	-0.09	0.23	-0.35					
Zr	-0.36	-0.38	-0.45	-0.40	-0.49	-0.49	0.08	0.25	0.24	0.38	0.37	0.11	0.09				
Ce	-0.12	-0.18	-0.24	-0.19	-0.33	-0.27	0.03	-0.13	-0.13	0.03	0.23	-0.12	0.47	0.27			
Ba	-0.46	-0.70	-0.69	-0.62	-0.55	-0.58	-0.36	0.56	0.52	0.48	0.47	0.29	-0.16	0.18	-0.06		
Rb	-0.49	-0.57	-0.50	-0.52	-0.38	-0.41	-0.36	0.47	0.39	0.32	0.03	0.44	-0.34	0.04	-0.14	0.56	
K	-0.53	-0.70	-0.65	-0.64	-0.55	-0.52	-0.39	0.58	0.52	0.45	0.29	0.34	-0.23	0.13	-0.11	0.83	0.79

Shear zone gneisses (N = 52)

	Ca	Fe(t)	Mn	Mg	Cr	Ni	Ti	Si	Al	Na	Sr	Pb	P	Zr	Ce	Ba	Rb
Ca																	
Fe(t)	0.77																
Mn	0.76	0.95															
Mg	0.68	0.90	0.90														
Cr	0.39	0.57	0.58	0.67													
Ni	0.55	0.59	0.67	0.65	0.69												
Ti	0.56	0.78	0.65	0.61	0.28	0.30											
Si	-0.47	-0.68	-0.69	-0.72	-0.57	-0.65	-0.45										
Al	-0.36	-0.67	-0.69	-0.76	-0.70	-0.67	-0.38	0.90									
Na	-0.48	-0.77	-0.75	-0.80	-0.68	-0.56	-0.56	0.80	0.89								
Sr	-0.13	-0.56	-0.56	-0.56	-0.69	-0.50	-0.34	0.36	0.59	0.66							
Pb	-0.07	-0.10	-0.12	-0.11	-0.11	-0.30	0.00	0.16	0.16	0.00	0.16						
P	-0.01	-0.28	-0.25	-0.31	-0.53	-0.14	-0.16	-0.12	0.13	0.30	0.65	-0.26					
Zr	-0.55	-0.69	-0.67	-0.63	-0.54	-0.54	-0.43	0.45	0.50	0.61	0.57	0.11	0.41				
Ce	-0.23	-0.31	-0.35	-0.33	-0.28	-0.26	-0.13	-0.06	0.01	0.09	0.29	-0.11	0.58	0.36			
Ba	-0.70	-0.69	-0.72	-0.66	-0.56	-0.64	-0.58	0.58	0.50	0.50	0.29	0.04	-0.01	0.30	0.08		
Rb	-0.75	-0.52	-0.52	-0.53	-0.31	-0.55	-0.40	0.55	0.37	0.32	-0.15	0.09	-0.33	0.19	-0.18	0.66	
K	-0.72	-0.56	-0.57	-0.55	-0.32	-0.56	-0.49	0.57	0.43	0.38	-0.06	-0.03	-0.30	0.11	-0.15	0.80	0.89

Table 2: crude correlation coefficients
Figures in large print are significant at the 99 % level

granulite facies gneisses (N = 39)

	Ca	Fe(t)	Mn	Mg	Cr	Ni	Ti	Si	Al	Na	Sr	Pb	P	Zr	Ce	Ba	Rb
Ca																	
Fe(t)	0.66																
Mn	0.70	0.94															
Mg	0.83	0.57	0.57														
Cr	0.67	0.32	0.39	0.74													
Ni	0.39	0.21	0.21	0.72	0.61												
Ti	0.11	0.61	0.57	-0.06	-0.22	-0.30											
Si	-0.89	-0.80	-0.79	-0.89	-0.60	-0.49	-0.25										
Al	0.03	-0.23	-0.17	0.07	-0.04	0.03	-0.34	-0.13									
Na	-0.78	-0.57	-0.56	-0.84	-0.60	-0.49	-0.02	0.78	-0.06								
Sr	-0.67	-0.54	-0.60	-0.57	-0.51	-0.30	-0.10	0.57	0.03	0.60							
Pb	-0.61	-0.30	-0.38	-0.55	-0.54	-0.43	0.12	0.45	0.05	0.68	0.56						
P	-0.50	-0.25	-0.30	-0.50	-0.47	-0.37	0.28	0.34	-0.04	0.47	0.78	0.53					
Zr	-0.71	-0.58	-0.61	-0.65	-0.48	-0.29	-0.11	0.69	-0.18	0.53	0.71	0.41	0.68				
Ce	-0.46	-0.24	-0.30	-0.42	-0.41	-0.40	0.31	0.30	-0.05	0.51	0.70	0.63	0.85	0.50			
Ba	-0.68	-0.49	-0.55	-0.48	-0.36	-0.21	-0.08	0.55	-0.07	0.44	0.74	0.51	0.57	0.63	0.60		
Rb	-0.12	-0.15	-0.16	0.17	0.14	0.21	-0.22	-0.05	0.26	-0.16	-0.05	0.11	0.02	-0.09	0.10	0.39	
K	-0.36	-0.32	-0.31	-0.01	0.04	0.19	-0.15	0.15	0.17	0.08	0.22	0.27	0.26	0.20	0.34	0.62	0.85

Undeformed retrograde gneisses in amphibolite facies (N = 145)

	Ca	Fe(t)	Mn	Mg	Cr	Ni	Ti	Si	Al	Na	Sr	Pb	P	Zr	Ce	Ba	Rb
Ca																	
Fe(t)	0.82																
Mn	0.82	0.95															
Mg	0.78	0.92	0.93														
Cr	0.59	0.70	0.66	0.79													
Ni	0.35	0.41	0.41	0.48	0.62												
Ti	0.56	0.68	0.63	0.56	0.32	0.17											
Si	-0.86	-0.91	-0.86	-0.85	-0.61	-0.35	-0.74										
Al	-0.26	-0.39	-0.45	-0.46	-0.47	-0.31	0.02	0.07									
Na	-0.57	-0.60	-0.62	-0.64	-0.55	-0.45	-0.27	0.41	0.63								
Sr	-0.12	-0.25	-0.32	-0.30	-0.34	-0.27	0.02	-0.02	0.61	0.56							
Pb	0.04	0.01	0.00	0.05	0.13	0.06	0.08	-0.02	-0.05	-0.12	-0.02						
P	0.25	0.13	0.08	0.09	-0.08	-0.05	0.33	-0.32	0.19	0.10	0.48	0.00					
Zr	-0.13	-0.14	-0.15	-0.14	-0.13	-0.09	0.25	0.06	0.08	0.15	0.20	0.17	0.15				
Ce	0.09	0.05	0.01	0.03	-0.06	-0.08	0.23	-0.13	-0.01	0.14	0.28	-0.01	0.57	0.43			
Ba	-0.38	-0.29	-0.30	-0.28	-0.18	-0.13	-0.01	0.27	0.10	0.06	0.18	0.11	0.04	0.31	-0.01		
Rb	-0.32	-0.19	-0.20	-0.19	-0.04	-0.08	-0.08	0.25	-0.09	-0.13	-0.24	0.40	-0.10	0.01	-0.06	0.27	
K	-0.44	-0.28	-0.29	-0.26	-0.15	-0.13	-0.02	0.27	0.07	-0.04	-0.02	0.28	-0.01	0.24	-0.05	0.62	0.73

Shear zone gneisses (N = 52)

	Ca	Fe(t)	Mn	Mg	Cr	Ni	Ti	Si	Al	Na	Sr	Pb	P	Zr	Ce	Ba	Rb
Ca																	
Fe(t)	0.90																
Mn	0.90	0.97															
Mg	0.89	0.94	0.95														
Cr	0.73	0.71	0.77	0.79													
Ni	0.65	0.64	0.73	0.73	0.88												
Ti	0.78	0.88	0.81	0.79	0.45	0.44											
Si	-0.91	-0.90	-0.88	-0.89	-0.67	-0.65	-0.81										
Al	-0.32	-0.46	-0.51	-0.48	-0.50	-0.40	-0.25	0.11									
Na	-0.58	-0.67	-0.66	-0.67	-0.49	-0.35	-0.56	0.40	0.72								
Sr	-0.08	-0.27	-0.33	-0.26	-0.35	-0.29	-0.11	-0.02	0.69	0.58							
Pb	0.32	0.34	0.32	0.37	0.22	0.16	0.34	-0.34	-0.04	-0.30	0.12						
P	0.09	-0.05	-0.10	-0.02	-0.16	-0.07	0.12	-0.21	0.50	0.36	0.75	0.06					
Zr	-0.25	-0.30	-0.33	-0.25	-0.25	-0.13	-0.15	0.12	0.44	0.46	0.58	0.02	0.50				
Ce	0.20	0.14	0.08	0.14	0.09	0.13	0.28	-0.27	0.18	0.07	0.41	0.24	0.73	0.37			
Ba	-0.52	-0.40	-0.45	-0.35	-0.32	-0.33	-0.36	0.41	0.06	0.13	0.09	-0.09	0.11	0.04	0.10		
Rb	-0.52	-0.32	-0.33	-0.30	-0.20	-0.30	-0.29	0.46	-0.15	-0.11	-0.38	-0.03	-0.28	-0.09	-0.18	0.51	
K	-0.49	-0.31	-0.32	-0.26	-0.17	-0.28	-0.34	0.39	-0.10	-0.04	-0.33	-0.16	-0.22	-0.16	-0.14	0.70	0.88

Table 3: differences of Fisher-transforms
Figures in large print are significant at the 99 % level

granulite facies gneisses (N = 39)

	Ca	Fe(t)	Mn	Mg	Cr	Ni	Ti	Si	Al	Na	Sr	Pb	P	Zr	Ce	Ba	Rb
Ca																	
Fe(t)	0.49																
Mn	0.58	0.27															
Mg	1.07	-0.02	0.09														
Cr	0.60	-0.12	-0.10	0.28													
Ni	0.44	-0.03	0.00	0.24	0.26												
Ti	0.00	-0.15	-0.10	-0.40	-0.38	-0.31											
Si	0.32	0.67	0.63	0.81	0.68	0.55	0.37										
Al	0.81	1.07	1.11	1.39	1.19	0.95	0.25	-0.95									
Na	-0.27	0.25	0.29	0.10	0.26	0.09	0.41	-0.75	-1.50								
Sr	-0.61	-0.28	-0.29	-0.18	0.02	0.05	-0.02	-0.46	-1.03	-0.34							
Pb	-0.52	0.15	-0.05	-0.12	-0.42	-0.36	0.35	-0.49	-0.67	0.39	0.34						
P	-0.90	-0.58	-0.59	-1.16	-0.86	-0.89	0.14	0.16	-0.47	0.57	0.40	0.87					
Zr	-0.70	-0.64	-0.65	-0.82	-0.28	-0.09	-0.21	-0.13	-0.75	0.29	0.58	0.35	0.78				
Ce	-0.75	-0.53	-0.46	-0.70	-0.52	-0.66	0.02	-0.17	-0.54	0.44	0.45	0.59	0.47	0.42			
Ba	-0.64	-0.23	-0.35	-0.15	0.01	0.19	0.07	-0.75	-1.15	-0.26	0.30	0.03	0.67	0.65	0.61		
Rb	0.60	0.55	0.49	0.82	0.59	0.65	-0.06	-0.47	-0.13	-0.84	-0.31	-0.72	-0.05	-0.40	-0.16	-0.54	
K	0.28	0.41	0.39	0.62	0.57	0.60	-0.06	-0.75	-0.62	-0.74	-0.28	-0.44	0.09	-0.12	0.08	-0.64	-0.09

Undeformed retrograde gneisses in amphibolite facies (N = 145)

	Ca	Fe(t)	Mn	Mg	Cr	Ni	Ti	Si	Al	Na	Sr	Pb	P	Zr	Ce	Ba	Rb
Ca																	
Fe(t)	0.04																
Mn	0.18	0.07															
Mg	0.58	-0.41	-0.36														
Cr	0.51	0.00	0.06	0.27													
Ni	0.07	-0.18	-0.26	0.13	0.30												
Ti	-0.52	-0.52	-0.37	-0.74	-0.48	-0.65											
Si	0.71	0.72	0.74	0.83	0.60	0.81	0.31										
Al	1.15	1.16	1.25	1.53	1.22	1.15	0.21	-0.66									
Na	0.00	0.47	0.43	0.29	0.18	0.08	0.62	-0.66	-1.49								
Sr	-0.70	-0.38	-0.45	-0.27	0.08	-0.01	0.00	-0.27	-0.97	-0.12							
Pb	-0.55	-0.32	-0.38	-0.53	-0.61	-0.35	-0.02	-0.11	-0.35	0.76	0.38						
P	-0.92	-0.27	-0.38	-0.81	-0.51	-0.84	0.46	-0.13	-0.86	0.17	-0.13	0.77					
Zr	-0.45	-0.21	-0.32	-0.50	-0.22	-0.03	0.33	-0.36	-1.06	-0.03	0.33	0.35	0.44				
Ce	-0.64	-0.40	-0.34	-0.55	-0.58	-0.68	0.17	-0.24	-0.69	0.38	0.39	0.58	0.32	0.32			
Ba	-0.27	-0.24	-0.28	-0.09	0.02	0.30	0.36	-0.78	-1.11	-0.28	0.51	0.29	0.51	0.52	0.48		
Rb	1.04	0.49	0.52	0.83	0.50	0.82	-0.01	-0.58	-0.11	-0.85	-0.13	-0.34	-0.06	-0.55	-0.12	-0.69	
K	0.59	0.16	0.27	0.49	0.29	0.65	0.07	-0.73	-0.50	-0.66	0.08	-0.07	0.16	-0.10	0.13	-0.55	-0.45

Shear zone gneisses (N = 52)

	Ca	Fe(t)	Mn	Mg	Cr	Ni	Ti	Si	Al	Na	Sr	Pb	P	Zr	Ce	Ba	Rb
Ca																	
Fe(t)	0.44																
Mn	0.40	0.20															
Mg	0.49	0.39	0.45														
Cr	0.09	-0.12	-0.17	0.01													
Ni	0.36	0.16	0.26	0.11	-0.04												
Ti	0.52	0.38	0.27	0.34	0.10	0.34											
Si	-0.39	-0.06	-0.11	-0.03	0.08	-0.25	0.06										
Al	-0.34	-0.09	-0.15	-0.14	-0.02	-0.20	0.04	-0.28									
Na	-0.27	-0.22	-0.14	-0.19	0.08	0.01	-0.21	-0.09	0.00								
Sr	0.09	0.10	0.16	0.09	-0.06	0.06	-0.03	-0.19	-0.06	-0.23							
Pb	0.02	0.48	0.34	0.41	0.19	-0.02	0.37	-0.39	-0.33	-0.37	-0.04						
P	0.03	-0.31	-0.21	-0.34	-0.35	-0.04	-0.33	0.29	0.39	0.40	0.54	0.10					
Zr	-0.25	-0.43	-0.33	-0.32	-0.06	-0.07	-0.54	0.23	0.30	0.31	0.26	0.00	0.34				
Ce	-0.11	-0.13	-0.12	-0.15	0.06	0.02	-0.15	0.07	0.15	0.05	0.05	0.06	0.00	0.16	0.11		
Ba	-0.37	0.02	-0.07	-0.06	-0.01	-0.10	-0.29	0.03	-0.04	0.03	-0.21	-0.26	0.15	0.12	0.13		
Rb	-0.44	0.07	-0.03	-0.01	0.09	-0.17	-0.05	0.11	-0.02	0.00	-0.18	-0.38	0.01	0.15	-0.05	0.15	
K	-0.31	0.25	0.12	0.13	0.28	-0.05	-0.13	-0.02	-0.12	-0.08	-0.36	-0.38	-0.07	-0.02	-0.04	-0.09	0.37

Appendix 4: source code of statistics computer program

The computer program for the log-ratio calculations was originally written in Turbopascal. What follows is a translation of this code into VBA. Due to the structure of VBA, the program is coupled to an EXCEL workbook. This makes data transfer easy. To do some log-ratio calculations, the EXCEL-workbook containing the following code (=makro) has to be loaded, as well as a workbook containing the data to be processed. Then the makro "LogRatio" is started (using the menus or Alt + F8). A dialog window appears where some entries can be made, viz. whether the data matrix has specimens as rows or as columns, which calculations shall be done, and parameters for some of the calculations. It is assumed that the data are in one continuous block on the active worksheet and that element/oxide and specimen designations are directly adjacent to this block.

Both program versions are available from the author.

```
' declaration of global variables:
Option Explicit
Public gWhichWay As Integer
Dim locUpperLeft(1 To 2) As Integer '1:Row, 2:column
Dim locLowerRight(1 To 2) As Integer '1:Row, 2:column
Public gArrayOfValues As Variant
Public gArrayOfValues2 As Variant
Public gListOfElements As Variant
Public gListOfSpecimens As Variant
Dim gNumberOfColumns As Integer
Dim gNumberOfRows As Integer
Dim gNumberOfElements As Integer
Dim gNumberOfSpecimens As Integer
Dim i As Integer, j As Integer
Dim d As Variant, c As Variant, v As Variant
Dim anz1 As Variant
Dim ar As Variant
Dim ew As Variant, gox As Variant, kk As Variant
Dim ordn As Variant
Dim it As Integer, p As Integer, k As Integer, l As Integer
Dim eps As Double, ma As Double, mao As Double, h As Double, sum As Double, z As Double
Dim ab1 As Boolean, ab2 As Boolean
Dim gStr As Variant, locName As String * 7
Dim gRange As Range
Dim gIterations As Integer
Dim gExactitude As Double
Dim cluster1 As Boolean
Dim cluster2 As Boolean
Dim RelFact As Integer
Dim locBvarimax As Boolean
Dim Scat As Boolean
Dim Pca As Boolean

Public Sub LogRatio()
```

```
Dim locString As String
Dim locPosition As Integer
Dim locPosition2 As Integer
Dim locBeginning As String
Dim locStartingColumn As Integer
Dim locStrExactitude As String

gWhichWay = 1
With UserForm1
    .Chk1.Value = True
    .ChkCluster1.Value = False
    .ChkCluster2.Value = False
    .ChkVarimax.Value = False
    .ChkPCA.Value = True
    .ChkScat.Value = True
    .txtExactitude = 0.00001
    .txtIterations = 20
    .txtRelF = 3
    .RefEdit1.SetFocus
    .Show

    If .Tag = -1 Then
        Exit Sub
    Else
        gIterations = .txtIterations
        locStrExactitude = .txtExactitude
        gExactitude = Val(locStrExactitude)
        RelFact = .txtRelF
        cluster1 = .ChkCluster1
        cluster2 = .ChkCluster2
        Scat = .ChkScat
        Pca = .ChkPCA
        locBvarimax = .ChkVarimax
        locString = .RefEdit1
        If locString = vbNullString Then
            MsgBox "mark data array!"
            Exit Sub
        End If
        locPosition = InStr(1, locString, "!$", vbBinaryCompare)
        locPosition2 = InStr(locPosition + 2, locString, "$", vbBinaryCompare)
        If locPosition2 - locPosition = 3 Then
            locBeginning = Mid(locString, locPosition + 2, 1)
            locUpperLeft(2) = Asc(locBeginning) - 64
        Else
            locBeginning = Mid(locString, locPosition + 2, 2)
            locUpperLeft(2) = Asc(Left(locBeginning, 1)) - 64
            locUpperLeft(2) = locUpperLeft (2) * 26 + Asc(Right(locBeginning, 1)) - 64
        End If
        locUpperLeft(1) = Val(Mid(locString, locPosition2 + 1, 5))
        locPosition = InStr(locPosition2, locString, ":", vbBinaryCompare)
        locPosition2 = InStr(locPosition2, locString, ";", vbBinaryCompare)
        If locPosition2 = 0 Then
        If locPosition = 0 Then
            MsgBox "lower right hand corner not marked! try again"
            Exit Sub
        Else
            locPosition = InStr(locPosition, locString, "$", vbBinaryCompare)
            locPosition2 = InStr(locPosition + 2, locString, "$", vbBinaryCompare)
            If locPosition2 - locPosition = 2 Then
```

```
                            locBeginning = Mid(locString, locPosition + 1, 1)
                            locLowerRight(2) = Asc(locBeginning) - 64
                    Else
                            locBeginning = Mid(locString, locPosition + 1, 2)
                            locLowerRight(2) = Asc(Left(locBeginning, 1)) - 64
                            locLowerRight(2) = locLowerRight(2) * 26 + Asc(Right(locBeginning, 1)) - 64
                    End If
                    locLowerRight(1) = Val(Mid(locString, locPosition2 + 1, 5))
                    End If
            Else
                    locPosition = InStr(locPosition2, locString, "!$", vbBinaryCompare)
                    locPosition2 = InStr(locPosition + 2, locString, "$", vbBinaryCompare)
                    If locPosition2 - locPosition = 3 Then
                            locBeginning = Mid(locString, locPosition + 2, 1)
                            locLowerRight(2) = Asc(locBeginning) - 64
                    Else
                            locBeginning = Mid(locString, locPosition + 2, 2)
                            locLowerRight(2) = Asc(Left(locBeginning, 1)) - 64
                            locLowerRight(2) = locLowerRight(2) * 26 + Asc(Right(locBeginning, 1)) - 64
                    End If
                    locLowerRight(1) = Val(Mid(locString, locPosition2 + 1, 5))
            End If
        End If
    End With

    gNumberOfRows = locLowerRight(1) - locUpperLeft(1) + 1
    gNumberOfColumns = locLowerRight(2) - locUpperLeft(2) + 1
    ReDim gArrayOfValues(1 To gNumberOfColumns, 1 To gNumberOfRows) As String
    gArrayOfValues = Range(Cells(locUpperLeft(1), locUpperLeft(2)), Cells(locLowerRight(1), locLowerRight(2))).Value

    Select Case gWhichWay
    Case 1
        ReDim gArrayOfValues2(1 To gNumberOfColumns, 1 To gNumberOfColumns) As Double

        Call ZeroReplacement

        gArrayOfValues2 = gArrayOfValues
        ReDim gListOfSpecimens(1 To gNumberOfColumns) As String
        ReDim gListOfElements(1 To gNumberOfRows) As String
        gNumberOfElements = gNumberOfRows
        gNumberOfSpecimens = gNumberOfColumns
        For i = 1 To gNumberOfColumns
            gListOfSpecimens(i) = Cells(locUpperLeft(1) - 1, locUpperLeft(2) - 1 + i)
        Next i
        For i = 1 To gNumberOfRows
            gListOfElements(i) = Cells(locUpperLeft(1) - 1 + i, locUpperLeft(2) - 1)
        Next i
    Case 2
        ReDim gArrayOfValues2(1 To gNumberOfColumns, 1 To gNumberOfRows) As Double

        Call ZeroReplacement

        For i = 1 To gNumberOfColumns
            For j = 1 To gNumberOfRows
                gArrayOfValues2(i, j) = gArrayOfValues(j, i)
            Next j
        Next i
        ReDim gListOfSpecimens(1 To gNumberOfRows) As String
        ReDim gListOfElements(1 To gNumberOfColumns) As String
```

```
        gNumberOfElements = gNumberOfColumns
        gNumberOfSpecimens = gNumberOfRows
        For i = 1 To gNumberOfRows
            gListOfSpecimens(i) = Cells(locUpperLeft(1) - 1 + i, locUpperLeft(2) - 1)
        Next i
        For i = 1 To gNumberOfColumns
            gListOfElements(i) = Cells(locUpperLeft(1) - 1, locUpperLeft(2) - 1 + i)
        Next i
End Select

Call CalculateLogRatios

End Sub
```

```
Private Sub ZeroReplacement()
'Replaces empty or not numeric (e.g. "b.d.") cells with zeros
If gWhichWay = 1 Then
    For i = 1 To gNumberOfRows
        For j = 1 To gNumberOfColumns
            If gArrayOfValues(i, j) = vbNullString Then
                gArrayOfValues(i, j) = 0
            End If
            If Not IsNumeric(gArrayOfValues(i, j)) Then
                gArrayOfValues(i, j) = 0
            End If
        Next j
    Next i
Else
    For i = 1 To gNumberOfColumns
        For j = 1 To gNumberOfRows
            If gArrayOfValues(i, j) = vbNullString Then
                gArrayOfValues(i, j) = 0
            End If
            If Not IsNumeric(gArrayOfValues(i, j)) Then
                gArrayOfValues(i, j) = 0
            End If
        Next j
    Next i
End If
End Sub
```

```
Private Function min(ByVal a As Integer, ByVal b As Integer) As Integer
If a < b Then min = a Else min = b
End Function
```

```
Function max(ByVal a As Integer, ByVal b As Integer) As Integer
 If a > b Then max = a Else max = b
End Function
```

```
Private Sub varimax(n As Integer, s As Integer, a As Variant)
'This follows Marcus (1993) and Harmann
Dim i As Integer, j As Integer, k As Integer, l As Integer, o As Integer, p As Integer, q As Integer
Dim m As Integer, r As Integer, it As Integer
Dim psi As Variant
Dim h As Variant, aa As Variant
```

```
Dim bb As Variant, cc As Variant
Dim dd As Variant
Dim v As Double, v0 As Double, a1 As Double, b As Double, d As Double
Dim c1 As Double, phi As Double, num As Double, den As Double
Dim locName As String * 7
Const pi = 3.14159265358
Err.Clear
On Error Resume Next
ReDim Preserve a(1 To gNumberOfElements, 1 To gNumberOfElements) As Double
ReDim ar(1 To gNumberOfElements) As Double
ReDim psi(1 To gNumberOfElements, 1 To gNumberOfSpecimens) As Double
ReDim h(1 To gNumberOfElements) As Double
ReDim aa(1 To gNumberOfElements) As Double
ReDim bb(1 To gNumberOfElements) As Double
ReDim cc(1 To gNumberOfElements) As Double
ReDim dd(1 To gNumberOfElements) As Double
v0 = 0
For i = 1 To gNumberOfElements
    For j = 1 To gNumberOfSpecimens
        psi(i, j) = 0
        Next j
Next i
o = min(n - 1, min(s - 1, 10))
m = RelFact
r = 0
For i = 1 To n
    h(i) = 0
    For j = 1 To n
        h(i) = h(i) + (a(i, j)) * (a(i, j))
    Next j
    h(i) = Sqr(h(i))
    If (i > m) And (h(i) = 0) Then
        MsgBox ("error as communality(" & i & ") = 0")
        GoTo Label_7
    End If
Next i
For i = 1 To n
    For j = 1 To m
        a(i, j) = a(i, j) / h(i) '(*normalizing by communalities*)
    Next j
Next i
ar = a
Do
    r = r + 1
    For i = 1 To m - 1
        For j = i + 1 To m

            For k = 1 To n
                bb(k) = (a(k, i) * (a(k, i)) - (a(k, j)) * (a(k, j)))
                cc(k) = (2 * a(k, i) * a(k, j))
                aa(k) = 2 * bb(k) * cc(k)
                dd(k) = (bb(k)) * (bb(k)) - (cc(k)) * (cc(k))
            Next k
            a1 = 0
            b = 0
            c1 = 0
            d = 0
            For k = 1 To n
                a1 = a1 + aa(k)
```

```
                b = b + bb(k)
                c1 = c1 + cc(k)
                d = d + dd(k)
            Next k
            num = (n * a1 - 2 * b * c1)
            den = n * d - (b * b - c1 * c1)
            phi = Atn(num / den) / 4 'arctan
            If num > 0 Then p = 1 Else p = -1
            If den > 0 Then q = 1 Else q = -1
            If (q = -1) Then
                If (p = -1) Then
                        phi = -phi - pi / 8
                Else
                        phi = phi + pi / 8
                End If
            End If
            psi(i, j) = phi
            For k = 1 To n
                If Abs(phi) > 0.0001 Then
                        ar(k, i) = a(k, i) * Cos(phi) + a(k, j) * Sin(phi)
                        ar(k, j) = -a(k, i) * Sin(phi) + a(k, j) * Cos(phi)
                End If
            Next k
            a = ar
        Next j
    Next i
    l = 1
    a1 = 0
    b = 0
    For j = 1 To m
        bb(j) = 0
        For i = 1 To n
            a1 = a1 + (a(i, j)) * (a(i, j)) * (a(i, j)) * (a(i, j))
            bb(j) = (a(i, j)) * (a(i, j)) + bb(j)
        Next i
        b = b + bb(j) * bb(j)
    Next j
    v = n * a1 - b
    c1 = v - v0
    v0 = v
    If c1 > 0 Then a = ar Else c1 = 0.0001
Loop Until (c1 < 0.0002) Or (r = 100)
Label_6:
'MsgBox ("it: " & r)

For i = 1 To n
    For j = 1 To m
        a(i, j) = a(i, j) * h(i)
    Next j
Next i

ReDim gStr(1 To gNumberOfElements) As String

On Error Resume Next
Err.Clear
ActiveWorkbook.Sheets("Varimax-rotation").Select
If Err.Number = 0 Then
    If MsgBox("the sheet 'Varimax-rotation' already exists. Overwrite?", vbYesNo) = vbYes Then
        Cells.Select
```

```
        Selection.ClearContents
        Range("A1").Select
     End If
Else
     ActiveWorkbook.Sheets.Add after:=ActiveSheet
     ActiveSheet.name = "Varimax-rotation"
End If

Cells(1, 1) = "varimax rotation applied to " & m & " factors, iterations: " & r
For i = 1 To n
     Cells(2, i) = gStr(i)
     For j = 1 To m
          Cells(i + 3, j + 1) = a(i, j)
     Next j
Next i
For j = 1 To n
     Cells(j + 3, 1) = gListOfElements(j)
Next j

For i = 1 To m
     h(i) = 0
     For j = 1 To n
          h(i) = h(i) + (a(j, i)) * (a(j, i))
     Next j
Next i
a1 = 0
For j = 1 To m
     a1 = a1 + h(j)
Next j
Cells(n + 4, 1) = "sum/sq "
For j = 1 To m
     h(j) = h(j) * 100 / a1
     Cells(n + 5, j + 1) = h(j)
Next j
Label_7:
End Sub
```

```
Private Sub jakob(n As Integer, w As Variant, v As Variant, ab1 As Boolean, ab2 As Boolean, _
ma As Double, mao As Double, it As Integer)
'Jakobi-procedure for extraction of eigenvalues and eigenvectors*)
Dim p As Integer, q As Integer, i As Integer, j As Integer
Dim sign As Integer, rot As Integer, it As Integer
Dim schr As Double, d As Double, t As Double, c As Double
Dim u As Double, r As Double, eps As Double
ReDim Preserve w(1 To gNumberOfElements, 1 To gNumberOfElements) As Double
ReDim Preserve v(1 To gNumberOfElements, 1 To gNumberOfElements) As Double
eps = 0.00001
eps = gExactitude
'(*outer norm*)
ma = 0
For p = 1 To n
     For q = 1 To n
     If p <> q Then
          ma = ma + Abs(w(p, q)) * Abs(w(p, q))
     End If
     Next q
Next p
mao = ma
```

```
'(*starting values*)
For p = 1 To n
    For q = 1 To n
        v(p, q) = 0
    Next q
Next p
For p = 1 To n
    v(p, p) = 1
Next p
rot = 1
j = 0
'(*iteration*)
Do
    If rot > 0 Then schr = Sqr(ma) / n Else schr = Sqr(ma) / (n * n)
    rot = 0
    j = j + 1
    '(*seek matrix elements > schr*)
    For p = 1 To n - 1
        For q = p + 1 To n
            '(*rotation*)
            If Abs(w(p, q)) > schr Then
                ma = ma - 2 * (w(p, q)) * (w(p, q))
                d = (w(q, q) - w(p, p)) / (2 * w(p, q))
                If d < 0 Then sign = -1 Else sign = 1
                t = sign / (Abs(d) + Sqr(1 + d * d))
                c = 1 / Sqr(1 + t * t)
                u = t * c
                r = u / (1 + c)
                For i = 1 To n
                    If (i <> q) And (i <> p) Then
                        w(i, p) = w(i, p) - u * (w(i, q) + r * w(i, p))
                        w(p, i) = w(i, p)
                        w(i, q) = w(i, q) + u * ((w(i, p) + u * w(i, q)) / (1 - r * u) - r * w(i, q))
                        w(q, i) = w(i, q)
                    End If
                Next i
                w(q, q) = w(q, q) + t * w(p, q)
                w(p, p) = w(p, p) - t * w(p, q)
                w(p, q) = 0
                w(q, p) = 0
                For i = 1 To n
                    v(i, p) = v(i, p) - u * (v(i, q) + r * v(i, p))
                    v(i, q) = v(i, q) + u * ((v(i, p) + u * v(i, q)) / (1 - r * u) - r * v(i, q))
                Next i '(*rotation*)
            End If
        Next q
        rot = 1
    Next p
    ab1 = (j = it)
    ab2 = (ma < mao * eps * eps)
Loop Until ab1 Or ab2
End Sub
```

```
Private Sub cluster(b As Variant, n As Integer, it As Integer, locName As String)
'(*Does cluster analysis; output not very sophisticated *)
Dim m As Integer, k As Integer, p As Integer, q As Integer
Dim i As Integer, j As Integer, l As Integer
Dim a1 As Double
```

```
Dim a As Variant, c As Variant
Dim m1 As Variant, m2 As Variant
Dim mt1 As Variant, mt2 As Variant
Dim hilf As Variant, e3 As Variant
Dim e4 As Variant
Dim mt3 As Variant, e1 As Variant
Dim e2 As Variant
Dim baum As Variant
ReDim Preserve b(1 To gNumberOfElements, 1 To gNumberOfElements) As Double
ReDim a(0 To gNumberOfElements, 1 To gNumberOfElements) As Integer
ReDim c(0 To locNumberOfElements + 1) As Integer
ReDim m1(0 To gNumberOfElements) As Integer
ReDim m2(0 To gNumberOfElements) As Integer
ReDim mt1(0 To gNumberOfElements) As Integer
ReDim mt2(0 To gNumberOfElements) As Integer
ReDim hilf(0 To gNumberOfElements + 1) As Integer
ReDim e3(0 To gNumberOfElements + 1) As Integer
ReDim e4(0 To gNumberOfElements + 1) As Integer
ReDim mt3(0 To gNumberOfElements) As Double
ReDim e1(0 To gNumberOfElements + 1) As Double
ReDim e2(0 To gNumberOfElements + 1) As Double
ReDim baum(1 To gNumberOfElements, 1 To gNumberOfElements) As String
For i = 0 To n + 1
    hilf(i) = i
    e1(i) = 0
    e3(i) = 0
Next i
e2 = e1
e4 = e3
For l = 1 To n
    '(*seeking smallest*)
    If it = 1 Then
        a1 = 1000
        For j = 1 To n
            For i = j + 1 To n
                If ((a1 >= b(i, j)) And (hilf(i) <> hilf(j))) Then
                    a1 = b(i, j)
                    p = j
                    q = i
                End If
            Next i
        Next j
    End If
    If it = 2 Then '(*seeking largest*)
        a1 = -1
        For j = 1 To n
            For i = j + 1 To n
                If ((a1 <= b(i, j)) And (hilf(i) <> hilf(j))) Then
                    a1 = b(i, j)
                    p = j
                    q = i
                End If
            Next i
        Next j
    End If
    mt3(l) = a1
    mt1(l) = hilf(p)
    m1(l) = p
    mt2(l) = hilf(q)
```

```
        m2(l) = q
        k = min(hilf(q), hilf(p))
        For i = 1 To n
            If ((hilf(i) = mt2(l)) Or (hilf(i) = mt1(l))) Then hilf(i) = k
        Next i
    Next l


    On Error Resume Next
    Err.Clear
    ActiveWorkbook.Sheets("cluster (" & locName & ") neighbors").Select
    If Err.Number = 0 Then
        If MsgBox("the sheet 'cluster (" & locName & ") neighbors' already exists. Overwrite?", vbYesNo) = vbYes Then
            Cells.Select
            Selection.ClearContents
            Range("A1").Select
        End If
    Else
        ActiveWorkbook.Sheets.Add after:=ActiveSheet
        ActiveSheet.name = "cluster (" & locName & ") neighbors"
    End If
    For i = 1 To locNumberOfElements - 1
        Cells(i, 1) = gListOfElements(m1(i))
        Cells(i, 2) = gListOfElements(m2(i))
        Cells(i, 3) = mt3(i)
    Next i
    For i = 1 To locNumberOfElements - 1
        Cells(i, 1) = gListOfElements(m1(i))
        Cells(i, 2) = gListOfElements(m2(i))
        Cells(i, 3) = mt3(i)
    Next i

    '(*ordering for the tree*)
    c(1) = min(mt1(n - 1), mt2(n - 1))
    c(2) = max(mt2(n - 1), mt1(n - 1))
    For i = 2 To n - 1
        k = 0
        Do
            k = k + 1
        Loop Until ((mt1(n - i) = c(k)) Or (mt2(n - i) = c(k)) Or (k > n))
        For j = i + 2 To k + 2 Step -1
            c(j) = c(j - 1)
        Next j
        c(k + 1) = max(mt1(n - i), mt2(n - i))
        c(k) = min(mt1(n - i), mt2(n - i))
    Next i
    For l = 1 To n - 1
        i = 0
        Do
            i = i + 1
        Loop Until (c(i) = min(mt1(l), mt2(l))) Or (i = n + 1)
        j = i - 1
        Do
            j = j + 1
        Loop Until (c(j) = max(mt1(l), mt2(l))) Or (j = n + 1)
        e3(l) = i
        e4(l) = j
    Next l
    For q = 1 To n
        For i = 1 To n
```

```
            baum(i, q) = " "
        Next i
    Next q
    For i = 1 To n - 1
        e1(i) = e3(i)
        e2(i) = e4(i)
    Next i
    '(*midpoints for where the strokes start*)
    For i = 1 To n - 1
        a1 = e2(i) - e1(i)
        For j = i + 1 To n - 1
            If e1(j) = e1(i) Then e1(j) = e1(i) + a1 / 2
            If e2(j) = e1(i) Then e2(j) = e1(i) + a1 / 2
        Next j
    Next i
    '(*"basic lines"*)
    For i = 1 To n
        k = 0
        Do
            k = k + 1
        Loop Until ((e3(k) = i) Or (e4(k) = i) Or (k = n + 1))
        For p = 1 To min(k, n)
            baum(i, p) = "-"
        Next p
    Next i
    '(*"follow-up lines"*)
    i = 1
    m = 2
    Do
        i = i + 1
        If i > locNumberOfElements Then
            GoTo Label_8
        End If
        k = i
        q = i
        j = i
        l = i
        If i > n - 1 Then GoTo Label_8
        Do
            k = k - 1
        Loop Until ((e3(k) = e4(i)) Or (k <= 0))
        If Abs(Fix((e2(k)) + e1(k))) / 2 >= 0.5 Then
            For p = k + 1 To l
                baum(Fix((e2(k) + e1(k)) / 2), p) = "_"
            Next p
        Else
            For p = k + 1 To l
                baum(Fix((e2(k) + e1(k)) / 2), p) = "-"
            Next p
        End If
        Do
            q = q - 1
        Loop Until ((e4(l) = e3(q)) Or (q <= 0))
        If (q = 0) Or (e4(l) > e3(l) + 1) Then
            Do
                j = j - 1
            Loop Until ((e3(j) = e3(l)) Or (j <= 0))
        End If
        If Abs(Fix(e1(l)) - e1(l)) / 2 >= 0.25 Then
```

```
            For p = j + 1 To l
                    baum(Fix(e1(l)), p) = "_"
            Next p
        Else
            For p = j + 1 To l
                    baum(Fix(e1(l)), p) = "-"
            Next p
        End If
 Loop Until i = n - 2


 i = n - 1
 k = i
 q = i
 j = i
 l = i
 Do
        q = q - 1
 Loop Until ((e4(l) = e3(q)) Or (q <= 0))
 If (q < 0) Or (q > gNumberOfElements) Then GoTo Label_7
 If (q = 0) Or (e4(l) > e3(l) + 1) Then
        Do
                j = j - 1
        Loop Until ((e3(j) = e3(l)) Or (j <= 0))
        If (j < 0) Or (j > gNumberOfElements) Then GoTo Label_7
 End If
 If Abs(Fix(e1(l)) - e1(l)) / 2 >= 0.25 Then
        For p = j + 1 To l
                baum(Fix(e1(l)), p) = "_"
                If p > l Then
                        GoTo Label_8
                End If
        Next p
 Else
        For p = j + 1 To l
                baum(Fix(e1(l)), p) = "-"
                If p > l Then
                        GoTo Label_8
                End If
        Next p
 End If
 Label_7:
 i = n - 1
 k = i
 q = i
 j = i
 l = i
 Do
        k = k - 1
 Loop Until ((e3(k) = e4(i)) Or (k <= 0))
 If k <= 0 Then GoTo Label_8
 If Abs(Fix(e2(k)) + e1(k)) / 2 >= 0.5 Then
        For p = k + 1 To l
                baum(Fix((e2(k) + e1(k)) / 2), p) = "_"
        Next p
 Else
        For p = k + 1 To l
                baum(Fix((e2(k) + e1(k)) / 2), p) = "-"
        Next p
 End If
```

220

```
Label_8: '(*top level line*)
If Abs(Fix(e1(n - 1) + e2(n - 1))) >= 0.5 Then
    baum(Fix((e1(n - 1) + e2(n - 1)) / 2), n) = "-"
Else
    baum(Fix((e1(n - 1) + e2(n - 1)) / 2), n) = "_"
End If


On Error Resume Next
Err.Clear
ActiveWorkbook.Sheets("Baum (" & locName & ")").Select
If Err.Number = 0 Then
    If MsgBox("the sheet 'Baum (" & locName & ")' already exists. Overwrite?", vbYesNo) = vbYes Then
        Cells.Select
        Selection.ClearContents
        Range("A1").Select
    End If
Else
    ActiveWorkbook.Sheets.Add after:=ActiveSheet
    ActiveSheet.name = "Baum (" & locName & ")"
End If


For i = 1 To gNumberOfElements
    For j = 1 To gNumberOfElements
        Cells(i, 1) = gListOfElements(c(i))
        Cells(i, 2) = "-"
        Cells(i, j + 2) = baum(i, j)
    Next j
Next i
Cells.Select
Selection.Columns.AutoFit

End Sub
```

```
Private Sub CalculateLogRatios()
Dim locZeroCounter As Integer
Dim locScaling As Integer
ReDim anz1(1 To gNumberOfElements, 1 To gNumberOfElements) As Integer
ReDim ar(1 To gNumberOfElements, 1 To gNumberOfSpecimens) As Double
ReDim d(1 To gNumberOfElements, 1 To gNumberOfElements) As Double
ReDim c(1 To gNumberOfElements, 1 To gNumberOfElements) As Double
ReDim v(1 To gNumberOfElements, 1 To gNumberOfElements) As Double
ReDim gox(1 To gNumberOfElements) As Double
ReDim ew(1 To gNumberOfElements) As Double
ReDim kk(1 To gNumberOfSpecimens) As Double

ar = gArrayOfValues2

'(*variance matrix*)
For i = 1 To gNumberOfElements
    For j = 1 To gNumberOfElements
        anz1(i, j) = 0
    Next j
Next i
For k = 1 To gNumberOfElements
    For i = 1 To gNumberOfSpecimens
        For j = 1 To gNumberOfElements
            If ((gArrayOfValues2(k, i) > 0) And (gArrayOfValues2(j, i) > 0)) Then
```

```
                    anz1(k, j) = anz1(k, j) + 1
                End If
            Next j
        Next i
Next k
locZeroCounter = 5
For i = 1 To gNumberOfElements
    For j = 1 To gNumberOfSpecimens
        If gArrayOfValues2(i, j) = 0 Then
            If j > 24 Then
                Cells(gNumberOfElements + locZeroCounter, 1) = "Value = 0 at " & Chr((j + 1) \ 26 + 64) & _
                Chr(((j + 1) Mod 26) + 64) & i + 1
            Else
                Cells(gNumberOfElements + locZeroCounter, 1) = "Value = 0 at " & Chr(j + 1 + 64) & i + 1
            End If
            locZeroCounter = locZeroCounter + 1
        End If
        If gArrayOfValues2(i, j) > 0 Then
            gArrayOfValues2(i, j) = Log(gArrayOfValues2(i, j))
        Else
            gArrayOfValues2(i, j) = 0
        End If
    Next j
Next i


For i = 1 To gNumberOfElements
    For j = 1 To gNumberOfElements
        For k = 1 To gNumberOfSpecimens
            If ((ar(i, k) > 0) And (ar(j, k) > 0)) Then
                c(i, j) = c(i, j) + gArrayOfValues2(i, k) - gArrayOfValues2(j, k)
            End If
        Next k
        If anz1(i, j) > 0 Then
            c(i, j) = c(i, j) / anz1(i, j)
        Else
            MsgBox ("division by 0, probably data matrix contains too many zeros")
            GoTo Label_3
        End If
    Next j
Next i
For i = 1 To gNumberOfElements
    For j = 1 To gNumberOfElements
        For k = 1 To gNumberOfSpecimens
            If ((ar(i, k) > 0) And (ar(j, k) > 0)) Then
                d(i, j) = d(i, j) + (-c(i, j) + gArrayOfValues2(i, k) - gArrayOfValues2(j, k)) * _
                (-c(i, j) + gArrayOfValues2(i, k) - gArrayOfValues2(j, k))
            End If
        Next k
        If anz1(i, j) - 1 > 0 Then
            d(i, j) = d(i, j) / (anz1(i, j) - 1)
        Else
            MsgBox ("division by zero, data matrix probably contains too many zeros")
            GoTo Label_3
        End If
    Next j
Next i

On Error Resume Next
Err.Clear
```

```
ActiveWorkbook.Sheets("variance matrix").Select
If Err.Number = 0 Then
     If MsgBox("the sheet 'variance matrix' already exists. Overwrite?", vbYesNo) = vbYes Then
          Cells.Select
          Selection.ClearContents
          Range("A1").Select
     End If
Else
     ActiveWorkbook.Sheets.Add after:=ActiveSheet
     ActiveSheet.name = "variance matrix"
End If
For i = 1 To gNumberOfElements
     Cells(1 + i, 1) = gListOfElements(i)
     Cells(1, 1 + i) = gListOfElements(i)
     For j = 1 To gNumberOfElements
          Cells(i + 1, j + 1) = d(i, j)
     Next j
Next i

If cluster1 Then
     Call cluster(d, gNumberOfElements, 1, "Var")
End If

'(*covariance matrix*)
ReDim gox(1 To gNumberOfElements) As Double
For i = 1 To gNumberOfElements
     gox(i) = 0
     For k = 1 To gNumberOfElements
          gox(i) = gox(i) + d(k, i)
     Next k
Next i
z = 0
For i = 1 To gNumberOfElements
     gox(i) = gox(i) / gNumberOfElements
     z = z + gox(i)
Next i
z = z / gNumberOfElements

For i = 1 To gNumberOfElements
     For j = 1 To gNumberOfElements
          v(i, j) = (gox(i) + gox(j) - d(i, j) - z) / 2
     Next j
Next i

On Error Resume Next
Err.Clear
ActiveWorkbook.Sheets("covariance matrix").Select
If Err.Number = 0 Then
     If MsgBox("the sheet 'covariance matrix' already exists. Overwrite?", vbYesNo) = vbYes Then
          Cells.Select
          Selection.ClearContents
          Range("A1").Select
     End If
Else
     ActiveWorkbook.Sheets.Add after:=ActiveSheet
     ActiveSheet.name = "covariance matrix"
End If
For i = 1 To gNumberOfElements
     Cells(1 + i, 1) = gListOfElements(i)
```

```
        Cells(1, 1 + i) = gListOfElements(i)
        For j = 1 To gNumberOfElements
            Cells(i + 1, j + 1) = v(i, j)
        Next j
    Next i

    c = v

    '(*correlation coefficients*)
    For j = 1 To gNumberOfElements
        For i = j + 1 To gNumberOfElements
            If ((c(i, i) <= 0) Or (c(j, j) <= 0)) Then
                MsgBox ("cannot do sqrt; probably too many zeros contained in data matrix")
                GoTo Label_11
            Else
                c(i, j) = v(i, j) / (Sqr(v(i, i)) * Sqr(v(j, j)))
                c(j, j) = 1
            End If
        Next i
    Next j
    For i = 1 To gNumberOfElements
        For j = 1 To gNumberOfElements
            If j > i Then c(i, j) = c(j, i)
        Next j
    Next i

    On Error Resume Next
    Err.Clear
    If Err.Number = 0 Then
        If MsgBox("the sheet 'correlation coefficients' already exists. Overwrite?", vbYesNo) = vbYes Then
            Cells.Select
            Selection.ClearContents
            Range("A1").Select
        End If
    Else
        ActiveWorkbook.Sheets.Add after:=ActiveSheet
        ActiveSheet.name = "correlation coefficients"
    End If

    For i = 1 To gNumberOfElements
        Cells(1 + i, 1) = gListOfElements(i)
        Cells(1, 1 + i) = gListOfElements(i)
        For j = 1 To gNumberOfElements
            Cells(i + 1, j + 1) = c(i, j)
        Next j
    Next i

    If cluster2 Then
        Call cluster(c, gNumberOfElements, 2, "Corr.")
    End If
    c = v
Label_11:
    If Pca Then
        For i = 1 To gNumberOfElements
            For j = 1 To gNumberOfElements
                v(i, j) = 0
            Next j
        Next I
        it = 30
```

```
it = gIterations
Call jakob(gNumberOfElements, c, v, ab1, ab2, ma, mao, it)
sum = 0
For i = 1 To gNumberOfElements
    ew(i) = c(i, i)
    sum = sum + ew(i)
Next i
If sum = 0 Then
    MsgBox ("sum of eigenvalues =0!, program aborted")
    GoTo Label_3
End If
ReDim ordn(1 To gNumberOfElements) As Integer
ordn(1) = 1
For i = 2 To gNumberOfElements
    ordn(i) = ordn(i - 1) + 1
Next i
For i = 1 To gNumberOfElements
    For j = i + 1 To gNumberOfElements
        If ew(i) < ew(j) Then
            h = ew(i)
            ew(i) = ew(j)
            ew(j) = h
            k = ordn(j)
            ordn(j) = ordn(i)
            ordn(i) = k
        End If
    Next j
Next i
For i = 1 To min(gNumberOfSpecimens, gNumberOfElements)
    ew(i) = ew(i) / sum
Next i
'(*there are locNumberOfElementseigenvectors
'min(0,gNumberOfElements - gNumberOfSpecimens) are degenerate*)
'(*displaying the 7 largest eigenvalues and eigenvectors

On Error Resume Next
    Err.Clear
    ActiveWorkbook.Sheets("pca").Select
    If Err.Number = 0 Then
        If MsgBox("the sheet 'pca' already exists. Overwrite?", vbYesNo) = vbYes Then
            Cells.Select
            Selection.ClearContents
            Range("A1").Select
        End If
    Else
        ActiveWorkbook.Sheets.Add after:=ActiveSheet
        ActiveSheet.name = "pca"
    End If

    If ab1 Then Cells(1, 1) = "reason for halting the algorithm: upper limit of iterations"
    If ab2 Then Cells(1, 1) = "reason for halting the algorithm: specified exactitude attained ; " & it & " iterations"
    it = min(gNumberOfSpecimens, gNumberOfElements)
    For i = 1 To gNumberOfElements
        Cells(8 + i, 1) = gListOfElements(i)
    Next i
    For i = 1 To it
        Cells(2, 1 + i) = "F" & i
    Next i
    Cells(3, 1) = "the " & it & " largest eigenvalues:"
```

```
    For i = 1 To it
        Cells(4, i + 1) = ew(i) * sum
    Next i
    Cells(5, 1) = "per cent of total variance:"
    For i = 1 To it
        Cells(6, i + 1) = ew(i) * 100
    Next i
    Cells(8, 1) = "corresponding eigenvectors:"
    For i = 1 To gNumberOfElements
        For k = 1 To it
            Cells(i + 8, k + 1) = v(i, ordn(k))
        Next k
    Next i
End If

ar = gArrayOfValues2
If Scat Then
    For i = 1 To gNumberOfSpecimens
        kk(i) = 0
        For j = 1 To gNumberOfElements
            gArrayOfValues2(j, i) = 0
        Next j
    Next i
    For i = 1 To gNumberOfElements
        For k = 1 To gNumberOfSpecimens
            kk(k) = kk(k) + ar(i, k)
        Next k
    Next i
    For k = 1 To gNumberOfSpecimens
        kk(k) = kk(k) / gNumberOfElements
    Next k
    For i = 1 To gNumberOfElements
        For j = 1 To gNumberOfSpecimens
            ar(i, j) = ar(i, j) - kk(j)
        Next j
    Next i

    For k = 1 To it
        For i = 1 To gNumberOfSpecimens
            For j = 1 To gNumberOfElements
                gArrayOfValues2(k, i) = gArrayOfValues2(k, i) + v(j, ordn(k)) * ar(j, i)
            Next j
        Next i
    Next k

    For i = 1 To gNumberOfElements
        gox(i) = 0 '(*centering*)
    Next i
    For j = 1 To gNumberOfSpecimens
        For i = 1 To it
            gox(i) = gox(i) + gArrayOfValues2(i, j)
        Next i
    Next j
    For i = 1 To it
        gox(i) = gox(i) / gNumberOfSpecimens
    Next i
    For i = 1 To it
        For j = 1 To gNumberOfSpecimens
            gArrayOfValues2(i, j) = gArrayOfValues2(i, j) - gox(i)
```

```
        Next j
        Next i

        On Error Resume Next
        Err.Clear
        ActiveWorkbook.Sheets("scattergram coordinates").Select
        If Err.Number = 0 Then
            If MsgBox("the sheet 'scattergram coordinates' already exists. Overwrite?", vbYesNo) = vbYes Then
                Cells.Select
                Selection.ClearContents
                Range("A1").Select
            End If
        Else
            ActiveWorkbook.Sheets.Add after:=ActiveSheet
            ActiveSheet.name = "scattergram coordinates"
        End If

        Cells(1, 1) = "coordinats of data points in the first " & it & " factors"
        For i = 1 To it
            Cells(2, 1 + i) = "F" & i
        Next i
        For k = 1 To gNumberOfSpecimens
            Cells(k + 2, 1) = gListOfSpecimens(k)
            For i = 1 To it
                Cells(k + 2, i + 1) = gArrayOfValues2(i, k)
            Next i
            locScaling = max(locScaling, Fix(Abs(gArrayOfValues2(1, k)) + 0.9999))
        Next k
        On Error Resume Next
        Err.Clear
        locName = Chr(1 + 65) & "3:" & Chr(2 + 65) & gNumberOfSpecimens + 2
        Call subScatterDiagram(locName, locScaling, 1, 2)
        locName = Chr(2 + 65) & "3:" & Chr(3 + 65) & gNumberOfSpecimens + 2
        Call subScatterDiagram(locName, locScaling, 2, 3)
        locName = Chr(1 + 65) & "3:" & Chr(3 + 65) & gNumberOfSpecimens + 2
        Call subScatterDiagram(locName, locScaling, 1, 3)
End If

If locBvarimax = True Then
    For i = 1 To gNumberOfElements
        For j = 1 To min(gNumberOfElements, gNumberOfSpecimens)
            d(i, j) = v(i, ordn(j))
        Next j
    Next i
    For i = 1 To gNumberOfElements
        If ew(i) < 0 Then
            ew(i) = 0
            If i > min(gNumberOfElements, gNumberOfSpecimens) Then MsgBox ("error: eigenvalue " & i & " < 0")
        End If
    Next i

    Call varimax(gNumberOfElements, gNumberOfSpecimens, d)
End If
Label_3:
End Sub
```

```
Private Sub subScatterDiagram(Bereich As String, maxSkalierung As Integer, _
locColumn1 As Integer, locColumn2 As Integer)
'for creating nice plots
Charts.Add
ActiveChart.ChartType = xlXYScatter
ActiveChart.SetSourceData Source:=Sheets("scattergram coordinates").Range( _
Bereich), PlotBy:=xlColumns
ActiveChart.SeriesCollection(1).Values = _
"='scattergram coordinates'!R3C" & locColumn2 + 1 & ":R" & gNumberOfSpecimens + 2 & "C" & locColumn2 + 1
ActiveChart.Location Where:=xlLocationAsObject, Name:= _
"scattergram coordinates"
With ActiveChart
     .HasTitle = True
     .ChartTitle.Characters.Text = "F" & locColumn1 & "-F" & locColumn2
     .Axes(xlCategory, xlPrimary).HasTitle = True
          .Axes(xlCategory, xlPrimary).AxisTitle.Characters.Text = "F" & locColumn1
     .Axes(xlValue, xlPrimary).HasTitle = True
     .Axes(xlValue, xlPrimary).AxisTitle.Characters.Text = "F" & locColumn2
End With
ActiveChart.HasLegend = False
With ActiveChart.Axes(xlValue)
     .MinimumScale = -maxSkalierung
     .MaximumScale = maxSkalierung
End With
With ActiveChart.Axes(xlCategory)
     .MinimumScale = -maxSkalierung
     .MaximumScale = maxSkalierung
End With
ActiveChart.Legend.Delete
ActiveChart.Axes(xlValue).MajorGridlines.Delete
ActiveChart.Axes(xlValue).Select
With Selection.Border
     .Weight = xlHairline
     .LineStyle = xlAutomatic
End With
With Selection
     .MajorTickMark = xlNone
     .MinorTickMark = xlNone
End With
ActiveChart.Axes(xlCategory).Select
With Selection.Border
     .Weight = xlHairline
     .LineStyle = xlAutomatic
End With
With Selection
     .MajorTickMark = xlNone
     .MinorTickMark = xlNone
End With
ActiveChart.Axes(xlCategory).Select
With Selection.Border
     .Weight = xlHairline
     .LineStyle = xlAutomatic
End With
With Selection
.MajorTickMark = xlNone
     .MinorTickMark = xlNone
     .TickLabelPosition = xlNone
End With
ActiveChart.Axes(xlValue).Select
```

```
With Selection.Border
    .Weight = xlHairline
    .LineStyle = xlAutomatic
End With
With Selection
    .MajorTickMark = xlNone
    .MinorTickMark = xlNone
    .TickLabelPosition = xlNone
End With
End Sub
```

The next part belongs to the UserForm:

```
Option Explicit


Private Sub cmdExit_Click()
Me.Tag = -1
Me.Hide
End Sub


Private Sub cmdOK_Click()
Me.Tag = 0
Me.Hide
End Sub


Private Sub chk1_Click()
gWhichWay= Me.chk1.Tag
Me.RefEdit1.SetFocus
End Sub


Private Sub opt2_Click()
gWhichWay= Me.opt2.Tag
Me.RefEdit1.SetFocus
End Sub


Private Sub chkCluster1_Click()
Me.RefEdit1.SetFocus
End Sub


Private Sub chkCluster2_Click()
Me.RefEdit1.SetFocus
End Sub


Private Sub chkPCA_Click()
If Me.chkPCA = True Then
    Me.txtExactitude.Visible = True
    Me.txtIterations.Visible = True
    Me.chkScat.Visible = True
    Me.Label4.Visible = True
    Me.Label5.Visible = True
    Me.chkScat = True
Else
```

UserForm1

mark your data array (without headings) and press OK

(•) Specimens = columns, elements = rows

( ) the other way round

Tabelle1!$B$3:$E$15

☐ cluster analysis by log-ratios

☐ cluster analysis by log-ratio correlation coefficients

☑ pca   20   max. iterations

0.00001   exactitude

☑ scattergram

☐ varimax rotation   3   number of relevant factors (max. 10)

&OK   &Exit

```
        Me.txtExactitude.Visible = False
        Me.txtIterations.Visible = False
        Me.chkScat.Visible = False
        Me.chkScat = False
        Me.Label4.Visible = False
        Me.Label5.Visible = False
End If
Me.RefEdit1.SetFocus
End Sub
```

```
Private Sub chkScat_Click()
Me.RefEdit1.SetFocus
End Sub
```

```
Private Sub chkVarimax_Click()
Me.RefEdit1.SetFocus
End Sub
```

```
Private Sub txtRelF_Change()
Me.RefEdit1.SetFocus
End Sub
```

References

Aitchison J (1984) The statistical analysis of geochemical compositions. Mathematical Geology 16, 531-564

Aitchison J (1986) The statistical analysis of compositional data. New York (Methuen)

Aitchison J (1997) The one-hour course in compositional data analysis or compositional data analysis is simple. In: Pawlowsky-Glahn V (ed), IAMG'97 Proceedings of the 3rd annual conference of the international association for mathematical geology, Barcelona (CIMNE), 2 vols, p. 3-35

Amato JM, Johnson CM, Baumgartner LP, Beard BC (1999) Rapid exhumation of the Zermatt-Saas ophiolite deduced from high-precision Sm-Nd and Rb-Sr geochronology. Earth Planet Sci Lett 171, 425-438

Argand (1911) Les grands plis couchés des Alpes pennines. Mat Carte géol Suisse NS 31

Ballèvre M, Merle O (1993) The Combin fault: compressional reactivation of a Late Cretaceous-Early Tertiary detatchement fault in the Western Alps. Schweiz mineral petrogr Mitt 73 205-227

Barnicoat AC (1988) Zoned high-pressure assemblages in pillow lavas of the Zermatt-Saas ophiolite zone, Switzerland. Lithos 21, 227-236

Barnicoat AC, Bowtell SA (1995) Sea-floor hydrothermal alteration in metabasites from the high-pressure ophiolites of the Zermatt-Aosta area of the Western Alps. Bolletino. Museo regionale di Scienze Naturali di Torino 13 (Suppl), 191-220

Barnicoat AC, Bowtell SA, Cliff RA (1991) Timing of high-pressure metamorphism in the Zermatt-Saas zone, Switzerland. Terra Abstracts 3, 89. Ditto in: Abstr Geol Soc Austr 27 (1990), 13

Barnicoat AC, Cartwright I (1995) Focussed fluid flow during subduction: oxygen isotope data from high-pressure ophiolites of the Western Alps. Earth Planet Sci Lett 132, 53-61

Barnicoat AC, Cartwright I (1997) The gabbro-eclogite transformation: an oxygen isotope and petrographic study of the West Alpine ophiolites. J Metam Geol 15, 93-104

Barnicoat AC, Fry N (1986) High-pressure metamorphism of the Zermatt-Saas ophiolite zone, Switzerland. J Geol Soc London 143, 607-618

Basaltic volcanism study project (1981) Basaltic volcanism on terrestrial planets. New York; 1286 pp

Baumgartner PO (1987) Age and genesis of Thetyan Jurassic radiolarites. Eclogae Geol Helv 80, 831-879

Beach A (1980) Retrogressive metamorphic processes in shear zones with special reference to the Lewisian complex. Journal of Structural Geology 2, 257–263

Beach A, Tarney J (1978) Major and trace element patterns established during retrogressive metamorphism of granulite-facies gneisses, NW Scotland. Precambrian research 7, 325-348

Bearth P (1953) Geologischer Atlas der Schweiz 1:25000, Blatt 29 Zermatt

Bearth P (1954a) Geologischer Atlas der Schweiz 1:25000, Blatt 30 Monte Moro

Bearth P (1954b) Geologischer Atlas der Schweiz 1:25000, Blatt 31 Saas

Bearth P (1963) Geologischer Atlas der Schweiz 1:25000, Blatt 43 Randa

Bearth P (1959) Über Eklogite, Glaukophanschiefer und metamorphe Pillowlaven. Schweiz mineral petrogr Mitt 39, 267-286

Bearth P (1952) Geologie und Petrographie des Monte Rosa. Beiträge geol Karte Schweiz NF, 96. 94 pp, xiv plates

Bearth P (1967) Die Ophiolite der Zone von Zermatt-Saas Fee. Beiträge geol Karte Schweiz NF, 132

Bearth P (1973) Gesteins- und Mineralparagenesen aus den Ophioliten von Zermatt. Schweiz mineral petrogr Mitt 53, 299-334

Bearth P, Schwander (1981) The post-triassic sediments of the ophiolite zone of Zermaatt-Saas Fee and the associated manganese mineralisations. Eclogae geol Helv 74 189-205

Bearth P, Stern W (1971) Zum Chemismus der Eklogite und Glaukophanite von Zermatt. Schweiz mineral petrogr Mitt 51, 349-359

Beattie P (1994) Systematics and energetics of trace-element partitioning between olivine and silicate melts: Implications for the nature of mineral/melt partitioning. Chem Geol 117, 57-71

Beccaluva L, DalPiaz GV, Macciotta G (1984) Transitional to normal MORB affinities in ophiolitic metabasites from Zermatt-Saas, Combin and Antrona units, Western Alps: Implications for the paleogeographic evolution of the western Thetyan basin. Geol Mijnbouw 63, 165-177

Bédard JH (1994) A procedure for calculating the equilibrium distribution of trace elements among the minerals of cumulate rocks, and the concentration of trace elements in the coexisting liquids. Chem Geol 118, 143-153

Bédard JH (2001) Parental magmas of the Nain plutonic suite anorthosites and mafic cumulates: a trace element modelling approach. Contrib Mineral Petrol 141, 747-771

Bowtell SA, Cliff RA, Barnicoat AC (1994) Sm-Nd isotopic evidence on the age of eclogitisation in the Zermatt-Saas ophiolite. J Metam Geol 12, 187-196

Boyd FR (1989) Compositional distinction between oceanic and cratonic lithosphere. Eart Planet Sci Letters 96, 15-26

Boynton WV (1984) Geochemistry of the Rare Earth elements: meteorite studies. In Rare Earth element geochemistry (ed Henderson P), pp 62-114, Elsevier

Bucher K & Frey M (1994) Petrogenesis of metamorphic rocks. Berlin etc (Springer), 318 pp.

Burkhard DJM & O'Neil JR (1988) Contrasting serpentinization processes in the eastern Central Alps. Contrib Mineral Petrol 99, 498-506

Burton JD, Culkin F (1972) Ga. Abundance in rock-forming minerals. In Wedepohl KH (ed) (1969ff) Handbook of Geochemistry. Berlin Heidelberg New York (Springer)

BVSP see Basaltiv Volcanism Study Project

Casey JF (1997) Comparison of major- and trace element geochemistry of abyssal peridotites and mafic plutonic rocks with basalts from the MARK region of the Mid-Atlantic Ridge. Proceedings Ocean Drilling Program, Scientific Results 153, 181-241

Castello P (1981) Inventario delle mineralizazzioni a magnetite, ferro-rame e manganese del complesso piemontese dei calcescisti con pietre verde in valle d'Aosta. Ofioliti 6, 5-46

Charlou JL, Donval JP, Douville E, Jean-Baptiste P, Radford-Knoery J, Y. Fouquet Y, Dapoigny A, Stievenard M (2000) Compared geochemical signatures and the evolution of Menez Gwen (37 50´N)

and Lucky Strike (37 17′N) hydrothermal fluids, south of the Azores Triple Junction on the Mid-Atlantic Ridge. Chem Geol 171, 49-75

Coleman RG (1971) Petrologic and geophysical nature of serpentinites. Geol Soc Am Bull 82, 897-918

Coleman RG (1977) Ophiolites. Berlin ets (Springer)

Coleman RG, Keith TE (1971) A chemical study of serpentinization – Burro Mountain, California. J Petrol 12, 311-328

Correns CW (1978) Ti. Abundance in rock-forming minerals. in Wedepohl KH (ed) (1969ff) Handbook of geochemistry. Berlin Heidelberg New York (Springer)

Cullars RL & Medaris LG (1973) Experimental studies of the distribution of Rare Earts as trace elements among silicate minerals and liquids and water. Geochim Cosmochim Acta 37, 1499-1512

DalPiaz GV & Ernst WG (1978) Areal geology and petrology of eclogites and associated metabasites of the Piemont ophiolitic nappe, Breuil - St. Jaques area, Italian Western Alps. Tectonophysics 51, 99-126

Dercourt J, Zonenshain LP, Ricou LE, Kazmin VE, Le Pichon X, Knipper AL, Grandjacquet C, Sbortshikov LM, Geyssant J, Lepvrier C, Pechersky DH, Boulin J, Sibuet J-C, Savostin LA, Sorokhtin O, Westphal M, Bazhenov MI, Lauer JP, Biju-Duval B (1986) Geological evolution of the Thetys Belt from the Atlantic to the Pamirs since the Lias. Tectonophysics 123, 241-315

Dunn T, Sen C (1994) Mineral/matrix partition coefficients for orthopyroxene, plagioclase, and olivine in basaltic to andesitic systems: A combined analytical and experimental study. Geochim Cosmochim Acta 58, 717-733

Escher J, Masson H, Steck A (1993) Nappe geometry in the Western Swiss Alps. J Struct Geol 15, 501-509

Flynn RT & Burnham CW (1978) An experimental determination of Rare Earth partition coefficients between a chloride containig vapour phase and silicate melts. Geochim Cosmochim Acta 42, 685-701

Frey FA (1969) rare earth abundances in a high-temperature peridotite intrusion. Geochim Cosmochim Acta 33, 1429-1447

Frondel C (1970) Sc. Abundance in rock-forming minerals. In Wedepohl KH (ed) (1969ff) Handbook of Geochemistry. Berlin Heidelberg New York (Springer)

Fry N, Barnicoat AC (1987) The tectonic implications of high-pressure metamorphism in the western Alps J Geol Soc London 144, 653-659

Fujimaki H, Tatsumoto M (1984) Partition coefficients of Hf, Zr and REE between phenocrysts and groundmass. Proc 14th Lunar Planet Sci Conf, Part 2, J Geophys Res 89 (suppl), B662-B672

Ganguin J (1988) Contribution a la charactérisation du métamorphisme polyphasé de la zone de Zermatt-Saas Fee. PhD thesis No 8731, ETH Zürich

Gariépy C, Ludden JN, Brooks CK (1983) Isotopic and trace element constraints on the genesis of the Faeroe lava pile. Earth Planet Sci Lett 63, 257-272

Gealey WK (1988) Plate tectonic evolution of the Mediterranean - Middle East region. Tectonophysics 155, 285-306

GEOROC Database (without year) http://georoc.mpch-mainz.gwdg.de/Start.asp)

GERM (without year) The Geochemical Earth Reference Model. A grassroots initiative. http://earthref.org/GERM/main.htm

Gernuks M (1999) Geochemische Analyse der Ophiolite der Tsaté Decke westlich von Zermatt (Schweiz): Petrogenese und geotektonische Position. Unpubl. Dipl. Thesis, University of Götttingen

Gieskes JM (1978) Sr. Abundance in natural waters. In Wedepohl KH (ed) (1969ff) Handbook of Geochemistry. Berlin Heidelberg New York (Springer)

Gopel C, Allègre CJ, Xu R-H (1984) Lead isotopic study of the Xigaze ophiolite (Tibet): the problem of the relationship between magmatites (gabbros, dolerites, lavas) and tectonites (harzburgites. Earth Planet Sci Lett 69, 301-310

Grant JA (1986) The isocon diagram – a simple solution to Gresens' equation for metasomatic alteration. Economic Geology 81, 1976-1982

Green TH (1981) experimental evidence for the role of accessory phases in magma genesis. J volcanol Res 10, 405-422

Gresens RL (1967) Composition volume relations of metasomatism. Chem. Geol 2, 47-65

Gruau G, Bernard-Griffiths J. Lecuyer C, Henin O, Macè J, Cannat M (1995) Extreme Nd isotopic variation of the Trinity Ophiolite Complex and the role of melt/rock reactions in the oceanic lithosphere. Contrib Mineral Petrol 121, 337-350

Hajash A Jr (1975) Hydrothermal processes along mid-ocean ridges: an experimental investigation. Contrib Miner Petrol 53, 205-226

Hajash A Jr (1984) Rare Earth Element abundances and distribution patterns in hydrothermally altered basalts: experimental results. Contrib Miner Petrol 85, 409-412

Hall A (1990) Geochemistry of spilites from South-West England: a statistical approach. Mineralogy and Petrology 41, 185-197

Hart SR, Brooks C (1974) Clinopyroxene-matrix partitioning of K, Rb, Cs, Sr and Ba. Geochim Cosmochim Acta 38, 1799-1806

Hart SR, Dunn T (1993) Experimental cpx/melt partitioning of 24 trace elements. Contrib Mineral Petrol 113, 1-8.

Heinrichs H, Herrmann AG (1990) Praktikum der analytischen Geochemie. Berlin etc (Springer), 669pp

Himmelheber W (1996) Geochemische Untersuchungen an Metamafiten und -ultramafiten aus der Zone von Zermatt-Saas Fee, Schweiz. Unpublished Diplomarbeit, University of Göttingen, 136 pp

Himmelheber W, Sheraton JW (in prep) A statistical method for studying element mobility revised

Holm PM, Hald N, Nielsen TFD (1992) Contrasts in composition and evolution of Tertiary CFBs between West and East Greenland and their relations to the establishment of the Icelandic mantle plume. In: Storey BC, Alabaster T, Pankhurst RJ (eds), Magmatism and the cause of continental break-up. Gelogical Society Special Publication 68, 349-362

Horn I, Foley SF, Jackson SE and Jenner GA (1994) Experimentally determined partitioning of high field strength- and selected transition elements between spinel and basaltic melt. Chem Geol 117, 193-218.

Humphris SE, Thompson G (1978) Hydrothermal alteration of oceanic basalts by seawater. Geochim Cosmochim Acta 42, 107-125

Humphris SE, Thompson G, Schilling J-G, Kingsley RH (1984) Petrological and geochemical variations along the Mid-Atlantic Ridge between 46 °S and 32 °S: Influence of the Tristan da Cunha mantle plume. Geochim Cosmochim Acta 49, 1445-1464

Hunziker JC (1974) Rb-Sr and K-Ar age determination and the Alpine tectonic history of the Western Alps. Memorie degli Istituti di Geologia e Mineralogia dell' Universitá di Padova 31, 1-55

Irber W, Bau M (1995) Fractionation of Zr/Hf, Y/Ho and the REE (tetrad effect) and its significance as a geochemical tool for the characterisation of granite evolution. Berichte der Dt Mineralog Ges 1, 112f

Irvine TN & Baragar WRA (1971) A guide to the chemical classificatin of the common volcanic rocks. Canad J Earth Sci 8, 523-548

Jagoutz E, Palme H, Baddenhausen H, Blum K, Cendales M, Dreibus G, Spettel B, Lorenz V, Wanke H (1979) The abundance of major, minor and trace elements in the earth's mantle as derived from primitive ultramafic nodules. Proc 10th Lunar Planet Sci Conf, vol 2, 2031-2050

Janetzky DR, Seyfried WE Jr (1986) Hydrothermal serpentinization of peridotite within the oceanic crust: experimental investigations of mineralogy and major element chemistry. Geochim Cosmochim Acta 50, 1357-1378

Johnson KTM, Dick HJB, Shimizu N (1990) Melting in the oceanic upper mantle: an ion microprobe study of diopsides in abyssal peridotites. J Geopys Res 95, 2551-2678

Kennedy AK, Lofgren GE, Wasserburg GJ (1993) An experimental study of trace element partitioning between olivine, orthopyroxene and melt in chondrules: equilibrium balues and kinetic effects. Earth Planet Sci Lett 115, 177-195

Kerr AC (1995) The geochemistry of the Mull-Morvern Tertiary lava succession, NW Scotland: an assessment of mantle sources during plume related volcanism. Chem Geol 122, 43-58

Kerr AC, Kent RW, Thomson BA, Seedhous JK, Donaldson CH (1999) Geochemical evolution of the Tertiary Mull volcano, Western Scotland. J Petrol 40, 873-908

Kinzler RJ (1997) Melting of mantle peridotite at pressures approaching the spinel to garnet transition: Application to mid-ocean ridge basalt petrogenesis. J Geophys Res 102, 853-874

Komor SC, Elthon D, Casey JF (1985a) Serpentinization of cumulate ultramafic rocks from the North American mountain massif of the Bay of Islands ophiolite. Geochim Cosmochim Acta 49, 2331-2338

Komor SC, Elthon D, Casey JF (1985b) Mineralogic variation in a layered utlramafic cumulate sequence at the North Arm Mountain Massif, Bay of islands ophiolite, New Foundland. J Geophys Res 90B, 7705-7736

Kretz R (1983) Symbols for rock-forming minerals. Am Mineralogist 68, 277-279

Landergren S (1974) V. Abundance in rock-forming minerals. In Wedepohl KH (ed) (1969ff) Handbook of Geochemistry. Berlin Heidelberg New York (Springer)

Lemoine M, Tricart P, Baillot G (1987) Ultramafic and gabbroic ocean floor of Ligurian Tethys (Alps, Corsica, Appennines) in search of a genetic model. Geology 15, 622-625

Maaløe S, Aoki K (1977) The major element composition of upper mantle estimated from the composition of lherzoliltes. Contrib. Miner. Petrol. 63, 161-173

Mahood GA, Hildreth EW (1983) Large partition coefficients for trace elements in high-silica rhyolites. Geochim Cosmochim Acta 47, 11-30.

Marcus LF (1993) Program appendix to Reyment & Jöreskog (1996)

Masuda A, Ikeuchi Y (1979) Lanthanide tetrad effects observed in marine environment. Geochem J 13, 19-22

McDonough WF (1990) Constraints on the composition of the continental lithospheric mantle. Earth Planet Sci Lett 101, 1-18

McDonough WF, Sun SS (1995) The composition of the earth. Chem Geol 120, 223-253

McGetchin TR, Silver LT, Chodos AA (1970) Titanclinohumite: a possible mineralogical site for water in the upper mantle. J Geophys Res 75, 255-259

McKenzie D, O'Nions RK (1991) Paritial melt distributions from inversion of rare earth element concentrations. J Petrol 32, 1021-1091

McLennan SM (1994) Rare earth element geochemistry and the "tetrad" effect. Geochim Cosmochim Acta 58, 2025-2035

Meyer J (1983a) Mineralogie und Petrologie des Allalingabbros. Unpubl doctoral thesis. Univ Basel, 329 pp

Meyer J (1983b) Mineralogie und Petrologie des Allalingabbros Terra Cognita 3, 187 (Abstract of the above)

Milnes AG (1974) Structure of the Pennine zone (Central Alps) a new working hypothesis. Bull Soc Geol Am 85, 1727-1732

Milnes AG, Greller M, Müller R (1981) Sequence and style of major post-nappe structures, Simplon-Pennine Alps. J Struct Geol 3, 411-421

Miyashiro (1973) The Troodos ophiolitic complex was probably formed in an island arc. Earth Planet Sci Lett 19, 21-224

Monié P (1985) La méthode ^{39}Ar-^{40}Ar appliqué au métamorphisme alpin dans la massif du Mont-Rose (Alpes Occidentales). Chronologie de haillée depuis 110 Ma. Eclogae geol Helv 78, 487-516

Mottl MJ, Holland HD (1978) Chemical exchange during hydrothermal alteration of basalts by seawater. I. Experimental results for major and minor componenets of seawater. Geochim Cosmochim Acta 42, 1103-1105

Muan A (1955) Phase equilibria in the system FeO–Fe_2O_3–SiO_2. Trans AIME 203, 965-976

Mueller C (1989) Albitization in the Zermatt area, Westen Alps. PhD-thesis Basel, 227 pp

Mysen BO (1983) Rare Earth Element partitioning between (H_2O + CO_2) vapor and upper mantle minerals: experimental data bearing on the conditions of formation of alkali basalt and kimberlite. Neues Jb Miner Abh 146, 41-65

Nicolas, A (1969) Serpentinisation d'une lherzolite: bilan chimique, implication tectonique. Bulletin Volcanologique 32, 499-508

Nielsen RL, Gallahan WE, Newberger F (1992) Experimentally determined mineral-melt partition coefficients for Sc, Y and REE for olivine, orthopyroxene, pigeonite, magnetite and ilmenite. Contrib Mineral Petrol 110, 488-499

Oberhänsli R (1980) P-T Bestimmungen anhand von Mineralparagenesen in Eklogiten und Glaukokphaniten der Ophiolite von Zermatt. Schweiz mineral petrogr Mitt 60, 215-235

Oberhänsli R (1982) The P-T history of some pillow lavas from Zermatt. Ofioliti 7, 431-436

Okay AI (1993) Sapphirine and Ti-clinohumite in ultra-high-pressure garnet-pyroxenite and eclogite from Dabie Shan, China. Contrib Mineral Petrol 116, 145-155

Ottonello G, Piccardo GB, Ernst WG (1979) Petrogenesis of some Ligurian peridotites - II. Rare earth element chemistry. Geochim Cosmochim Acta 43, 1273-1284

Pearce JA (1980) Geochemical evidence for the genesis and eruptive setting of lavas from Thetyan ophiolites. In Panayiotou A (ed) Ophiolites. Proc Int ophiolite Symp Cyprus 1979. Nicosia 1980

Pearce JA (1983) Role of the sub-continental lithosphere in magma genesis at sctive continental margins. In Hawkesworth CJ & Norry MJ (eds) Continental basalts and mantle xenoliths. Nantwich, 230-249

Pearce JA, Norry MJ (1979) Petrogenetic implications of Ti, Zr, Y and Nb variations in volcanic rocks. Contrib Mineral Petrol 69, 33-47

Pearce JA, ParkinsonIJ (1993) Trace element models for mantle melting: application to volcanic arc petrogenesis. In: Prichard HM, Alabaster T, Harris NBW, Neary CR (eds) Magmatic processes and plate tectonics. Geol Soc Lond Spec Publ 76, 373-403

Pearson K (1896) Mathematical contributions to the theory of evolution. On a form of spurious correlation which may arise when indices are used in the measurement of organs. Proc R Soc 60, 489-498

Pfeifer H-R, Colombi A, Ganguin J (1989) Zermatt-Saas and Antrona zone: a petrographic and geochemical comparison of polyphase metamorphic ophiolites of the West-Central Alps Schweiz mineral petrogr Mitt 69, 217-236

Pfiffner A (1992) Alpine orogeny. In: Blundell D, Freeman R, Mueller S, A continent revealed. Teh European geotraverse. Cambridge, 275 pp

Philpotts AR (1990) Principles of igneous and metamorphic petrology. Englewood Cliffs (Prentice Hall), 498 pp

Philpotts JA, Schnetzler CC (1970) Phenocryst-marix partition coefficients for K, Rb, Sr, and Ba, with applications to anorthosite and basalt genesis. Geochim Cosmochim Acta 34, 307-322

Piccardo GB, Rampone E, Vannucci R (1992) Ligurian peridotites and ophiolites: from rift to ocean formation in the Jurassic Ligurian-Piemonte basin. Acta Vulcanologica. Marinelli Volume. 2, 313-325

Platt JP (1986) Dynamics of orogenic wedges and the uplift of high-pressure metamorphic rocks. Geol Soc Am Bulletin 97, 1037-1053

Prinzhofer A, Allègre CJ (1985) esidual peridotites and the mechanisms of partial melting. Earth Planet Sci Lett 74, 251-265

Rahn M, Bucher K (1998) Titanian clinohumite formation in the Zermatt-Saas ophiolites, Central Alps. Mineralogy and Petrology 64, 1-4

Rampone E, Hofmann AW, Raczek I (1998) Isotopic contrasts within the Internal Liguride ophiolite (N. Italy): the lack of a genetic mantle-crust link. Eart Planet Sci Lett 163, 175-189

Rampone E, Hofmann AW, Piccardo GB, Vanucci R, Bottazzi P, Ottolini L (1995) Petrology, mineral and isotope geochemistry of hte external Liguride peridotites (Northern Appenines, Italy). J Petrol 36, 81-105

Rampone E, Hofmann AW, Piccardo GB, Vanucci R, Bottazzi P, Ottolini L (1996) Trace element and isotope geochemistry of depleted peridotites from a N-MORB type ophiolite (Internal Ligurides, N. Italy). Contrib Mineral Petrol 123, 61 76

Reinecke T (1991) Very-high pressure metamorphism and uplift of coesite-bearing metasediments from the Zermatt-Saas zone, western Alps. European Journal of Mineralogy 3, 7-17

Reinecke T (1998) Prograde high- to ultrahigh-pressure metamorphism and exhumation of oceanic sediments at Lago di Cignana, Zermatt-Saas Zone, western Alps. Lithos 42, 147-189

Reyment RA (1989) Compositional data analysis with Aitchison's log-ratios. Terra Nova 1, 29-34

Reyment RA, Jöreskog KG (1996) Applied factor analysis in the natural sciences. Cambridge (University press), second edition 1993

Ringwood AE (1970) Petrogenesis of Apollo 11 basalts and implications for lunar origin. J Geophys Res 75, 6453-6479

Rollinson HR (1992) Another look at the constant sum problem in geochemistry. Mineral Mag 56, 469-475

Rollinson H (1993) Using geochemical data: evaluation, presentation, interpretation. Harlow (Longman), 352 pp

Rubatto D, Gebauer D, Fanning M (1997) Dating the UHP/HP metamorphism in the Western Alps (Sesia-Lanzo and Zermatt-Saas Fee) evidence for subduction events at the cretaceous-tertiary boundary and in the middle eocene. Terra Abstracts 9, suppl 1, 30-31

Rubatto D, Gebauer D, Fanning M (1998) Jurassic foramtion and eocene subduction of the Zermatt–Saas-Fee ophiolites: implications for the geodynamic evolution of the Central and Western Alps. Contrib Mineral Petrol 132, 269-287

Ruppel C, Royden L, Hodges KV (1988) Thermal modelling of extensional tectonics: application to pressure - temperature - time histiories of metamorphic rocks. Tectonics 7, 947-957

Ryerson FJ, Watson EB (1987) Rutile saturation im magmas: implications for Ti-Nb-Ta depletion in island-arc basalts.Earth Planet Sci Lett 86, 225-239

Sartori M (1987) Structure de la zone du Combin entre les Diablons et Zermatt (Valais). Eclogae Geol Helv 80,789-814

Scambelluri M, Müntener O, Hermann J, Piccardo GB, Trommsdorff V (1995) Subduction of water into the mantle: history of an Alpine peridotite. Geology 23, 459-462

Scarrow JH, Cox KG (1988) Basalts generated by decompressive adiabatic melting of a matle plume: a case study from the Isle of Skye. J Petrol 36, 3-22

Schliestedt M (1986) Eclogite-blueschist relationship as evidenced by mineral equiligria in the high-pressure metabasic rocks of Sifnos (Cycladis Islands), Greece. J Petrol 27, 1437-1459

Schilling JG, Zajac M, Evans R, Johnston T, White W, Devine JD, Kingsley R (1983) Petrologic and geochemical variations along the mid-atlantic ridge from 29° N to 73° N. Am J Sci 283, 510-586

Schnetger B (1994) Partial melting during the evolution of the amphibolite- to granulite-facies gneisses of the Ivrea Zone, northern Italy. Chemical Geology 113, 71–101

Seyfried & Bischoff (1979) Low temperature basalt alteration by seawater: an experimental study at 70° and 150°. Geochim Cosmochim Acta 43, 1937-1949

238

Seyfried WE & Bischoff JL (1981) Experimental seawater-basalt interaction at 300°, 500 bars, chemical exchange, secondary mineral formation and implications for the transport of heavy metals. Geochim Cosmochim Acta 45, 135-147

Seyfried WE, Dibble WE (1980) Seawater-peridotite interaction at 300 °C and 500 bar: implications for the origin of oceanic serpentinites. Geochim Cosmochim Acta 44, 309-321

Sheraton JW (1969) The geology and geochemistry of the Lewisian rocks of the Drumbeg area, Sutherland. Unpubublished PhD Thesis, University of Birmingham

Sheraton JW (1970) The origin of the Lewisian gneisses of northwest Scotland, with particular reference to the Drumbeg area, Sutherland. Earth Planet Sci Lett 8, 301–310

Sheraton JW, Skinner AC, Tarney J (1973) The geochemistry of the Scourian gneisses of the Assynt district. In: Park RG, Tarney J (eds), The early Precambrian of Scotland and related rocks of Greenland. University of Keele, 13–30

Sills JD (1983) Mineralogical changes occurring during the retrogression of Archaean gneisses from the Lewisian complex of NW Scotland. Lithos 16, 113–124

Sokolov SV (1977) Distribution of trace elements in magnetites. Zapiski Vsesoyusnogo Mineralogicheskogo Obshchestva 106, 281-290

Stampfli GM (1993) Le Briançonais, terrane exotique dans les Alpes? Eclogae Geol Helv 86, 1-45

Stampfli G, Marthaler M (1990) Divergent and convergent margins in the north-western Alps. Confrontation to actualistic models. Geodinamica Acta 4, 159-184

Stampfli GM, Marchant RH (1997) Geodynamic evolution of the Thetyan margins of the Western Alps. In: Pfiffner OA Lehner P Heithmann P Müller S Steck A (eds) Deep structure of the Swiss Alps. Results of NRP 20. Basel etc (Birkhäuser), 380 pp, 223-239

Steck A (1989) Structures des déformations alpines dans la région de Zermatt. Schweiz mineral petrogr Mitt 69, 205-210

Steck A (1990) Une carte des zones de cisaillement ductil des Alpes centrales. Eclogae GeolHelv 83, 603-627

Steck A, Epard J-L, Escher A, Marchand R, Masson H, Spring L (1989) Coupe tectonique horizontale des Alpes centrales. Mémoires de Géologie (Lausanne) 5

Sun S, Nesbitt RW, Sharaskin AYa (1979) Geochemical characteristics of mid-ocean ridge basalts. Earth Planet Sci Lett 44,0119-138

Takahashi E (1986) Melting of a dry peridotite KLB-1 up to 14 GPa: implications on the origin of peridotitic upper mantle. J Geophys. Res 91, 9367-9382

Tarney J, Skinner AC, Sheraton JW, (1972) A geochemical comparison of major Archaean gneiss units from northwest Scotland and East Greenland. 24th International Geological Congress, Section 1, 162–174

Tarney J, Weaver B, Drury SA (1979) Geochemistry of Archaean trondhjemitic and tonalitic gneisses from Scotland and East Greenland. In Barker F (ed), Trondhjemites, dacites, and related rocks. Elsevier, Amsterdam, 275–299

Teagle DAH, Alt JC, Chiba H, Humphris SE, Halliday AN (1998) Strontium and oxygen isotopic constraints on fluid mixing, alteration and mineralization in the TAG hydrothermal deposit. Chem Geol 149, 1-24

Thompson RN (1982) British Teriary volcanic province. Scott J Geol 18, 49-102

Thompson RN, Morrison MA (1988) Asthenospheric and lower-lithospheric mantle contributions to continental extensional magmatism: an example from the British Tertiary volcanic province. Chem Geol 68, 1-15

Thompson RN, Morrison NA, Hendry GL, Parry SJ (1984) An assessment of the relative roles of crust andd mantle in magma genesis: an elemental approach. Phil Tans R Soc London A310, 549-590

Trommsdorff V, Piccardo GB, Montrasio A (1993) From magmatism through metamorphism to seafloor emplacement of subcontinental Adria lithosphere during pre-Alpine rifting (Malenco, Italy). Schweiz Mineral Petrol Mitt 73, 191-203

Trümpy R (1998) Die Entwicklung der Alpen: Eine kurze Übersicht. Z dt geol Ges 149, 165-182

Watkins KP, (1983) Petrogenesis of Dalradian albite porphyroblast schists. Journal of the Geological Society of London 140, 601–618

Wayte GJ, Worden RH, Rubie DC, Droop GTR (1989) A TEM study of disequilibrium plagioclase breakdown at high pressure: the role of the infiltration fluid. Contrib Mineral Petrol 101, 426-437

Weaver BL, Tarney J (1980) Rare earth geochemistry of Lewisian granulite-facies gneisses, northwest Scotland: implications for the petrogenesis of the Archaean lower continental crust. Earth Planet Sci Lett 51, 279–296

Weaver BL, Tarney J, Windley B (1981) Geochemistry and petrogenesis of the Fiskenaesset anorthosite complex, southern West Greenland: nature of the parent magma. Geochim Cosmochim Acta 45, 711–725

Wedepohl KH (1969) Composition and abundance of common igneous rocks. Chapter 7 in Wedepohl KH (ed) (1969ff) Handbook of Geochemistry, vol 1. Berlin Heidelberg New York (Springer)

Wedepohl KH (ed) (1969ff) Handbook of geochemistry. 2 vols, Berlin Heidelberg New York (Springer)

Wedepohl KH (1988) Spilitisation of the ocean crust and seawater balances. Fortschr Mineral 66, 129-146

Weiss M (1997) Clinohumites: a field and experimental study. PhD-thesis ETH Zürich. 168 + 130 pp

Weiss M, Müntener O (1996), Crystal chemistry of titanian-clinohumite: implications for storage of HFSE in the mantle. J Conf Abs 1 (6th VM Goldschmidt Conference)

Wernicke B (1981) Low-angle normal faults in the Basin and Range province – nappe tectonics in an extending orogen. Nature 291, 645-648

Wernicke B (1985) Uniform sense normal simple shear of the continental lithosphere. Canadian Journal of Earth Sciences 22, 108-125

Widmer TW (1996) Entwässerung ozeanisch alterierter Basalte in Subduktionszonen (Zone von Zermatt-Saas Fee). PhD-thesis ETH Zürich

Widmer TW, Ganguin J, Thompson AB (2000) Ocean floor hydrothermal veins in eclogite facies rocks of the Zermatt-Saas zone, Switzerland. Schweiz Mineral Petrogr Mitt 80, 63-73

Wood DA, Gibson IL, Thompson RN (1976) Elemental mobility during Zeolite facies metamorphism of the tertiary basalts of East Iceland. Contrib Miner Petrol 55, 241-254

Woronow A (1997a) The elusive benefits of log-ratios. In: Pawlosky-Glahn V (ed), IAMG'97 Proceedings of the 3rd annual conference of the international association for mathematical geology, Barcelona (CIMNE), 2 vols, p 97-101

Woronow A (1997b) Regression and discriminant analysis using raw data – is it really a problem? In: Pawlosky-Glahn V (ed), IAMG'97 Proceedings of the 3rd annual conference of the international association for mathematical geology, Barcelona (CIMNE), 2 vols, p 157-162

Woronow A, Love KM, Butler JC (1989) Reply to interpreting and testing compositional data, Mathematical Geology 21, 65-71

Lebenslauf
Wendelin Himmelheber

Ich wurde am 23.12.1958 als das zweite Kind meiner Eltern, der Kunsthistoriker Dr. Georg und Dr. Jo Himmelheber in Stuttgart geboren. 1964 wurde ich in Karlsruhe eingeschult. 1965 zogen meine Eltern nach München, wo ich die Volksschule und das humanistische Wilhelmsgymnasium besuchte. Nach dem Abitur 1978 schrieb ich mich an der Ludwigs-Maximilians-Universität in München für Philosophie mit den Nebenfächern Mathematik und Linguistik ein. Nach 3 Semestern wurde ich im September 1979 zum Zivildienst eingezogen, den ich teils in der Bayerischen Landesschule für Blinde, teils im Kreiskrankenhaus Perlach bis Ende 1980 ableistete. Danach ging ich zurück an die Uni, wo ich nach weiteren 2 Semestern 1982 das Vordiplom in Mathematik ablegte. Da mir dies alles zunehmend lebensfern und perspektivlos erschien, nahm ich mir ein Jahr frei, um über meine Zukunft nachzudenken. In diesem Jahr reiste ich 4 Monate in den USA, verbrachte einige Zeit in einer Landkommune und veröffentlichte in der Zeitschrift Transatlantik einen Essay über das Flippern. 1983 zog ich nach Göttingen, um eine Lehre als Drucker zu beginnen. Ich lernte 3 Jahre bei der Firma Musterschmidt und arbeitete danach noch bis 1995 als Drucker, davon die letzten 7 Jahre im Kollektivbetrieb Aktivdruck. 1991 begann ich Geologie zu studieren, was mir aufgrund der günstigen Arbeitsbedingungen in meiner Druckerei (3-Tage-Woche) möglich war. Im Februar 1996 legte ich nach 10semestrigem Studium mein Diplom ab; die Diplomarbeit befasste sich mit der Geochemie von Gesteinen aus der Zone von Zermatt-Saas Fee, die Diplomkartierung wurde im Altkristallin Osttirols durchgeführt. Ein Jahr vorher war ich nach Kassel zu meiner Frau gezogen, die ich im Februar 1996 heiratete. Nach dem Studium wollte ich als Geologe arbeiten; an Promotion war nicht gedacht. Jedoch war ich zunächst arbeitslos. Deshalb reichte ich nach einem halben Jahr in Zusammenarbeit mit Dr. A. Schneider und Prof. Dr. J. Hoefs einen entsprechenden Antrag bei der DFG ein und fuhr zur Probennahme ins Gelände. Der Antrag wurde jedoch abgelehnt, dafür aber bekam ich Arbeit als Scheinselbständiger in der Baugrundbegutachtung, insbesondere von Eisenbahnstrecken. Diese Arbeit übte ich 14 Monate lang aus bis Sommer 1998. Danach hatte ich bis Ende 1998 für 5 Monate 2 Stellen als Wissenschaftliche Hilfkraft an der Göttinger Uni. Nachdem wir zu Januar 1999 nach Bielefeld gezogen waren, da meine Frau dort Arbeit gefunden hatte, bewarb ich mich in der Softwareindustrie und lebe seither als Programmierer.